T0319532

DRUG DISCOVERY FOR THE TREATMENT OF ADDICTION

DRUG DISCOVERY FOR THE TREATMENT OF ADDICTION

Medicinal Chemistry Strategies

BRIAN S. FULTON
Visiting Lecturer
Department of Chemistry and Chemical Biology
Northeastern University

For general information on our other products and services or for technical support, please contact our
Customer Care Department within the United States at (800) 762-2974, outside the United States at
(317) 572-3993 or fax (317) 572-4002.

Wiley also publishes its books in a variety of electronic formats. Some content that appears in print may
not be available in electronic formats. For more information about Wiley products, visit our web site at
www.wiley.com.

Library of Congress Cataloging-in-Publication Data:

Fulton, Brian S., author.
  Drug discovery for the treatment of addiction : medicinal chemistry strategies / Brian S. Fulton.
     p. ; cm.
  Includes bibliographical references and index.
  ISBN 978-0-470-61416-7 (cloth)
  I. Title.
  [DNLM: 1. Substance-Related Disorders–drug therapy.   2. Drug Discovery.   3. Neurotransmitter
Agents.   WM 270]
  RM301.25
  615.1'9–dc23

                                                                                    2014012670

Printed in the United States of America

For those who want, but can't.

CONTENTS

PREFACE

This book arose from a review article I wrote in 2008 for Annual Reports in Medicinal Chemistry. Following publication, Jonathan Rose of Wiley and Sons contacted me to see if I would be interested in writing a book on the subject. Sure, I thought, how hard can that be? That will take only of couple of years. Five long years later it finally became reality. I am grateful then to the patience of Jonathan with my, I am sure it seemed, perpetual "only 3 more months."

My interest in addiction developed as a NIDA funded Research Fellow at McLean Hospital from 2005 to 2009. In 2005, I was working as a contract chemist at Polaroid, considering a career switch from industry to academic; though, at the then age of 47, I was not sure how feasible that might be. Although I had never met John Neumeyer I was aware of him and noticed he ran a medicinal chemistry group at McLean Hospital. John had been a professor at Northeastern University and had also started Research Biochemicals International, so I thought as one who had lived in both worlds, he might be able to offer some sound advice. Upon meeting with John I was surprised when he said he had a position open in his lab that would be funded by a NIDA Training Grant under Jack Mendelson and Nancy Mello. Though I knew very little about addiction it seemed like an ideal opportunity to fulfill my dream of conducting CNS research, so I accepted. Little did I realize that I was joining a research center started by pioneers in the study of addiction. During my stay at McLean I attended weekly research meetings where I was introduced to the arcane (at least to me) world of behavioral pharmacology. Luckily, I was surrounded by leaders in the study of rodent and primate behavior and addiction; Jack Bergman, Barak Caine, Steven Negus, and Nancy Mello, who were all very patient with explaining behavioral pharmacology to a simple organic chemist. It was probably the most interesting time of my scientific career, and I will be forever grateful for their guidance and patience. My biggest regret is that Jack Mendelson passed away shortly after I joined, so I never really got to know him. Unfortunately, Nancy also passed away in 2013, and so the torch has been passed.

In this book, I will attempt to convey my understanding of addiction to the general medicinal chemistry community. Primarily, this is a book written by an organic chemist for organic chemists. Addiction is a fascinating field of research with very real therapeutic outcomes that deserves more attention by medicinal chemists. As we will see, addiction research relies heavily on the use of animal models that mimic the different stages of addiction. A close working relationship between chemists and behavioral pharmacologists is therefore critical. To aid chemists interested in addiction, I have tried to reduce a complex subject to where it is understandable to those not fluent in the languages of human and nonhuman behavior and the structure and function of the brain. As such, I have taken some liberties during this reduction that experts in the different subjects may find too simplistic, and they will be right. My defense is that it is probably not necessary for a medicinal chemist to expertly understand the controversies and intricacies of self-administration versus conditioned place preference. It won't help making molecules and it is probably a more productive use of time and intellectual energy to have behavioral pharmacologists explain it over a cup of coffee. Nonetheless, some level of understanding is required if one is to correctly interpret pharmacology data in order to direct your efforts in the right direction.

The book is divided into two broad sections. The first section of Chapters 1–4 deals with general aspects of addiction, neuropharmacology, behavioral pharmacology, and drug development. The second section of Chapters 5–10 dives more deeply into medication development. Chapter 1 is a general discussion on the effects of addiction in society. It presents questions of what is addiction and how is it described. Chapter 2 covers the neurobiology and neurochemistry of addiction. This chapter looks at the important neurotransmitter and receptor systems involved in the development of addiction. The goal of the chapter is to provide a solid neurochemistry mechanistic understanding of how addictive drugs work and potential targets to treat addiction. The neurobiology of addiction is very complex and is beyond the scope of this book, and myself, to present it with the accuracy and depth it deserves. Fortunately, it is well covered in other books, most notably by Koob and Le Moal in *Neurobiology of Addiction*. I have concentrated on presenting it more from of a "systems biology" viewpoint with concise discussions on the important cellular and anatomical changes that occur in addiction. In order to help the reader fully understand results discussed in the subsequent chapters, a description of common behavioral pharmacology testing methods is presented in Chapter 3. It will be written with the assumption that the average reader has limited exposure to this area. Topics covered are animal models of the different stages of addiction, interpreting results, some pros and cons of rodent versus nonhuman primate models, and extrapolation of animal models to the human disease state. As an introduction to Chapters 5–10, Chapter 4 covers general approaches to drug development for the treatment of addiction. Special areas of concern relative to the treatment of CNS diseases such as the blood–brain barrier are discussed. While the majority of content in this chapter will be known to medicinal chemists, the non-chemist will hopefully find it informative.

In Chapters 5–10, we more extensively study each drug of abuse and the development of medications to treat its addictive properties. General themes in each chapter are some discussion on the chemistry and pharmacology of each drug of abuse, what drugs are currently approved and the drug's properties, and then finally the current medicinal chemistry strategies being conducted on medication development for the treatment of addiction. It needs to be emphasized that I have focused on drugs that have been tested in a clinical setting. This will exclude many interesting and important preclinical animal studies and the compounds that were developed to be used in those studies. I do not want to diminish the importance of this work; fortunately, it has already been amply reviewed, and I have tried to direct the reader to recent reviews covering the subjects. My emphasis on clinical studies is to show the reader what is known to actually work, or not work.

Some general comments on data and information in the book; first, the primary literature was used as much as possible. However, if not referenced then binding data and functional activity are taken from the PubChem or the NIMH Psychoactive Drug Screening Program databases. Drug properties, especially clinical ones, are taken from the National Library of Medicine database. I have also relied heavily on public information from the National Institute of Drug Abuse, Drug Enforcement Agency, and the United Nations Drug Abuse websites. A special acknowledgement is given to the individuals in each government agency who supply this valuable information to the public. Lastly, if a synthesis of a drug is not referenced, then it was taken from the book *Pharmaceutical Substances: Syntheses, Patents and Applications of the most relevant AIPs*, 5th edition.

On a more personal note, I would like to thank my parents and brothers for their patience and understanding for the missed Christmases, Ozark float trips, and High Sierra climbing as I tried to complete this book during semester breaks. Special thanks goes to my psychological consultants Sylvia Halperin, Ph.D., and Elissa Klienman, M.D., as well as to Anna Sole for the encouragement.

BRIAN S. FULTON

Somerville, MA
2014

1

WHAT IS DRUG ADDICTION?

I can resist everything except temptation

(Oscar Wilde)

It is a simple question with complicated answers. First, and foremost, drug addiction is a medical condition and should be viewed as such. Gone are the days when drug addiction, as with all mental illness, was simplistically viewed as a problem of "free will." A simple answer to the question is when a person cannot stop using a substance (drug) even though they are fully aware the substance is destroying them. We will look at more specific descriptions of addiction later. We also will discuss the difference between addiction, abuse, and dependence. In the categorization of addiction, the user can be classified as being addicted to a single drug or to multiple drugs (e.g., alcohol and nicotine).

Complicating the situation is the fairly common phenomena of comorbidity. The term "comorbidity" describes two or more disorders occurring in the same person such as addiction comorbid with depression or schizophrenia comorbid with addiction.[1] This will complicate the treatment strategy, for example, which disorder to treat first? Are they separate or linked? Did one precede the other? The clinician must take into account these factors. It may also be of importance to the medicinal chemist, especially if there is an underlying physiological commonality.

In this chapter, we will look at some of the societal effects of addiction and then look more closely at the distinct stages of addiction. Unless otherwise mentioned, all statistics in the upcoming discussion are taken from the National Institutes of Drug

Drug Discovery for the Treatment of Addiction: Medicinal Chemistry Strategies, First Edition. Brian S. Fulton.
© 2014 John Wiley & Sons, Inc. Published 2014 by John Wiley & Sons, Inc.

Abuse (NIDA) web site or from the 2012 NSDUH (National Survey on Drug Use and Health) study by the US Department of Health and Human Services.

In most literature addressed for law enforcement agencies, the medical profession, and for the general public, distinctions are often made between illicit and legal drugs. The illicit drugs are those we commonly associate with substance abuse: morphine or heroin, cocaine, methamphetamine, and marijuana. Legal drugs are alcohol, nicotine, prescription medications, and now in some states, marijuana. In this book, I will not make a distinction, that is, when the term "drug" or "drug addiction" is used, it can refer to both illicit and/or legal drugs. With regard to the practicing medicinal chemist who is developing medications for the treatment of addiction, the distinction is irrelevant.

1.1   DEFINITIONS

Before we start, let us examine some basic terminology in the field of substance addiction. As with all mental illness, objective laboratory analytical methods that can be used to diagnose the disease do not yet exist. For example, it is not possible to say take a blood sample, analyze it, and declare that an individual is addicted to a drug. One certainly can analyze for the presence of drugs in blood but that simply shows use, it does not automatically imply addiction. As such, medical personnel in the field of mental illness such as psychiatrists gather and agree on what criteria is required to declare that a person suffers from a mental illness. The consensus is then published in the *Diagnostic and Statistical Manual of Mental Disorders (DSM)*. We are currently at the 5th edition of the DSM, which was released in May 2013.[2] The DSM-V codes agreed upon are designed as guidelines to assist psychiatrists in the diagnosis of mental disorders. The diagnosis of a mental disorder is thus based on a subjective examination of a patient by a psychiatrist. As one might imagine, then there can be some disagreement on what criterion should be used. This is certainly true in the field of drug addiction. Three terms in particular can be confusing: drug abuse, drug dependence, and drug addiction.[3]

In brief, drug abuse refers to the use of a drug in such a way that normal functioning is impaired. Note that one can abuse a drug without being addicted to it. The over consumption of alcohol readily comes to mind. The term "dependence" originally represented purely observable physiological effects of drug use such as withdrawal. The term "addiction" more accurately describes both the observable physiological effects and the more psychological effects of craving. The DSM-IV used the term "dependence" while the DSM-V completely avoids the use of dependence and addiction. NIDA uses the term "addiction", which is what will be used in this book.

The DSM-V lists the criteria for the diagnosis of addiction under substance-related and addictive disorders. Substance-related disorders are divided into substance use disorders and substance-induced disorders. The theme of this book will be addressed toward the development of medication for substance use disorders. Diagnostic criteria are given for 10 separate classes of drugs: alcohol, caffeine, cannabis, depressants, hallucinogens, inhalants, opioids, stimulants, tobacco, and other drugs. As the criteria were just released and the new criteria and guidelines will be debated for some time, let us also examine the criteria in the DSM-IV.

DSM-IV criteria for substance dependence are:

A maladaptive pattern of substance use, leading to clinically significant impairment or distress, as manifested by three (or more) of the following, occurring at any time in the same 12-month period:

(1) Tolerance, as defined by either of the following:
   (a) A need for markedly increased amounts of the substance to achieve intoxication or desired effect.
   (b) Markedly diminished effect with continued use of the same amount of the substance.

(2) Withdrawal, as manifested by either of the following:
   (a) The characteristic withdrawal syndrome for the substance (refer to Criteria A and B of the criteria sets for withdrawal from the specific substances).
   (b) The same (or a closely related) substance is taken to relieve or avoid withdrawal symptoms.

(3) The substance is often taken in larger amounts or over a longer period than was intended.

(4) There is a persistent desire or unsuccessful efforts to cut down or control substance use.

(5) A great deal of time is spent in activities necessary to obtain the substance (e.g., visiting multiple doctors or driving long distances), use the substance (e.g., chain smoking), or recover from its effects.

(6) Important social, occupational, or recreational activities are given up or reduced because of substance use.

(7) The substance use is continued despite knowledge of having a persistent or recurrent physical or psychological problem that is likely to have been caused or exacerbated by the substance (e.g., current cocaine use despite recognition of cocaine-induced depression, or continued drinking despite recognition that an ulcer was made worse by alcohol consumption).

The DSM-V now does not separate between abuse and dependence. They view addiction, or as called in the DSM-V—substance use disorder, as a single disorder measured on a continuum from mild to severe. Each substance will now be addressed as a separate disorder, and drug craving will now be a symptom. Psychiatrics and psychologist specialized in addiction will need to further define this topic.

A second source of diagnostic criteria is available from the World Health Organization. The World Health Organization has developed an international system of disease classification that can be used as a standard diagnostic tool for epidemiology, health management, and clinical purposes. More than 100 countries use the system to report mortality data that is a primary indicator of health status. This system helps to monitor death and disease rates worldwide and measure progress toward the millennium development goals. About 70% of the world's health expenditures (USD \$3.5 billion) are allocated using International Classification of Diseases (ICD) for reimbursement and resource allocation. The criteria are listed in ICD that is in the 10th revision. ICD-10 diagnostic codes for Mental and Behavioral Disorders are listed in Chapter 5, F00-F99. Specific codes for addiction are listed under: Mental and Behavioral Disorders due to Psychoactive Substance Use, blocks F10-F19. The ICD uses the term "dependence" that is defined as: "a cluster of behavioral, cognitive, and physiological phenomena that develop after repeated substance use and that typically include a strong desire to take the drug, difficulties in controlling its use, persisting in its use despite harmful consequences, a higher priority given to drug use than to other activities and obligations, increased tolerance, and sometimes a physical withdrawal state."

1.2   THE DRUGS OF ABUSE

The structures and Chemical Abstract Services registry numbers of the drugs of abuse, which we will discuss, are shown in Figure 1.1. The structure of ethanol is simply CH_3CH_2OH. Broadly speaking, the drugs consist of the naturally occurring opioid narcotic morphine that is present in poppies and its synthetic analog heroin, the naturally occurring stimulant cocaine that is derived from the coca plant and the synthetic stimulants amphetamine and methamphetamine, the mild stimulant/anxiolytic compound nicotine that is present in tobacco plants, and the hallucinogen Δ^9-THC that is present in marijuana. Morphine, cocaine, nicotine, and marijuana are all natural products that are produced in plants. Their existence has been known for thousands of years and a rich literature exists concerning their history. As we will see, all these drugs, and ethanol, are abused and can result in addiction.

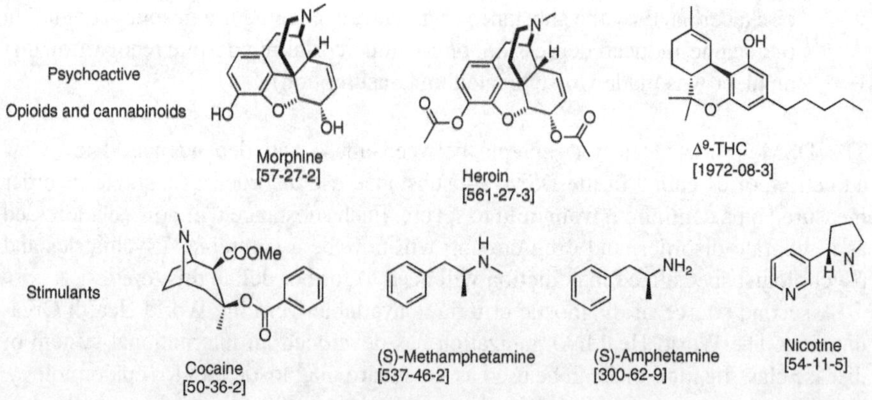

FIGURE 1.1   Structures of the most common drugs of abuse

1.3 SCHEDULE OF CONTROLLED SUBSTANCES

Controlled substances are most closely associated with one's thoughts of drug addiction. The Drug Enforcement Agency (DEA) is responsible for control of illegal drugs in the United States. The mission of the Drug Enforcement Administration is to "enforce the controlled substances laws and regulations of the United States and bring to the criminal and civil justice system of the United States, or any other competent jurisdiction, those organizations and principal members of organizations, involved in the growing, manufacture, or distribution of controlled substances appearing in or destined for illicit traffic in the United States; and to recommend and support non-enforcement programs aimed at reducing the availability of illicit controlled substances on the domestic and international markets."

To assist in their mission, drugs are classified as controlled substances according to the Controlled Substance Act. A listing of the substances and their schedules is found in the DEA regulations, 21 C.F.R. Sections 1308.11 through 1308.15. A controlled substance is placed in its respective schedule based on whether it has a currently accepted medical use in treatment in the United States and its relative abuse potential and likelihood of causing addiction. The Office of the Attorney General assumes responsibility for drug scheduling. There are five classifications from Schedule I to V and they are defined as follows.

1.3.1 Schedule I Controlled Substances

Substances in this schedule have a high potential for abuse, have no currently accepted medical use in treatment in the United States, and there is a lack of accepted safety for use of the drug or other substance under medical supervision. Some examples of substances listed in Schedule I are: heroin, lysergic acid diethylamide, marijuana (cannabis), peyote, methaqualone, and 3,4-methylenedioxymethamphetamine ("ecstasy").

1.3.2 Schedule II Controlled Substances

Substances in this schedule have a high potential for abuse that may lead to severe psychological or physical dependence. Note that included in this list are many drugs that have been approved as medication for the treatment of pain and CNS disorders. Examples of single entity Schedule II narcotics include morphine and opium. Other Schedule II narcotic substances and their common name brand products include: hydromorphone (Dilaudid®), methadone (Dolophine®), meperidine (Demerol®), oxycodone (OxyContin®), and fentanyl (Sublimaze® or Duragesic®). Examples of Schedule II stimulants include: amphetamine (Dexedrine®, Adderall®), methamphetamine (Desoxyn®), and methylphenidate (Ritalin®). Other Schedule II substances include: cocaine, amobarbital, glutethimide, and pentobarbital.

1.3.3   Schedule III Controlled Substances

Substances in this schedule have a potential for abuse less than substances in Schedules I and II and abuse may lead to moderate or low physical dependence or high psychological dependence. Examples of Schedule III narcotics include combination products containing less than 15 mg of hydrocodone per dosage unit (Vicodin®) and products containing not more than 90 mg of codeine per dosage unit (Tylenol with codeine®). Also included are buprenorphine products (Suboxone® and Subutex®) used to treat opioid addiction. Examples of Schedule III non-narcotics include benzphetamine (Didrex®), phendimetrazine, ketamine, and anabolic steroids such as oxandrolone (Oxandrin®).

1.3.4   Schedule IV Controlled Substances

Substances in this schedule have a low potential for abuse relative to substances in Schedule III. An example of a Schedule IV narcotic is propoxyphene (Darvon® and Darvocet-N 100®). Other Schedule IV substances include many of the benzodiazepines: alprazolam (Xanax®), clonazepam (Klonopin®), clorazepate (Tranxene®), diazepam (Valium®), lorazepam (Ativan®), midazolam (Versed®), temazepam (Restoril®), and triazolam (Halcion®).

1.3.5   Schedule V Controlled Substances

Substances in this schedule have a low potential for abuse relative to substances listed in Schedule IV and consist primarily of preparations containing limited quantities of certain narcotics. These are generally used for antitussive, antidiarrheal, and analgesic purposes. Examples include cough preparations containing not more than 200 mg of codeine per 100 mL or per 100 grams (Robitussin AC®).

1.4   SOME FACTS FROM 2012 NSDUH STUDY

The NSDUH study is an annual survey sponsored by the Substance Abuse and Mental Health Services Administration. The survey is the primary source of information on the use of illicit drugs, alcohol, and tobacco in the civilian, noninstitutionalized population of the United States aged 12 years old or older. The survey interviews approximately 67,500 persons each year. The reports can be obtained from http://store.samhsa.gov/home.

Substance abuse is a worldwide problem costing an estimated $600 billion per year in the United States alone. This includes about $181 billion for illicit drugs, $193 billion for tobacco, and $235 billion for alcohol.[4]

Highlights from the 2012 NSDUH study for *illicit drug* use are:

In 2012, an estimated 23.9 million Americans aged 12 or older were current (past month) *illicit drug users*, meaning they had used an illicit drug during the month prior to the survey interview. This estimate represents 9.2% of the population aged 12 or older. This is an increase from levels recorded in 2002–2011. Illicit drugs include marijuana and hashish, cocaine, heroin, hallucinogens, inhalants, or prescription-type psychotherapeutics used for nonmedical purposes.

Marijuana was the most commonly used illicit drug with 18.9 million past month users in 2012. Daily or almost daily use of marijuana (used on 20 or more days in the past month) increased from 5.1 million persons in 2007 to 7.6 million persons in 2012. The second most illicit drug used was cocaine with 1.6 million current cocaine users aged 12 or older.

Cocaine use was quickly followed by the use of hallucinogens with 1.1 million individuals aged 12 or older having used a hallucinogen in the past month. The use of heroin and methamphetamine is less with an estimated 669,000 users of heroin and 440,000 users of methamphetamine in the past month. Some good news with regard to methamphetamine is that the number of past year initiates of methamphetamine was 133,000 in 2012. This number was lower than the estimates in 2002–2004, which ranged from 260,000 to 318,000. There appears to be an increase in the number of heroin users, though.

Some items of special note are, that of illicit drugs marijuana is the most abused drug but of both illicit and legal (alcohol, nicotine) drugs, alcohol by far is the most commonly abused drug. Slightly more than half of Americans aged 12 or older reported being current drinkers of alcohol (51.8%). This translates to an estimated 135.3 million people.

Of these individuals, nearly one quarter (23.1%) participated in binge drinking at least once in the 30 days prior to the survey. This translates to about 59.7 million people. Binge drinking is defined as having five or more drinks on the same occasion on at least one day in the past month. Heavy drinking, which is defined as consuming five or more drinks on the same occasion on at least five different days in the month, was reported by 6.5% of the population or 17 million people.

Tobacco use is still high with 69.5 million Americans (26.7% of the population) who were current users of tobacco. Of these, 57.5 (22.1% of population) smoked cigarettes while 13.4 million (5.2%) smoked cigars, 9.0 million (3.5%) used smoke-less tobacco, and 2.5 million (1.0%) smoked tobacco in pipes. There was a high correlation with tobacco use and illicit drug use where 54.6% of illicit drug users also smoked cigarettes.

The above data refer to the use of drugs, not necessarily substance addiction or abuse. However, in 2012, an estimated 22.2 million persons (8.5% of the population aged 12 or older) were classified with substance dependence or abuse in the past year based on criteria specified in the *Diagnostic and Statistical Manual of Mental Disorders*, 4th edition (DSM-IV). Of these 22.1 million persons, 14.9 million were addicted to or abused alcohol alone, 2.8 million were classified with addiction of or abuse of both alcohol and illicit drugs, and 4.5 million were addicted to or abused illicit drugs but not alcohol.

In terms of use/abuse potential, opioid use is very dangerous in the sense that 58% of the heroin users were classified as addicted to or abusers of heroin, whereas only 12% of those using alcohol were classified as addicted to or abusers of alcohol.

There are also 4.9 million persons who were nonmedical users of pain relievers, for example, 358,000 persons reported the nonmedical use of OxyContin within the past year.

As mentioned, substance abuse is now recognized as a medical condition. The DSM-V divides substance abuse into two categories: substance use disorders (substance dependence and abuse) and substance-induced disorders (intoxication, withdrawal, dementia, etc.). This book will focus on drug development for the treatment of addiction, abuse, and withdrawal.

1.5 THE ADDICTIVE STATE

Addiction is described as "a chronic, often relapsing brain disease that causes compulsive drug seeking and use despite harmful consequences to the individual that is addicted and those around them."[5] The use of the word "chronic" is important and needs to be differentiated from acute use. Chronic use will be associated with addiction; acute use is more associated with the concept of social drug use (Stage 1). This distinction is important as animal models used to study the effects of new drugs as therapeutic agents for the treatment of addiction will need to address chronic use of the addictive drug. Animal models that study acute drug use may give results that cannot be extrapolated toward chronic drug use. By the time dependence is reached neurophysiological changes have occurred that may not be fully represented in animal models that study only acute drug use.

The addiction process tends to follow a progression from casual social use followed by routine compulsive use to finally substance addiction with the possible development of tolerance. A withdrawal response from substance addiction can occur with the degree of severity being dependent on the drug of abuse. Although it is possible to complete a withdrawal program for all drugs of abuse to the desired end point of abstinence, relapse to the addictive state is common and is, in fact, not unexpected. The most effective treatment paradigms involve a combination of psychopharmacological treatment (when available) and behavioral therapy (counseling).

The reasons why one individual can become addicted while another does not involves a complex interaction of genetic, environmental, emotional, neurobiological, and social conditioning factors. Despite the great structural diversity of drugs of abuse, they all have several similar effects. They are all acutely rewarding and result in negative emotional reinforcement upon withdrawal of the drug. Expressed differently in the terminology of behavioral pharmacological, we can speak of addictive drugs as positive reinforcers or negative reinforcers. Reinforcement in general refers to a stimulus that increases the frequency of a behavior. A drug is labeled as a positive reinforcer if it increases the frequency of a behavior it is contingent on. It is something that is presented to the subject. In animal models, this is investigated by determining whether the frequency of a response such as lever pressing will increase if it is followed by an infusion of the drug. Likewise, lever pressing will decrease if pressing the lever does not result in an infusion of the drug. A positive reinforcer usually affects results by generating feelings of pleasure (hedonism) and will thus increase the frequency of behavior induced by the drug.

A negative reinforcer, as defined by B.F. Skinner, is something that when removed increases the probability of a response. An example is the drug Antabus for the treatment of alcoholism. When a patient is taking Antabus, the ingestion of alcohol causes the person to become sick, mimicking some of the same symptoms as withdrawal. In this case, Antabus could be considered a negative reinforcer and alcohol an adverse stimulus. The aversive event is becoming sick. The removal of the aversive stimulus, alcohol, will increase the probability of not be sick, thus preventing drinking. We will see how the field of behavioral pharmacology plays an extremely vital role in the study of substance addiction. Many of the terms used in behavioral pharmacology have their origins in the work of B.F. Skinner.[6]

Drugs of abuse act on neurocircuits that have evolved to insure survival of the organism. The drugs perturb areas of the brain involved on choice (free will) and emotions. This is a finely tuned system that controls our impulses and allows us to analyze and recognize dangerous situations. Drugs of addiction act in part by removing this braking mechanism.

There is initially a behavior choice; I choose to take this drug. Once the drug is on board, physical changes (plasticity) in the brain rapidly occur. Pathological and pharmacological commonalities seen among all drugs of substance abuse are long-term neurobiological and neuroanatomical changes and the effect of the drug (directly or indirectly) upon the mesolimbic dopamine system of the brain with modulation of dopamine transmission and levels.[7] Increases in dopamine concentration in the ventral tegmental area of the brain overstimulate the reward system resulting in feelings of euphoria, a positive rewarding effect. In some individuals, the braking system malfunctions and they spin into abuse and/or addiction. Why some and not all? That questions remains to be answered.

1.5.1   Stages of Addiction

The different stages of addiction have been defined as:[8]

- Acute reinforcement/social drug taking/impulsive use
- Escalating/compulsive use
- Dependence
- Withdrawal
- Protracted abstinence

Let us look at these stages in more detail. The consumption of alcohol will be used as an example of a drug as its use is most relatable to the general public.

Acute Reinforcement/Social Drug Taking/Impulsive Use   The first stage is characterized by what psychiatrists call impulse control disorder. There will be some sense of stress or anticipation of a positive reinforcement before the drug is taken.

Using the drinking analogy, one looks forward to "relaxing" at the end of a stressful day by enjoying a drink of alcohol. The alcohol is acting as the stimulus and the contingent behavior is feeling relaxed. There may not be any sense of committing "bad" behavior. In this case, alcohol can be classified as a positive reinforcer. An important component of this stage is the development of behavioral cues. Cues could be a favorite chair, room, restaurant, tavern, etc., where one might particularly enjoy the drink. Of later importance is that the brain begins to associate the location with drinking and pleasure.

Escalating/Compulsive Use   During the second stage, one will feel compelled to have a drink after work, or that it is necessary to have a drink to feel good. The key distinction is that to feel good one must have a drink. Consumption of alcohol then starts to become repetitive. If consumption is linked to anxiety or stress then the alcohol starts to act as a negative reinforcer. Drinking is seen to *prevent* the negative feelings of anxiety or stress. One can be said to be self-medicating oneself to relieve the negative feeling of stress and anxiety. Note during Stages 1 and 2, no physical signs of drug taking may be present. However, it is becoming clear that the CNS is undergoing changes (plasticity) as it adapts to the recurring presence of alcohol.

A classic example of escalating use is binge drinking. Different definitions exist but in general it can be classified as women having four or more drinks in a sitting and men drinking five or more (Centers for Disease Control and Prevention) and in a period of 2 hours or less (National Institute on Alcohol Abuse and Alcoholism). The Centers for Disease Control (CDC) recently reported that an estimated 38 million adults binge drink an average of four times each month. This has clearly progressed beyond "social" drinking. We are entering the realm of addiction, the user clearly understands that binge drinking will lead to negative consequences, such as a hangover, yet continues despite this knowledge. The underlying neurological reasons why one individual may progress to compulsive use while another does not are not clear. One study suggests that approximately 20% of people who use addictive drugs will switch from controlled to compulsive use.[9]

Dependence   At this stage, the patient is addicted to the drug and can be considered to be abusing the drug with harmful effects upon themselves. Physical signs of drug use will become apparent. Addiction has been described as a state where discontinuation of the drug causes withdrawal symptoms and the person compulsively takes the drug.[10] Physiological tolerance to the drug can develop meaning that increasing doses of the drug are required to achieve the same effect. The development of tolerance can develop due to either a decrease in the drug concentration, by increased metabolism of the drug, or by decreased sensitivity to the drug. Decreased sensitivity to a drug often results due to the reduced expression of receptors that the drug acts upon. Drug tolerance should be viewed as occurring due to the effect of the drug on the body rather than to the drug itself.[10] Cross-tolerance to similar acting drugs can occur. Addiction and tolerance to prescription narcotic pain pills can result in tolerance to heroin. Sensations of craving will also be present during the addictive stage.

Withdrawal   Withdrawal occurs when the drug is no longer available. Withdrawal can occur by voluntary cessation of use or can be induced (e.g., the opioid antagonist naltrexone will induce immediate withdrawal symptoms in narcotic users). The negative emotional and physiological effects that can occur upon withdrawal will often lead the substance abuser to resume drug use. The feelings of craving will develop during withdrawal. Each drug has its unique withdrawal symptoms. Some may be mild as with marijuana to severe symptoms as with heroin. Withdrawal starts some hours after the last dosage of the drug has worn off.

At this stage, the term "psychological dependence" or "addiction" is often used, suggesting that the addiction is only a "state of mind." This of course is nonsense. During the different stages of addiction, physiological changes such as fluctuations in receptor levels and neurotransmitter levels are continuously occurring. New homeostatic states are reached. The withdrawal symptoms are a result of the body acting to the lack of the drug at these new homeostatic set points.

Medication that can alleviate these negative effects will assist in the recovery process. A well-known example is the use of the μ-opioid receptor partial agonist methadone for the treatment of heroin use. A more recent example is the use of the α4β2 nAChR partial agonist varenicline for smoking cessation.[11]

Protracted Abstinence   The most difficult goal to achieve is abstinence. It can be said that one is never "cured" of addiction. It can always return with one being just a casual drink or stressful episode away from relapse. For the individual recovering from addiction, staying clean will be a never-ending task. A good support network is important as well as an understanding of what may be the root causes of their addiction or need for drugs. Psychopharmacology intervention can play an important role. However, patient compliance is always a critical issue. After all, who wants to take a drug everyday for the rest of their life? Especially, when there are no obvious external or physical signs of a disorder. Can medication promote protracted abstinence? Perhaps, but at this time it would appear that the patient, medical personnel, and society must recognize addiction as a chronic disorder that may require the use of medication for the rest of the individual's life. Just as individuals with schizophrenia and other serious mental illness may be required to take medication for their entire life.

Relapse is the resumption of drug taking following detoxification and abstinence. A distinction of relapse is that it occurs following chronic drug use and after protracted abstinence. Relapse can be triggered by different mechanisms. Consumption of even a small amount of the drug can result in drug-induced reinstatement of drug taking while association of an environmental cue such as a visual cue can result in the cue-induced reinstatement of drug taking. It is thought that different parts of the brain control these different mechanisms of relapse.

The treatment of addiction is complicated due to the different stages of addiction as defined above. Future points of entry for medicinal chemistry are at the stages of withdrawal and, most importantly, relapse.

1.6   THEORIES OF ADDICTION

At a cellular and molecular level, drugs of abuse interact with a variety of protein targets eliciting a series of responses. We will discuss this in much more detail but in brief opioids interact with opioid receptors, cocaine with monoamine transporters, methamphetamine with the vesicular transporter, ethanol and nicotine with ion channels, and marijuana with cannabinoid receptors. The primary responses the drugs induce will be via these interactions; however, the situation is much more complex than this. Again, the common theme will be an increase in synaptic dopamine levels followed by changes in cellular plasticity.

In an attempt to develop global models to explain addiction, different unified theories of what might cause addiction have been proposed. At his level, the theories deal more with behavior and have been proposed by researchers specialized in studies of human behavior. For the sake of completeness, they are briefly presented but the interested reader should consult more detailed and specialized descriptions of the theories of addiction.[8,12] Of note is that each theory involves different neurocircuitry dysfunctions but in general there is a breakdown of the ability of the cortex to process information from the limbic areas of the brain that can prevent compulsive drug use.

Before this is discussed, however, we should be aware of two competing theories of how an individual becomes involved with psychoactive drugs. These are the gateway hypothesis (aka steppingstone) and the common liability to addiction hypothesis. The foundation of the gateway hypothesis is that the use of "soft" drugs such as marijuana leads one down a path toward "hard" drugs such as heroin. The order of drug use is therefore very important. The common liability to addiction hypothesis is based on a genetic liability to addiction and on common physiological mechanisms of addiction across different drugs. A 2012 supplemental issue of *Drug and Alcohol Dependence* explores these theories in detail.

The aberrant-learning theory of addiction proposes that repeated exposure to addictive drugs causes an over response to drug-associated cues despite the knowledge of adverse consequences. The frontostriatal-dysfunction theory of addiction proposes that repeated exposure to addictive drugs leads to dysfunction in the ability of the cortex to control impulse behavior and decision making. There is a lack of control to drug-associated cues despite adverse consequences. In the hedonic-allostasis theory of addiction, the initial positive rewarding effects of the drug are replaced by a new emotional state called the "hedonic allostatic" state. It asserts that a new homeostasis reward set point is achieved leading to a reduction in the drug-rewarding effects.

A more complex theory is the incentive-sensitization theory of addiction. Here, it is suggested that the sensation of craving (incentive salience) is controlled differently than drug "liking." It is proposed that drug taking within different rewarding contexts and cues is associated with an increase in mesocorticolimbic dopamine neurotransmission. Chronic drug taking will result in long-lasting neuronal plasticity that makes the individual hypersensitive to drug-associated cues. Finally, in the psychomotor-stimulant theory of addiction, the positive reinforcing activity of addictive drugs causes psychomotor activation. It is proposed that approach behaviors are affected.

1.7   COMORBIDITY

Drug addiction is often linked with other psychiatric disorders, particularly disorders of mood, depression, bipolar, anxiety, and schizophrenia. It is unknown whether one precipitates the other. In a recent study in the United States of 471 patients with bipolar disorder, it was found that use of cannabis was 6.8 times greater than a non-bipolar control group.[13] Certain populations of schizophrenic patients may be unusually susceptible to cannabis use and cannabis abuse has been associated with an earlier onset of schizophrenia.[14] An increase in motor cortex excitability in first-episode schizophrenia patients was found in those with chronic cannabis use (weekly consumption over a period of at least 12 weeks in the last 12 months) versus those with no cannabis use.[15]

Here, one must presume that the mental illness, for example, bipolar disorder or schizophrenia, preceded drug taking, even if the signs of the mental illness were not apparent.

As mentioned earlier, substance-induced mental disorders are recognized in the DSM-V. These will be manifested themselves as changes in mood, development of psychosis, anxiety, etc. Carefully note that these disorders are often observed during the course of intoxication or withdrawal. These changes are usually temporary and can disappear after a month or so following stoppage of drug taking or withdrawal.

1.8   GENETIC ASPECTS OF ADDICTION

Can you inherit addiction? A pressing question is if there exists common factors of liability to drug addiction. While there is clearly a level of genetic liability toward developing addiction the situation is complicated and to date no single part of the genome has been indentified that leads to addiction. There are different levels of genetic investigation. Some are quite broad in scope where they try to separate genetic factors from environmental factors, differences between sexes, age, and between drugs (substance specific). Others are more specific where specific regions of the chromosome or specific genes that are associated with addiction are investigated. Many studies have investigated whether gene linkage and association is present in addiction to drugs. Association studies investigate variations throughout the entire genome to identify single nucleotide polymorphs (SNPs) specifically identified with a disease. Single nucleotide polymorphs are changes to a single base pair in a DNA sequence that could result in the change in the resultant amino acid sequence of the protein produced upon translation. Linkage analyses meanwhile investigate whether certain regions of the genome are specifically associated with a disease.

We will not take a detailed look at this subject though it is worthwhile to review some basic concepts to more fully appreciate correlations of genetics and addiction. It is of great importance but at this time may not be of great help to chemists involved in drug discovery. For those interested in a more in-depth understanding, some excellent reviews are available.[16–18]

The human genome contains some 30,000 genes distributed among 23 pairs of chromosomes, 1–22 autosomes and an X or Y sex chromosome. For the discussion at hand, the main interest is in the production of proteins from genes. Genes of course are composed of DNA, which in turn is constructed from a variable sequence of the four nucleotides: adenine (A), cytosine (C), guanine (G), and thymine (T). The four nucleotides are grouped into codons consisting of three nucleotides each. Each codon will be translated into an amino acid. The general scheme for protein synthesis is the transcription of DNA into messenger RNA. There is a direct transcription of nucleotides, for example, G of DNA is transcribed into G of mRNA, from DNA to mRNA except for thymine that is transcribed into uracil (U). Coding regions (exons) within the mRNA are then translated into proteins. Sequence variations in the exon, due for example to SNPs, can alter the structure and function of the protein. Genes also contain promoter regions where transcription of the DNA into mRNA is initiated. Sequence variation in the promoter region may affect the rate of mRNA and protein production.

There are numerous studies on the pharmacogenetics of drug addiction. To give the reader some idea of what type of research is conducted in this area, some recent studies will be summarized.

A number of human studies indicate that there are both common and substance-specific genetic factors. A large study using the Virginia Adult Twin Study of Psychiatric and Substance Use Disorders database revealed a significant genetic contribution to the risk of addiction. They investigated the effects of genetics and environment on the use of five drugs: cannabis, cocaine, alcohol, caffeine, and nicotine. The genetic effect on risk was drug dependent with the highest correlation seen with cannabis and cocaine. Environmental factors influencing the risk of addiction appeared to be less significant. No common genetic factor was detected; caffeine and nicotine dependence in fact appeared to be influenced by genetic factors unique to these two drugs.[19]

In a study from the University of Colorado, twin modeling was used to determine the heritability of alcohol, tobacco, and cannabis dependence and the etiology of multiple drug problems. The participants consisted of 2484 registrants of the Center for Antisocial Drug Dependence. It was found that additive genetic factors explained more than 60% of the common liability to drug dependence and that the data suggested that both common and substance-specific genetic and environmental factors contribute to individual differences in the levels of alcohol, tobacco, and cannabis dependence symptoms.[20]

Gene linkage studies have identified regions of chromosomes 1, 3, 4, 9, 14, 17, and 18 and gene association studies have identified the genes CNR1, CB2, FAAH, MGLL, TRPV1, and GPR55 as being important in cannabis addiction.[21] These genes presumably then produce receptors and other proteins important in the psychopharmacological effects of marijuana.

A possible linkage between chromosomes 2 and 17 has been identified in subjects with opioid dependence.[22] The opioid receptors consist of three receptors, μ, δ, and κ, that are encoded by the OPRM1, OPRD1, and OPRK1 genes, respectively. The OPRM1 gene is located on chromosome 6, the OPRD1 gene on chromosome 1, and the OPRK1 gene on chromosome 8.

The most common OPRM1 SNP produces two alleles—A118G. The SNP occurs in exon 1 and alters an amino acid at position 40 from asparagine to aspartate in the μ-opioid receptor. Studies of the effect of this variant on receptor function have yielded contradictory results. Some suggest that receptors encoded by this OPRM1 variant have an increased affinity for β-endorphin and greater receptor activation upon binding, while others have found no change in receptor function, signaling, or binding affinities for various opioids. Although variation in OPRM1 has been found to contribute to the risk for heroin addiction in some populations, not all studies agree. The A118G SNP is also associated with an increased risk of alcoholism.[18] Three placebo-controlled trials have found that alcohol-dependent individuals with a G allele have better clinical responses, including lower relapse rates, on naltrexone than those with the A allele.

Variants in the nicotinic receptor CHRNA5, CHRNA3, CHRNB4 subunit genes, which are a cluster of genes on chromosome 15, have been associated with nicotine dependence. It was investigated whether these genetic variants were associated with comorbid disorders and whether comorbid psychiatric disorders modified the genetic risk of nicotine dependence. The link is important as individuals with nicotine dependence have a two- to threefold increased odds for having another psychiatric disorder. Genetic material was obtained from blood samples from individuals in St. Louis, MO, and Detroit, MI, who were part of the Collaborative Genetic Study of Nicotine Dependence. The blood samples were analyzed for the presence of two SNPs that are in the CHRNA5 gene. Along with nicotine dependence, evidence of the presence of alcohol dependence, cannabis dependence, major depressive disorder, panic attack, social phobia, posttraumatic stress disorder, attention deficit hyperactivity disorder, conduct disorder, and antisocial personality disorder was examined. Though, as mentioned above, comorbidity of nicotine dependence and other psychiatric disorders is common, in this study, no strong correlation between the genetic variants in CHRNA5-A3-B4 subunit genes and the presence of other psychiatric disorders was detected.[23]

Genetic variants in metabolism can affect the effectiveness of addiction treatment. For example, the cytochrome P450 enzyme 2A6 metabolizes nicotine. Individuals using transdermal nicotine replacement that have a faster nicotine metabolic system have been linked to a decreased probability of smoking cessation. CYP2A6 polymorphism leading to a reduction in CYP2A6 enzymatic activity, and hence, increased levels of nicotine when using nicotine replacement therapy are associated with a better outcome on nicotine replacement therapy.[16]

Another aspect of genetics and addiction is the effect drugs of abuse themselves have on transcription and translation. It is well known that drugs alter the transcriptional process leading to changes in protein levels such as receptors. This is achieved through the second messenger cascade affecting transcription factors in the nucleus. Transcription factors are proteins that bind to the promoter regions of a gene and initiate or prevent the expression of the gene. A detailed exploration of this topic can be found in the review by Robison and Nestler but in brief, it is proposed that changes in the transcriptional potential of genes through the actions of transcription factors (e.g., ΔFOSB, CREB, and NF-κB), chromatin modifications, and non-coding RNAs contribute substantially to many of the neuroadaptations that result from chronic exposure to drugs of abuse.[24]

At the time of writing, a Program Announcement from NIDA was released encouraging genetic research in addiction. As an illustration of what future research is of interest in addiction, part of the Program Announcement is shown below.

PA-14-025 Discovering Novel Targets: The Molecular Genetics of Drug Addiction and Related Co-Morbidities:

This FOA seeks investigator-initiated applications for genetic and genomic research projects that identify chromosomal loci and/or genetic variation and haplotypes that are associated with either increased or decreased vulnerability to drug addiction, including stimulants (e.g., cocaine, amphetamine, caffeine), narcotics (e.g., opiates), nicotine/tobacco products, benzodiazepines, barbiturates, cannabis, hallucinogens, and/or multiple drugs of abuse and, as needed, accounting for associated co-morbidities (e.g., HIV/AIDS, major depression, schizophrenia, bipolar disorder, alcoholism) in human beings or animal models.

The state of the science on genetics of addiction may be quite different for a given drug. For example, the literature is robust with genome-wide association studies (GWAS) on nicotine dependence, but replicated genetic findings are limited for certain other drugs of abuse. Similarly, there is interest in chromosomal loci and/or genetic variation and haplotypes that are associated with differences in responses to treatment for addiction to these drugs of abuse and related disorders.

This FOA encourages applications for research projects that identify and/or validate chromosomal loci and variations in genes that are associated with vulnerability to addiction and that inform the likelihood of responsiveness to treatment. Applications that propose to examine intermediate phenotypes or endophenotypes to assess the molecular genetics of drug addiction, addiction vulnerability and/or their associated co-morbidities and how they are related to drug addiction are especially encouraged. Also encouraged are genetic as well as computational and large-scale genomic approaches, which may include but are not limited to linkage, linkage disequilibrium, case-control or family-based studies, and integration of data from other databases that may supplement substance abuse genetics and genomics data. Data may be collected from the general population, special populations, recent admixed populations, and/or animal models. Secondary data analysis of data collected from the general population, special populations, recent admixed populations, and/or animal models are also appropriate for this announcement. Investigators are encouraged to include, as a component of their project and as appropriate, gene-gene interactions, gene-environment interactions, gene–environment-development interactions, pharmacogenetics, and non-human models.

1.9   APPROVED MEDICATIONS FOR THE TREATMENT OF SUBSTANCE ABUSE AND ADDICTION

Very few drugs have been approved for the treatment of substance addiction. In clinical practice, many drugs that are approved for other indications are used off-label for the treatment of addiction. They will be discussed later. Drugs approved for opioid addiction include: methadone (withdrawal and maintenance), naltrexone, buprenorphine (withdrawal and maintenance), and clonidine. Drugs approved for alcohol addiction include: naltrexone (maintenance), acamprosate (maintenance), and disulfiram (maintenance). Drugs approved for nicotine addiction include: varenicline (withdrawal and maintenance), bupropion (withdrawal and maintenance), and nicotine patch (withdrawal and maintenance). No drugs have been approved for the treatment of stimulant addiction (i.e., cocaine and methamphetamine) and marijuana.

2

PHYSIOLOGICAL BASIS OF ADDICTION—A CHEMIST'S INTERPRETATION

A proper discussion of the theories of addiction and the different strategies taken to develop medications for the treatment of addiction must focus on what is called the reward system of the brain. All drugs of abuse, and presumably medications to treat substance abuse, do, and will, interact on the reward system. The reward system can be viewed on a hierarchical level as composed of:

(1) Anatomical—neurocircuitry pathways and their interactions
(2) Cellular—role of individual neuronal cell types
(3) Molecular—role of neurotransmitters and drugs and their protein targets

This chapter will present a concise overview of the reward system and its role in substance abuse. The chapter will consist of two broad sections, though the goal of the discussion is to integrate them into a continuous theme. The two broad sections are the neurobiology of addiction and the neurochemistry of addiction. This subject is a vast area of information that cannot be fully covered in this book. The interested reader who desires a more detailed understanding can consult the many excellent texts devoted to this subject. Appendix A contains a list of books pertinent toward gaining a deeper understanding of the neurobiology and neurochemistry of addiction. In particular, the neurobiology of addiction is well covered by Koob and Le Moal in *Neurobiology of Addiction*.[8] Our goal will be to gain a basic understanding of neuroanatomical aspects of the brain and the reward system to more fully understand and interpret pharmacological studies of drugs of addiction.

In general, the physiological etiology of diseases of the central nervous system (CNS) can be ascribed to dysfunctions of certain regions of the brain. This dysfunction can be caused by degenerative loss of neurons and/or due to abnormal (nonhomeostatic) levels of neurotransmitters. Parkinson's disease, for example, is due to loss of dopamine-producing neurons in the basal ganglia while low levels of the monoamine neurotransmitters can explain depression. It is critical to always remember that neurotransmitter dysfunction is at the heart of all mood disorders. The ultimate goal of the medicinal chemist will be to discover a drug that restores neurotransmitter homeostasis.

Though each addiction is unique, in general, a fluctuation of neurotransmitter concentrations and cellular proteins is observed where their concentrations are increased during the rewarding addiction stage and then decreased during the withdrawal stage. It is also clear that each drug of abuse causes structural changes (plasticity) in parts of the brain. The regions affected and the plasticity incurred can vary with each drug and with acute versus chronic use. These changes will have an observable effect on behavior. An immense challenge is to accurately correlate neurotransmitter changes and plasticity changes with behavior. For example, it is known that opiates and stimulants regulate structural changes in the opposite directions, opiates decrease structural plasticity while stimulants increase it. Yet, both drugs can cause similar behavioral responses in rodents such as inducing locomotor activity.[25]

The extension of rodent behavioral pharmacology results to humans is a very difficult task that is the subject of intense research. Throughout this book, I will try to concentrate on using results obtained from human clinical trials. By necessity, there are many studies that cannot, and will not, be conducted on humans or non-human primates. A simple example is represented by intracerebroventricular (i.c.v.) drug injection. A common technique to determine localized sites of neuronal activity is to inject a drug directly into the brain of a rodent such that the drug may be localized in a particular region of the brain. The technique is also done to study the effect of drugs in the brain that may not be able to pass the blood-brain barrier. It would of course be unethical to do this in humans and other primates.

Experiments such as this though can be important as they allow the study of any drug on its effects in modulating specific neuronal circuits and receptors. Though the initial drug tested by i.c.v. injection may not be able to cross the blood-brain barrier and be bioavailable to the brain, if promising pharmacological results are obtained, it would validate a potential target and warrant future work on developing analogs that are bioavailable.

2.1   THE REWARD SYSTEM

The reward system was "discovered" by Olds and Milner in 1954.[26] Rats were placed in a test chamber called a Skinner Box, a more modern term is operant conditioning chamber, and electrodes were implanted in various parts of the brain. A modern version of an operant chamber is shown in Figure 2.1.

The rats were then trained to press a lever to obtain a dose of alternating current as long as the lever was repressed. It was found that rats would press a lever that

FIGURE 2.1   Modern operant conditioning chamber (skinner box) (Reprinted with permission from Thomsen and Caine.[168] Copyright © 2005 John Wiley & Sons, Inc.)

stimulated the brain when the electrodes were in the septal nuclei area of the brain. The septal nuclei are a set of neurons located in the limbic system as part of the cerebral hemisphere and interact with the hippocampus, amygdala, and hypothalamus. When the electricity was on, the rats would press the levers hundreds of times over a six-hour period. When the electricity was off, the rate of responding fell close to zero. They interpreted the results as "a phenomenon representing strong pursuit of a positive stimulus rather than escape from some negative condition." The rats were self-administering a rewarding stimulus.

Building on this and other research, Stein reported in 1958 the effects of cues, or as was called in this paper "secondary reinforcement."[27] Stein placed rats in a box that contained two levers. Pressing one lever produced an audible tone, pressing the other lever had no effect. After a period of training sessions, the rats had a slight preference to pressing the no-effect lever. The levers were removed and the rats were then stimulated with a low dose of electricity via electrodes implanted in the limbic area of the brain. Of importance, just before each shock, the audible tone was presented. The rats were placed back in the box with both levers as before. Upon presentation of the audible tone, there was now a dramatic change in lever preference and the rate of lever pressing. Even though upon pressing of the audible tone lever no stimulating shock was given, the rats greatly preferred pressing the audible tone lever. The rats had learned to associate the audible tone with a positive reinforcement (electrical stimulation). The audible tone was now acting as a cue.

The last piece of the puzzle was the discovery that dopamine was the neurotransmitter active in the reward process. In 1958, the presence of dopamine (3-hydroxytyramine) in the brain was reported.[28] A catecholamine hypothesis for reward mediation was developed in the late 1960s. This was a rich period for monoamine research. In 1965, Schildkraut proposed the catecholamine hypothesis of depression and in 1975 Yokel and Wise presented conclusive evidence that dopamine was the neurotransmitter involved in the mediation of the reward system.[29,30]

2.2   NEUROANATOMY OF THE REWARD SYSTEM

The reward system is considered to be located in the limbic system area of the brain. The neuronal components of the limbic system are not rigorously defined but for our purposes can be considered to consist of cellular structures between the hypothalamus and the neocortex, which include the amygdala, ventral tegmental area (VTA), nucleus accumbens, and the hippocampus.[31] An extremely detailed discussion of the primate reward circuit is available.[32] Throughout this book, a recurring theme will be important neuronal input and output connections with the VTA, nucleus accumbens, hypothalamus, and the cortex playing a central role in examining the etiology of addiction. The VTA is a component of the limbic midbrain nuclei system, while the nucleus accumbens is a component of the basal ganglia. The limbic system is considered to be an evolutional ancient part of the brain responsible for the control of emotional experiences, responses, learning, and memory. The amygdala receives direct sensory neuronal inputs from the sensory organs and also from the cortex and insula that are of a more general type related to feelings of physical and emotional comfort and discomfort.

As one might imagine, the neurocircuitry of addiction is exceedingly complex and a detailed analysis is quite beyond the scope of this book.[33] My goal will be to give a brief introduction with the average medicinal chemist as the audience. A simplified schematic of the brain will be used. The schematic in Figure 2.2 contains the major anatomical regions that will be discussed. It is important to appreciate that it is greatly simplified scheme. For example, the nucleus accumbens is composed of different shells, each thought to contribute to different aspects of the different stages of addiction. As we proceed, a central theme to remember is that all drugs of abuse are thought to exert their *reinforcing* effects by increasing dopamine levels in the nucleus accumbens by acting upon different molecular targets throughout the brain but in particular targets on the VTA.[34]

Figure 2.3 depicts a much simpler scheme of the addiction circuitry to keep in mind as we proceed.

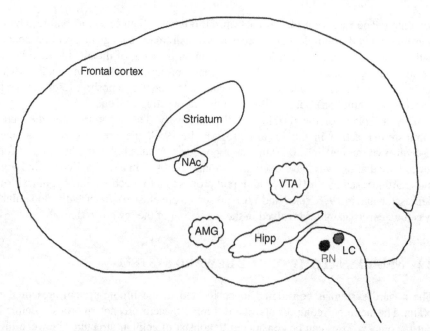

FIGURE 2.2 Addiction circuitry schematic. Abbreviations: AMG, amygdala; LC, locus ceruleus; Hipp, hippocampus; NAc, nucleus accumbens; RN, raphe nuclei; VTA, ventral tegmental area

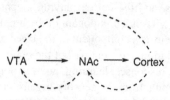

FIGURE 2.3 Addiction circuit

We can consider activity upon or by the VTA as causing dopamine release into the nucleus accumbens that in turn affects cognition in the cortex. All three areas further communicate (via neurotransmitters) less directly as represented by the hashed arrows.

2.3   BRIEF REVIEW OF THE CENTRAL NERVOUS SYSTEM AND ADDICTION

This section will focus on presenting a concise overview of the important cellular and anatomical changes that occur in addiction. A brief overview of the CNS is necessary though before we embark.[35] After this, we will discuss important components of the CNS that will be of particular relevance for our understanding of addiction.

The CNS consists of the brain and the spinal cord. The brain is composed of water, capillary blood vessels, and individual cell types called neurons and glia cells. Neuronal cells, or neurons, have been categorized by physiological function into three types, those that are excitatory, those that are inhibitory, and others that have a modulating influence. Neurons are unique in the body as they can interact with each other by the exchange of chemicals. The chemicals can be small organic compounds such as neurotransmitters, peptides, or inorganic compounds such as alkali and alkaline cations (Na^+, K^+, and Ca^{2+}) or halide anions (Cl^-). This exchange occurs through a small gap between neurons called a synapse. The neuron cell body has a wide variety of shapes and sizes but they all contain two appendages called dendrites and axons extending from the cell body. In general, we can view dendrites as receiving information from other neurons (afferent) and sending the information to the neuron cell body and axons as sending information (efferent) from the cell body to other neurons.

As mentioned above, the site of interaction between a dendrite of one neuron and the axon of another neuron is called the synaptic gap. There are considered to be two types of synapses: electrical and chemical. Electrical synapses contain bridging sections called gap junction channels that allow the movement of ions between neurons. The physical width of the electrical synapse is 3.5 nm. Contrast this with the chemical synapse, where a distance of 20-40 nm separates the two neurons. Neurotransmitters are the agent of transmission in a chemical synapse. The neuron on the axon side of the gap is called the presynaptic neuron and the neuron on the dendrite side is called the postsynaptic neuron. Multitudes of receptors, ion channels, and transporters are located on the surface of the cell membranes both presynaptic and postsynaptic. A unique component of the presynaptic neuron is synaptic vesicles of 40-50 nm in diameter that contain neurotransmitters. The majority of our efforts will be spent studying the interactions of drugs in these tiny areas of the brain and their resultant effects on addiction. The synapse as depicted in Figure 2.4 is often pictured as the presynaptic terminus represented as a triangle and the postsynaptic neuron as a line or other symbol.

The transport of chemicals into and out of the brain from the rest of the body occurs through the blood vessels that are intertwined with the cellular tissue. The physical interaction between blood vessels and neurons is the blood-brain barrier. The "barrier" is actually just unusually tightly packed endothelial cells of the capillary vessels that hinder passive diffusion of chemicals between the blood and the brain. This protective mechanism plays a major role in the bioavailability of drugs for the

FIGURE 2.4   Depiction of a synapse

treatment of CNS disorders and the development of any CNS active drugs must take this into account.

Though neurons are the main players in the transmission of information, the majority of cells in the CNS are glia cells. As with neurons, several types of glia cells exist such as astrocytes. Glia cells play an important structural role in the support of neurons but they also participate in the information process. Glia cells are densely packed such that the common distance between them is 15-20 nm. Glia cells contain receptors, release neurotransmitters and growth factors, control clearance of neurotransmitters from the synapse, and are involved in neuroplasticity. The main differences between glia cells and neurons are that glia cells do not contain typical synaptic contacts with other cells, cannot maintain an action potential, and can continuously divide.[36] Astrocytes can communicate via gap junctions.

2.3.1   Neuron Firing

Neurons can communicate via the transfer of chemicals initiated by an action potential. The cytoplasm of a neuron is more negatively charged relative to the exterior of the cell resulting in a potential difference across the membrane. The membrane potential is defined as the potential inside the neuron minus the potential outside the neuron where

$$V_{membrane} = V_{inside\ cell} - V_{outside\ cell}$$

and by convention the potential outside the cell is defined as 0. Neurons under homeostatic conditions have a resting membrane potential that can range from −40 to −90 mV but on average is about −70 mV. The cell membrane potentials are measured using different patch-clamp techniques.

Neurons are said to be depolarized when the cell potential becomes more positive, say by the influx of positive ions. When the cell potential becomes more negative, the neuron becomes hyperpolarized. The ions typically involved in this polarization/depolarization cycle are K^+, Na^+, and Cl^-. Because the ions are not distributed in equal concentrations across the cell membrane, a chemical gradient exists in the neuron. The interior of the neuronal cell has higher concentrations of K^+ ions and organic anions (e.g., amino acids) and lower concentrations of Na^+, Ca^{2+}, and Cl^- ions relative to the exterior. If ligand binding allows positive ions to enter the cell, the interior of the cell becomes more positively charged and the neuron is depolarized. When the cell reached a certain cell potential (approximately −40 mV) or threshold, the neuron will "fire an action potential" down the axon. When the potential (impulse) reaches the terminus of the axon, the change in the localized cell potential results in the release of neurotransmitters. These neurotransmitters diffuse across the synapse, and interact with another receptor or ion channel causing a second set of activities to occur.

Even though the membrane is not very permeable to ions there is, at the resting state of the neuron, a small level of exchange of ions across the membrane. For the neuron to become depolarized, much larger changes in ion concentrations must be achieved.

This is accomplished by the opening or closing of proteins called ion channels. Ion channels modulate cell potentials and neuron firing by controlling the concentration of anions and cations in the neurons, hence the membrane potential. The activity of the ion channels is caused by binding of extracellular ligands, neuronal voltage changes, or by intracellular second messengers. This opening and closing process is called gating.

2.4   NEUROTRANSMITTERS AND THEIR TARGETS

Information flow within the CNS occurs via what is called the neurocircuitry of the brain.

The active components of these neurocircuits are neurotransmitters. Neurotransmitters are chemicals that are synthesized in a neuron and stored in synaptic vesicles. They are released upon stimulation of the neuron and interact with another neuron through the synapse. By acting on receptors such as ion channels and G-protein-coupled receptors (GPCRs), neurotransmitters transmit and regulate information.

The chemistry, pharmacology, and cell biology of neurotransmitters and receptors have been extensively documented and their general concepts are usually well known to medicinal chemists. Before we launch into the details of the neuronal circuits, however, it is worth some time to review key molecular components of it; in particular, GPCRs, ion channels, and transporters. The bulk of the medicinal chemistry discussions will involve drugs interacting with one of these three receptors.

Neurotransmitters and their receptors are not always uniformly distributed throughout CNS, for example, receptors for cocaine are localized in the components of the reward system, whereas receptors for cannabinoids are expressed throughout the brain. An important fact going forward is to remember that any given neuron can contain several receptors imbedded in the cell membrane. In other words, a glutaminergic neuron may not only contain receptors for glutamate.

Drugs of abuse elicit their effects by binding to receptors and modulating neurotransmission. Likewise, drugs that treat psychiatric disorders also modulate neurotransmitter levels in the brain by binding to receptors. This modulation of neurotransmission can occur directly but drugs of abuse also do so by indirectly affecting the levels of one neurotransmitter by interacting first at a different neurotransmitter system. The interactions of drugs of abuse with receptors, therefore, follow complex patterns and they act upon different parts of the brain.

2.4.1   Neurotransmitters

Neurotransmitters are either small organic compounds or peptides. All are synthesized inside a neuron, stored in vesicles, and released from the cell into the synapse. A key difference though is that small organic or non-peptidic neurotransmitters are synthesized in the axon, while peptidic neurotransmitters are synthesized in the nucleus. The peptide neurotransmitters are then transported to the axon. The synthesis and transport of peptidic neurotransmitters occur in stages. First, normal translation produces a peptide called the precursor peptide. This precursor peptide is packaged into a vesicle for transport. During the transport process, peptidases cleave the precursor peptide into the neuroactive peptides. Note that regulation of the amounts of neuroactive peptides is a function of peptidase activity. There are said to be more than 100 neuroactive peptides. Regulation of neurotransmitter synthesis and release also occurs by different feedback mechanisms, for example, the release of neurotransmitters from synaptic vesicles is activated by the influx of Ca^{2+} ions. Detailed discussions of individual neurotransmitters will occur when needed.

Neurotransmitters are defined as follows:

(1) The chemical must be found in the presynaptic neuron and must be released when the neuron is stimulated.
(2) The chemical must be inactivated after it is released. This occurs by reuptake or metabolic degradation.
(3) If the chemical is applied exogenously at the postsynaptic membrane, the effect will be the same as when the presynaptic neuron is stimulated.
(4) The chemical applied to the synapse must be affected in a manner similar to that of the naturally occurring chemical.

For reference, the structures of the main non-peptidic neurotransmitters of particular importance in addiction are shown in Figure 2.5 along with their Chemical Abstract Service (CAS) numbers and PubChem identification numbers.

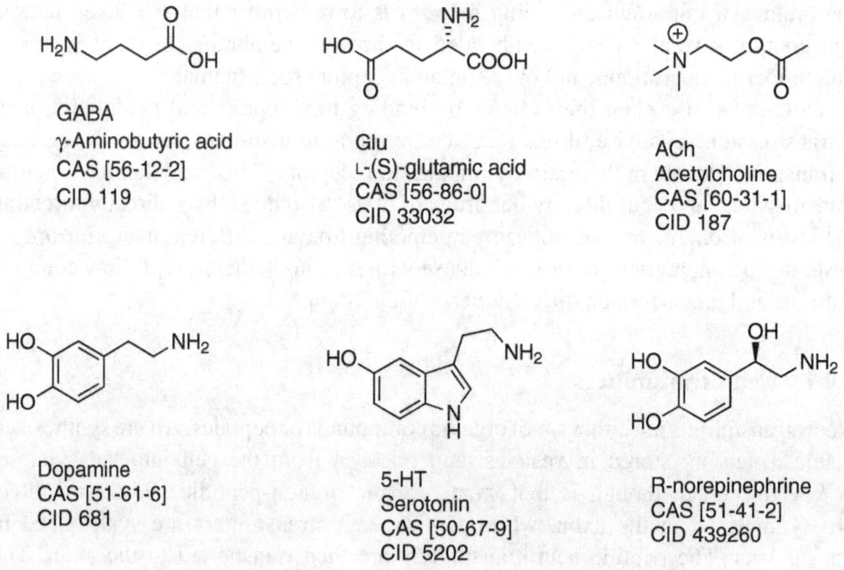

GABA
γ-Aminobutyric acid
CAS [56-12-2]
CID 119

Glu
L-(S)-glutamic acid
CAS [56-86-0]
CID 33032

ACh
Acetylcholine
CAS [60-31-1]
CID 187

Dopamine
CAS [51-61-6]
CID 681

5-HT
Serotonin
CAS [50-67-9]
CID 5202

R-norepinephrine
CAS [51-41-2]
CID 439260

FIGURE 2.5   Structures of neurotransmitters

The γ-aminobutyric acid (GABA) and glutamate neurotransmitters exist in the highest concentration (μM-mM) and are the major chemicals controlling brain activity. GABA attenuates or slows brain activity, while glutamate and acetylcholine potentiate or increase brain activity. The other neurotransmitters modulate brain activity to lesser degrees. While doing so, they can act in both an inhibitory or excitatory mode. All influence second messenger effects. Under "normal" conditions, a homeostatic concentration of neurotransmitters exist, over- or under-secretion of a neurotransmitter usually has secondary effects on the levels of other neurotransmitters along with protein synthesis due to disruptions of the second messenger cycles. Instead of discussing "a homeostatic concentration of neurotransmitters," the concept of tone can be introduced. The word "tone" is somewhat loosely used but in terms of neurochemistry, it can be interpreted as the mean concentration of a neurotransmitter within a synapse.

2.4.2   Receptors

Throughout this book, the term "receptor" is used as defined in IUPAC Glossary of terms used in Medicinal Chemistry (1998): *A receptor* is a molecule or a polymeric structure in or on a cell that specifically recognizes and binds a compound acting as a molecular messenger (*neurotransmitter*, hormone, lymphokine, lectin, *drug*, etc.).

The receptors that will most interest us are GPCRs, transporters, and ion channels. Unless otherwise specified, when the term "receptor" is used in isolation, it can signify any of the above.

Receptors can lie on the presynaptic or postsynaptic neurons. Their normal function is to interact with neurotransmitters to control the homeostatic operation of the CNS in response to stimuli. Modulation of these receptors usually results in a series of enzymatic processes within the cell called secondary messenger effects that ultimately regulate protein synthesis. Thus, the interaction of, say, morphine with a μ-opioid receptor can modulate the future synthesis of the μ-opioid receptor in the nucleus.

We will see how drugs of abuse often indirectly affect dopamine levels via interaction with a different neurotransmitter system. An understanding of neuroanatomy and neurocircuitry is important to appreciate the levels of interdependence of neurotransmitter systems and the actions of drugs of abuse. Each drug of abuse is different but commonalities can occur such that a drug useful for the treatment of narcotic addiction can be potentially useful for the treatment of alcoholism.

The drugs of abuse bind to their protein targets where they elicit a biological response, that is, they act as agonists. The pharmacology of ligand binding will be briefly reviewed in Chapter 4. As we proceed from here though it is important to remember that a drug acting as an agonist elicits some type of biological response, whereas an antagonist does not. An antagonist blocks the action of an agonist on its receptor. A nuance though is that an agonist can affect an antagonist response even though in cellular functional assays it is defined as an agonist. This is accomplished if the drug strongly binds to a receptor, elicits a weak agonist response, and competes with the endogenous neurotransmitter at the binding site. By preventing binding of the endogenous neurotransmitter, the full agonist effect of the neurotransmitter is attenuated and the drug then behaves as an antagonist. This approach has lead to the successful partial agonist approach that will be discussed more in Chapter 4.

2.4.3   G-Protein-Coupled Receptors

G-protein-coupled receptors consist of two components: the protein receptor sitting on the cell surface encased in the cell membrane, and a trimeric G-protein with an attached molecule of guanosine diphosphate (GDP), residing next to the cytoplasmic side of the cell membrane. The receptor component of GPCRs is composed of seven transmembrane (TM) domains with an extracellular N-terminus and an intracellular C-terminus. When a ligand (agonist) noncovalently binds to the receptor, it causes a conformational change in the receptor that can allow a G-protein to interact with the receptor through noncovalent intermolecular bonding interactions. Upon agonist binding, guanosine triphosphate (GTP) displaces GDP from the G-protein causing the G-protein to dissociate into two different components: an α-unit bound to the GTP and a $\beta\gamma$-unit. The α-GTP unit and the $\beta\gamma$-unit are responsible for cell signaling and their dissociation elicits second messenger effects to occur. The agonist will eventually dissociate from the membrane-bound receptor, dispersing into the extracellular milieu where it can then bind to a different receptor. The activation of G-proteins by an agonist-occupied GPCR is catalytic rather than stoichiometric. The α subunit contains some intrinsic GTPase activity; conversion of α-GTP to α-GDP by cleavage of a phosphate unit allows the dissociated G-protein units to again associate returning the system to equilibrium. Understanding how drugs of abuse perturb this process is an active area of addiction research.

In general, G-proteins and the receptors do not interact in the absence of an agonist. There are thought to be cases though when they do couple. In these situations, the receptor is said to be constitutively active and has been used to explain the phenomenon of inverse agonism.

There are four basic families of G-proteins. They can be broadly placed into two groups, ones that activate second messenger enzymes and others that inhibit secondary messenger enzymes. The G_s family activates the key enzyme adenylyl cyclase (AC) and the G_i family, which includes G_o proteins, inhibits the activity of adenylyl cyclase. The enzyme adenylyl cyclase hydrolyzes adenosine triphosphate (ATP) to form the important second messenger cyclic adenosine $3'$-$5'$ monophosphate (cAMP) as shown in Figure 2.6.

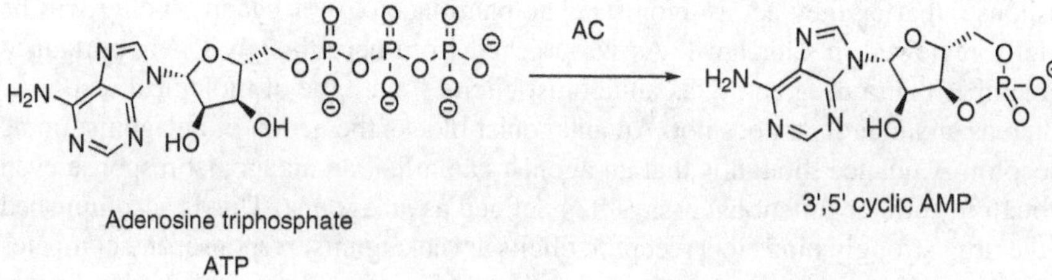

Adenosine triphosphate
ATP

3',5' cyclic AMP

FIGURE 2.6   Formation of cAMP from ATP

The brain utilizes large amounts of ATP that is largely generated through the metabolism of glucose via glycolysis, the tricarboxylic acid cycle, and oxidative phosphorylation. Approximately, 120 grams of glucose per day is required for the brain to meet its ATP requirements. Basal levels of cAMP in the cytosol of neurons are approximately $0.1–1$ µM while ATP is present at levels of $0.8–1.4$ mM.[37] Adenylyl cyclase is a membrane-bound enzyme, and nine different and unique membrane-bound forms, AC1–AC9, are known. Only one, AC1, is expressed solely in the brain. The enzyme contains two membrane-spanning regions each consisting of six domains. The ATP binding/catalytic domains are in the cytoplasm. One of the main targets of cAMP is cAMP-dependent protein kinase or protein kinase A (PKA). PKA is a tetrameric enzyme containing two catalytic and two regulatory domains. Upon binding of two molecules of cAMP to each regulatory domain, the tetramer dissociates to form the catalytically active enzyme. This protein kinase then phosphorylates a diverse set of proteins including other enzymes in the second messenger system along with ion channels and receptors. The kinase can also enter into the nucleus where it phosphorylates, thus activating, the transcription factor cAMP response element-binding protein (CREB), ultimately regulating protein synthesis. It is clear that modulation of adenylyl cyclase can have large effects on the activity of the neuron.

A large number of receptors in the brain are coupled with the G_q family of G-proteins. These G-proteins activate the second messenger's inositol triphosphate and diacylglycerol. Inositol triphosphate causes the release of Ca^{2+} stored in the endoplasmic reticulum, thereby increasing the levels of Ca^{2+} in the cytosol. Cellular Ca^{2+} levels play an important role in regulating neuronal firing and neurotransmitter release. Diacylglycerol binds to a regulatory region on protein kinase C resulting in activation of the enzyme (also activated by Ca^{2+}).

Upon ligand binding to the GPCR, the $G_{\alpha q}$ unit dissociates and activates the enzyme phosphoinositide-specific phospholipase C (PI-PLC) that hydrolyzes phosphoinositide (PIP2, PIP_2, PtdIns(4,5)P_2) to produce 1D-*myo*-inositol 1,4,5-trisphosphate [Ins(1,4,5)P_3 or IP_3) and a diacylglycerol as shown in Figure 2.7.

IP_3 is further metabolized to inositol for the regeneration of phosphatidylinositol. Lithium is still a first choice for the treatment of bipolar disease. It should be of interest that lithium's proposed mechanism of action, as depicted in Figure 2.8, is the inhibition of an enzyme involved in the regeneration of phosphatidylinositol.

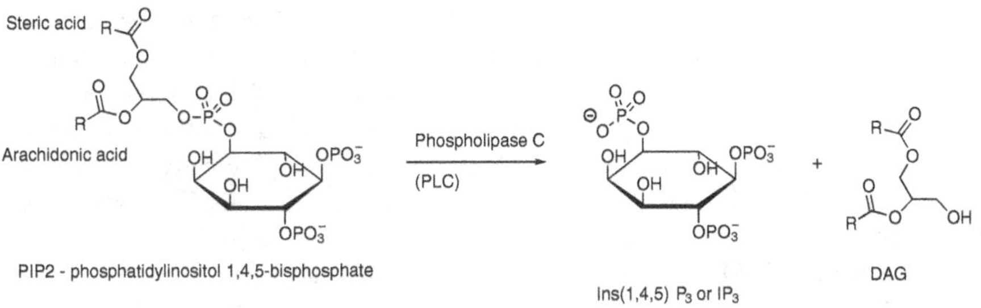

FIGURE 2.7   Function of PLC

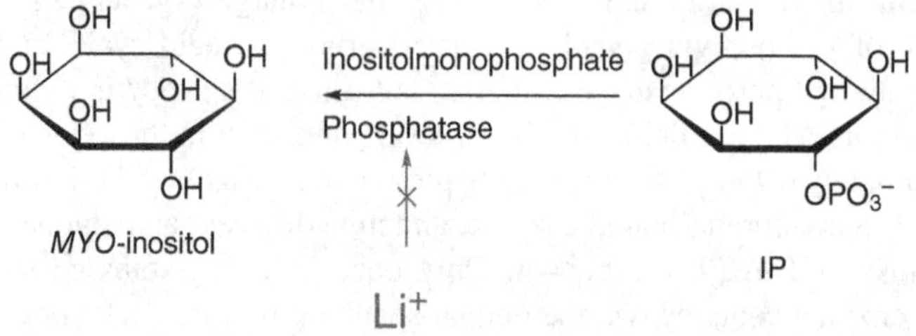

FIGURE 2.8   Inhibition of inositol monophosphate phosphatase by lithium

Finally, the G_{12} family activates guanine nucleotide exchange factors (Rho-GEF). Guanine nucleotide exchange factors are proteins that help facilitate the exchange of GDP by GTP, thus activating the dissociation of the G-protein.

Further discussion of GPCRs will be reserved for the specific medicinal chemistry discussions of drugs and their receptors. The chapter by Zachariou, Duman, and Nestler in *Basic Neurochemistry* is recommended for the readers who want a more detailed discussion of G-proteins specific to the CNS.[38]

Then, in general, we can view the GPCR as mediating the actions of a neurotransmitter, or drug, from the exterior of the cell on cellular processes within the cell. GPCRs play an intimate and a fundamental role in signal transduction. Another function of G-proteins in neurons includes depolarization and firing of an action potential. When agonist binding causes the α- and βγ-subunits of the G-protein to dissociate from the receptor, the α- and βγ-subunits can then bind to a membrane-bound ion channel causing it to open or close.

2.4.4   Ion Channels

Ion channels are cell surface proteins and like GPCRs span the membrane connecting the cytoplasm with the exterior of the cell. There are two types of ion channels, those gated by ion concentration changes (cell potential changes) called voltage-gated ion channels, and those gated by binding of a ligand to the ion channel and are called ligand-gated ion channels. Ion channels are more structurally complex than the protein receptor of GPCRs in that they exist as subunits bound together versus one continuous peptide chain as in GPCRs. The principle voltage-gated ion channels are the Na^+, K^+, and Ca^{2+} ion channels. Their main function is to control the excitability of a neuron. The structures of the voltage-gated ion channels are as follows. Sodium channels consist of three subunits: a large α subunit and smaller β_1- and β_2 subunits. Calcium channels are similar consisting of a large subunit called α_1, and smaller subunits α_2, β, and γ. There are three families of potassium channels consisting of two, four, or six TM domains. Their activity can be regulated by voltage, calcium ions, or neurotransmitters. The simplest potassium channels are the 2TM K_{ir} family. A wide variety of voltage-gated ion channels can exist. The potassium channels themselves consist of about 40 different types.

Of particular importance to us will be the ligand-gated ion channels that include the nicotinic acetylcholine receptors and the GABA$_A$ ion channel. These ligand-gated ion channels differ in structure from voltage-gated ion channels in that they consist of five subunits forming a pentagon shape. The diversity of the subunits is enormous. These will be discussed in more detail later. Ion channels are also referred to as ionotropic receptors.

2.4.5   G-Protein-Activated Inwardly Rectifying K$^+$ Channels

G-protein-activated inwardly rectifying potassium channels, also known as GIRK or K$_{ir}$3, are worth examining as they have been implicated in addiction, for example, ethanol is a K$_{ir}$3 agonist. Several interesting recent reviews have extensively reported on the structure, physiology and regulation of K$_{ir}$3, and role of K$_{ir}$3 in addiction.[39–41] A standardized nomenclature for potassium channels has been proposed and will be used in this discussion.[42]

The inwardly rectifying potassium channels are comprised of four different but homologous subunits labeled K$_{ir}$3.1-3.4. The subunits assemble into a tetrameric functional ion channel and as mentioned above each subunit consists of a 2TM domain.

In electronics, a rectifier is a device that converts alternating current to direct current. Direct current only flows in only one direction so when applied to ion channels it says that current, or K$^+$ ions in this case, will move more easily in one direction than in another direction. The process by which this occurs is called rectification. The primary physiological function of inwardly rectifying potassium channels is to regulate the resting membrane potential. They are called inward rectifiers because the current (K$^+$ ions) more readily flows through the channel into the neuron than out of the neuron. This is against the concentration gradient. There is a nonlinear change in ionic current through an ion channel pore as a function of the electrochemical driving force. Under physiological conditions, the resting membrane potential of a typical neuron, $E \cong -70$ mV, is positive to the equilibrium potential of potassium, $E_K \cong -90$ mV, and the small outward K$^+$ current through the K$_{ir}$3 channels decreases the excitability of a neuron.

When the neuron becomes hyperpolarized and the membrane potential becomes more negative, $E < -90$ mV, the K$^+$ inward rectifier opens allowing the resting state of -70 mV to be re-established by increasing the positive charge in the neuron. When the neuron becomes depolarized, the membrane potential is more positive than E_K and the K$^+$ inward rectifier passes very little charge out of the cell *relative* to when $E < -90$ mV. Charge flow is mainly one way—into the neuron. There are other K$^+$ channels that can pass K$^+$ ion charge out of the cell after depolarization to re-establish the membrane resting potential.

As the name implies, agonist binding to a GPCR can regulate the activity of these ion channels. Following dissociation of the G-protein, the βγ subunit can interact with membrane bound inwardly rectifying potassium channels. If the βγ subunit is from a G$_{i/o}$ protein, then binding with K$_{ir}$3 activates the channel, opening it, allowing potassium ions to leave the cell. The decrease in positive charge will cause the neuron

TABLE 2.1   Equilibrium Binding Constants (K_d, μM) for Monoamine Neurotransmitters

	DAT	NET	SERT
Dopamine	2.4	32	>100
Norepinephrine	16.2	2.2	>100
Serotonin	>100	>100	2.1

to become hyperpolarized reducing cell firing. If the βγ subunit is from a G_q protein, then binding with $K_{ir}3$ deactivates the channel. The later occurs as the activity of $K_{ir}3$ is also regulated by phosphoinositide. $K_{ir}3$ channel currents are enhanced in the presence of phosphoinositide and their depletion usually leads to loss of activity. As G_q proteins activate PLC for whom phosphoinositide is a substrate, levels of phosphoinositide will be reduced resulting in loss of $K_{ir}3$ activity.

2.4.6   Transporters

Transporters are proteins that can transport a neurotransmitter or drug from the extracellular space through the cell membrane into the neuron. This is an energy-driven process; transport does not occur by passive diffusion. The three most familiar transporters are those for the biogenic amines dopamine, norepinephrine, and serotonin. Transporters differ structurally from GPCRs in that they have 12 TM domains versus seven for GPCRs, and both the N-terminus and C-terminus are located in the cytoplasm of the cell. All three monoamine transporters are about the same size consisting of 617-630 amino acids. They share about 50% overall homology with the greatest homology being in TM domains 1 and 2, and 4-8. This homology can make it difficult to develop selective ligands. The intrinsic dissociation constants of the monoamine neurotransmitters at the dopamine transporter (DAT), the norepinephrine transporter (NET), and the serotonin transporter (SERT) are given in Table 2.1.[43] Of the three, serotonin is unique in that the serotonin transporter is particularly selective and serotonin does not bind strongly to DAT and NET. As we can see from their structures, this can be rationalized as dopamine and norepinephrine differ in structure only by a benzylic hydroxyl group.

2.5   NEUROCIRCUITRY AND NEUROTRANSMITTERS IN ADDICTION

Information flow throughout the CNS occurs via what is called the neurocircuitry of the brain. Another term gaining popularity is "connectome" that is meant to represent the totality of connections between the neurons in the CNS.[44] A complexity of CNS drug discovery is that all mental illnesses appear to involve the disruption of at least one neurocircuit. Even if only one circuit is disrupted, it invariably affects the activities of other neurocircuits. This is certainly true for drug addiction. To

TABLE 2.2   Neurocircuits

Neurocircuitry system	Neurotransmitter	Receptors
Glutaminergic (excitatory)	Glutamic acid	Ionotropic ion channels NMDA, AMPA (Na^+), KA (Na^+) Metabotropic glutamate receptors mGlu1–mGlu8
GABAergic (inhibitory)	γ-Aminobutyric acid	$GABA_A$ and $GABA_B$ ion channels
Cholinergic	Acetylcholine	Muscarinic GPCRs M_1–M_5 Nicotinic Na^+ channels (NAChR)
Adrenergic	Norepinephrine	α- and β-adrenergic receptors Norepinephrine transporter
Dopaminergic	Dopamine	Dopamine receptors D_1–D_5 Dopamine transporter
Serotonergic	Serotonin	Serotonin receptors $5\text{-}HT_1$–$5\text{-}HT_7$ Serotonin transporter

date, no magic target that can be hit with a magic bullet has been discovered. As such, an understanding of the neurocircuitry of the brain is critical to understanding the etiology of a mental illness. The neurocircuits of the brain are defined by the neurotransmitter that is released into the synapse and also by the anatomical location of the neurons. The major neurocircuitry systems are shown in Table 2.2.

We will now look at each major neurocircuitry system in the CNS specifically with regard to addiction. From this, an understanding of approaches toward the treatment of addiction will develop. The goal of this section is to provide a solid neurochemical mechanistic understanding of how addictive drugs work and the identification of potential targets to treat addiction.

2.5.1   Glutaminergic System in Drug Addiction

Glutamate-containing neurons are widely distributed throughout the brain. Important connections relative to addiction are shown in Figure 2.9. Of particular importance to addiction are a bundle of neurons between the cortex and the striatum called the corticostriatal circuit that is modulated by glutamate. From the cortex, two subcircuits have been described: the limbic and motor subcircuits. The limbic subcircuit comprises limbic brain regions such as the prefrontal cortex, the amygdala, the nucleus accumbens, and the VTA. The motor subcircuit connects the motor cortex, the dorsal striatum, and the substantia nigra. The glutaminergic neurons in the nucleus accumbens important in addiction originate from the prefrontal cortex, amygdala, and the hippocampus.

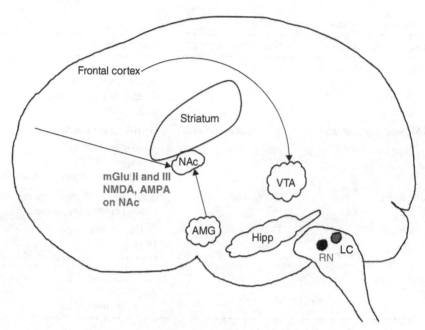

FIGURE 2.9   Glutaminergic system in drug addiction

This model has been described by Peter Kalivas to function as follows:[45]

"In this simplified corticostriatal circuit, the nucleus accumbens serves as a gateway through which information that has been processed in the limbic subcircuit gains access to the motor subcircuit. When a novel stimulus capable of motivating an adaptive behavioral response (for example, food) is encountered, the limbic subcircuit is engaged to attend to and process new and previously learnt information that is relevant to the stimulus. In this way, an animal can determine the adaptive value of implementing a previously learnt behavior and/or forming new behavioral strategies. If an established behavior continues to yield the desired outcome (for example, if pressing a lever successfully provides food), the influence of the limbic subcircuit progressively diminishes, and neuronal activity in the motor subcircuit becomes more organized around task performance. However, if the behavior fails to yield the adaptive outcome (for example, if a lever press no longer delivers food), then the limbic subcircuit is strongly engaged by the altered environmental contingencies and activity in the motor subcircuit becomes more disorganized, which in turn allows modification of the established behavior and maximizes the likelihood of an adaptive outcome. Once a behavior has been updated to a new environmental contingency (for example, if pressing a different lever provides food), over the course of repeated successful outcomes the limbic subcircuit progressively exerts less of an influence as the updated behavior becomes the preferred behavioral response."

Glutamic acid, or more specifically its salt glutamate, is the major excitatory neurotransmitter in the CNS. It is estimated that up to 90% of the neurons in the brain use glutamate as their neurotransmitter and that a significant portion of the ATP utilized by the brain is required for the repolarization of neurons that have been depolarized by glutamate. The brain contains huge amounts of glutamate (about 5–15 mmol/kg wet weight depending on the region), but only a fraction of this glutamate is normally present in extracellular fluid. The concentrations in the extracellular fluid, which represents 13–22% of brain tissue volume, and in the cerebrospinal fluid are normally around 3–4 μM and around 10 μM, respectively. Consequently, the concentration gradient of glutamate across the plasma membranes is several 1000-fold. The highest levels of glutamate are found inside nerve terminals where it is packaged in synaptic vesicles at concentrations of up to 210 mM per vesicle.[46]

Acting as an agonist on its receptors, glutamate induces a cell potential. Glutamate binds to both GPCRs and ion channels. The GPCRs are called metabotropic glutamate receptors of which there are eight known subtypes, designated as $mGlu_1$– $mGlu_8$. The mGlu receptors are all coupled to $G_{i/o}$ G-proteins except for $mGlu_1$ and $mGlu_5$ that are $G_{q/11}$ coupled. Important to addiction are $mGlu_2$, $mGlu_3$, and $mGlu_5$. The mGlu receptors are also activated by the endogenous amino acids: aspartate, serine-O-phosphate, N-acetylaspartylglutamate, and cysteine sulfinic acid. The role of metabotropic glutamate receptors in addiction has been recently reviewed.[47]

mGlu₂ and mGlu₃ Receptors   The $mGlu_2$ receptors are expressed only in neurons while the $mGlu_3$ receptors are thought to be present in both neurons and glia cells; both receptors are located presynaptic and postsynaptic. An important feature of mGlu receptors is that they do not mediate fast excitatory transmission. Rather, they appear to have a modulatory effect on glutaminergic transmission. The mGlu receptors are Class C GPCRs and exist as obligatory homodimers made by the association of two domains consisting of an extracellular ligand-binding region (LBR), described as a bilobate "venus flytrap" or "clamshell" domain, which is associated with the G-protein activating seven-TM domain. In most Class C GPCRs, a cysteine-rich domain consisting of 70 amino acids connects the extracellular domain and the seven-TM domain.

The research team of Jingami and Morikawa has been able to obtain crystal structures of the extracellular LBR of $mGlu_1$, $mGlu_3$, and $mGlu_7$ in both unbound and agonist bound forms.[48,49] The crystal structures reveal two molecules of the LBR (protomers) forming a homodimer related by a pseudo-twofold axis covalently connected by an interprotomer disulfide bridge. Each protomer, MOL1 and MOL2, consists of the amino-terminal and the carboxy-terminal domains forming two lobes that were labeled LB1 and LB2. These domains form a "clamshell"-like shape with glutamate bound in the "hinge" of the clamshell as shown in Figure 2.10 for the $mGlu_1$.

A diagram of full-length $mGlu_1$ receptor is depicted in Figure 2.10(a), where functional regions are boxed. The LB1 and LB2 domains, which constitute a LBR, are colored blue (dark) and red (gray), respectively. Numerical positions of amino-acid residues are indicated according to the primary amino-acid sequence and the open boxes indicate protein regions that were not determined in this study. In Figure 2.10(b), the spatial arrangements of the $mGlu_1$ receptor domains are shown with the MOL1 and MOL2 molecules of the m1-LBR dimer distinguished by dark and light

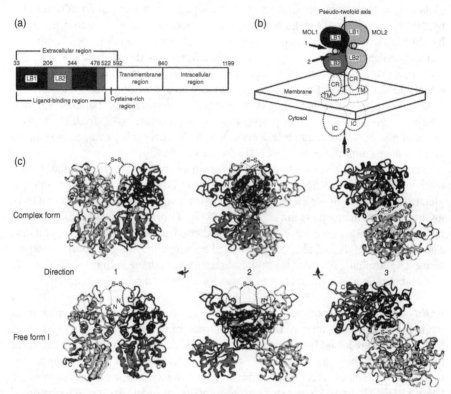

FIGURE 2.10   Crystal structures of the mGluR1 receptor (Adapted with permission from Kunishima et al.[49] Copyright © 2000 Macmillan Publishers Ltd.)

coloring, respectively. The ligand, glutamate, is shown as yellow spheres (e.g., arrow 1) bound in between LB1 and LB2 domains and within the membrane or attached to it are cysteine-rich regions, intracellular regions (IC), and the TM region. Figure 2.10(c) depicts different stereoviews of the dimeric structure when complexed to glutamate and when free. The interprotomer disulfide bridge is clearly visible.

We can see how glutamate binds between the two extracellular lobes. This will cause conformational changes in the lobes resulting, in turn, in conformational changes of the seven-TM domain leading to G-protein binding. Of great interest, then, is how and where drugs would bind and affect the functional activity of the mGlu receptors. In theory, there are two main domains where binding could occur: the extracellular domains or the seven-TM domain. Potentiating the effect of glutamate would generate positive allosteric modulators while attenuating the effect of glutamate would generate negative allosteric modulators.

During addiction, there is a reduction in the concentration of extracellular, or extrasynaptic, glutamate but an increase in synaptic glutamate in the nucleus accumbens. Under homeostatic conditions, synaptic levels of glutamate in the nucleus

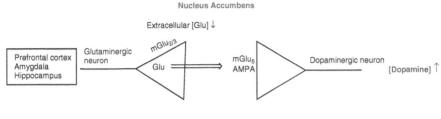

Nucleus Accumbens

Extracellular [Glu] ↓

Addiction decreases levels of extracellular glutamate,
disinhibiting mGlu$_{2/3}$. resulting in increased release of synaptic glutamate

FIGURE 2.11   Glutaminergic synapse during addiction

accumbens are controlled, in part, by binding of extracellular glutamate to presynaptic mGlu$_2$ and mGlu$_3$ receptors. It should be noted that these receptors are not located in the synaptic cleft but more on the exterior, or extrasynaptic, of the neuron distant from the synapse. Agonists binding to presynaptic mGlu receptors block excitatory glutaminergic synaptic transmission resulting in a *decrease* in synaptic glutamate. It is proposed that mGlu binding attenuates voltage-dependent Ca^{2+} channel activity thus inhibiting cell firing.

The mGlu$_2$ and mGlu$_3$ receptors are thus behaving as autoreceptors in a feedback-type mechanism of controlling levels of glutamate. Concentrations of glutamate required for mGlu$_2$ and mGlu$_3$ agonist activity have been measured as EC$_{50}$'s = 4-12 μM with mGlu$_3$ binding showing greater efficacy.[50,51] As homeostatic concentrations of extracellular glutamate are approximately 3-4 μM, mGlu$_3$ will be near full saturation while mGlu$_2$ will be partially saturated allowing the receptors to be constitutively active. This process is detailed in Figure 2.11.

During addiction though, the reduction in homeostatic levels of glutamate reduces the "tone" of the mGlu$_{2/3}$ receptors, they no longer act as a brake on synaptic glutamate release, resulting in a compensatory mechanism that *increases* the release of glutamate into the synapse.

During cocaine and heroin reinstatement (relapse), an increase in synaptic glutamate is observed. This is not only due to the lack of control by mGlu$_{2/3}$ receptors as just discussed, but also to lower levels of cystine-glutamate exchange that occurs on cocaine and nicotine self-administration.[52]

In all, an increase in glutamate levels corresponds with drug reinstatement (relapse). Therefore, restoring glutamate levels to the basal (i.e., homeostatic) levels should help prevent relapse or drug-seeking behavior. This has in fact observed with mGlu$_{2/3}$ agonists such as a drug from Lilly, LY379268.[53] This will be discussed in more detail in Chapter 6.

Beyond treatment of drug addiction, other potential clinical indications for mGluR$_{2/3}$ direct agonists and potentiators are in generalized anxiety disorder, panic disorder, depression, schizophrenia, neuropathic pain, Parkinson's disease, seizure disorders, and stroke and other neurodegenerative disorders.[54]

mGlu₅ Receptor   The $mGlu_5$ receptor is located postsynaptic and concentrations of glutamate required for $mGlu_5$ agonist activity have been measured as $EC_{50} = 11$ µM. Blocking the receptor with an antagonist or deleting the gene producing the receptor has been shown to attenuate cocaine and also ethanol- and nicotine-seeking behavior.[55,56] As will be discussed in Chapter 6, an large amount of effort has gone into the development of negative allosteric modulators of $mGlu_5$ for addiction. Positive allosteric modulators of $mGlu_5$ are being developed for the treatment of schizophrenia.

NMDA, AMPA, and Kainate Ion Channels   Glutamate also binds to postsynaptic ionotropic receptors that consist of the NMDA (*N*-methyl-D-aspartate), AMPA (α-amino-3-hydroxy-5-methyl-4-isoxazoleproprionic acid), kainate (KA), and δ ligand-gated ion channels. The glutamate ion channels have been the subject of a recent extensive review from which the following discussion is taken.[57]

The ion channels are composed of four large subunits that form a central ion pore of a diameter of approximately 0.7-0.8 nm. The preferred IUPHAR names are GluA for AMPA, GluN for NMDA, GluK for kainate, and GluD for the δ ion channels. Several subtypes exist for each. A crystal structure at 3.6 Å resolution is available for the GluA2 subtype bound to an antagonist. The GluA receptor is a tetramer with an overall twofold symmetry perpendicular to the membrane plane; the extracellular amino-terminal domains and the extracellular ligand-binding domains are organized as dimers of dimers, and the ion channel domain exhibits a fourfold symmetry.[58]

Glutamate receptors show limited selectivity for alkali metal cations. In terms of selectivity at a synapse (K^+ vs. Na^+), the differences are minimal with relative permeability's of $P_K/P_{Na} = 1.14$ for NMDA receptors and $P_K/P_{Na} = 1.25$ for AMPA and kainate receptors. Thus, the current carried through most glutamate receptor channels is a mixture of the monovalent cations Na^+ and K^+ plus the divalent Ca^{2+}. The kainate receptor is less permeable to Ca^{2+} than are AMPA and NMDA receptors.

Modulation of these ion channels has not received a lot of attention in addiction medication development. As will be discussed, the NMDA antagonist memantine that is approved for Alzheimer's disease and the multimodal drug topiramate ($GABA_A$ agonist/AMPA antagonist) have been studied for treatment in addiction. Dysfunction of the NMDA and AMPA receptors has been linked to major depressive disorder and bipolar disorder.[59]

Glutamate Regulation   As mentioned earlier, neuronal cells are constantly awash in extracellular, non-synaptic glutamate. This extracellular glutamate appears to be produced mainly by glia cells. Levels of extracellular glutamate are modulated via their uptake by excitatory amino-acid transporters (EAAT) located on glial cells and neurons. Five EAAT subtypes have been identified with EAAT1 and EAAT2 expressed predominantly in glial cells and astrocytes and EAAT3 and EAAT4 expressed almost exclusively on neurons. In terms of glutamate uptake, the EAAT1 and EAAT2 are particularly important with EAAT2 expressed throughout the brain and have been estimated to be responsible for 90% of the glutamate uptake from the extracellular fluid into the glia cells.[60] Glutamate has moderate binding affinity to EAAT1-3 with

K_m's of 22, 25, and 42 µM at EAAT1, EEAT2, and EAAT3, respectively.[61] The EAATs are very stereoselective for the transport of L-(S)-glutamate with K_m's for D-glutamate > 1000 µM.

Glutamate levels in the extracellular space are also regulated by a cystine-glutamate exchange mechanism present in glia cells. Cystine is the disulfide of cysteine. This mechanism has been less examined but may prove to be important. Glia cells uptake extracellular cystine by a protein called a cystine-glutamate antiporter or system x_c^- and exchanges it for intracellular glutamate in a 1:1 stoichiometry. This exchange of the two amino acids occurs across a concentration gradient, as the intracellular concentration of glutamate is well in excess of cystine in the neuron.

Cocaine and nicotine are known to reduce cystine-glutamate exchange resulting in reduced concentrations of extracellular glutamate in the nucleus accumbens. How this occurs is not clear but repeated cocaine use does increase the K_m of [^{35}S]cystine uptake as measured in nucleus accumbens tissue slices from 2.1 µM in saline-treated rats to 4.2 µM with no noticeable change in V_{max}.[62] We will return to this topic in Chapter 6 in the discussion of *N*-acetylcysteine for the treatment of cocaine addiction.

Looking postsynaptic, levels of the ion channel glutamate receptor AMPA are increased following administration of cocaine.[63] Glutamate acts as an agonist on binding to AMPA receptors, so AMPA antagonists, by blocking glutamate activity, could attenuate cocaine or heroin reinstatement.

While the above discussion has focused mainly on the role of glutamate in the nucleus accumbens, there is evidence that glutaminergic afferents from the prefrontal cortex and other regions into the VTA are important in the addiction process. The importance of the VTA appears secondary to the nucleus accumbens with regard to glutamate and addiction.[52] In the VTA, AMPA ionotropic receptors appear to play a major role. The increase in dopamine levels in the VTA is linked to the release of glutamate. A complex mechanism has been proposed where stress causes the release of corticotropin-releasing factor (CRF) in the VTA, which in turn causes the local release of glutamate that then activates the dopaminergic system in the VTA increasing dopamine levels.[64,65] The VTA studies were conducted under cocaine-seeking conditions with rodents, whether this holds true for other drugs of addiction remains to be seen. In fact, it has been shown that the anesthetic propofol (Diprivan) at concentrations as low as 1 nM increases glutamate transmission in the VTA leading to the increased release of dopamine in the VTA.[66] This would occur by glutamate binding to glutamate receptors on dopaminergic neurons causing an action potential to develop resulting in dopamine release. It is suggested that second-hand exposure to anesthetics such as propofol may contribute to the unusually high drug abuse rate of anesthesiologists relative to other physicians.

2.5.2   GABAergic System in Drug Addiction

Contra to glutamate is the neurotransmitter GABA. GABA is the major inhibitory neurotransmitter in the CNS and exerts its action via the ionotropic $GABA_A$ and the metabotropic $GABA_B$ receptors. GABA neurons are widely distributed throughout

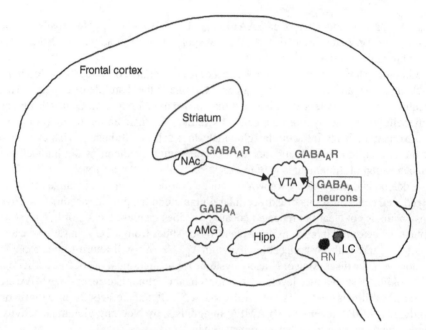

FIGURE 2.12   GABAergic circuitry in addiction

the brain including the striatum and the cortex. Nucleus accumbens neurons use GABA as their main neurotransmitter and GABAergic neurons also originate from the VTA. Important connections relative to addiction are shown in Figure 2.12.

Eugene Roberts at Washington University in St. Louis, MO, first discovered the presence of GABA in the brain in 1949. While looking at the components of mouse brain extracts by the novel method of 2D paper chromatography, a new compound was observed that reacted with ninhydrin signifying it contained amino groups. This compound was isolated from the paper chromatograms and was shown to be GABA.[67] GABA and glutamate have a symbiotic relationship as GABA is biosynthetically generated from glutamate via the enzyme glutamic acid decarboxylase. GABA is removed from the synapse by the transporters SCL1A1 (EAAT3) located in presynaptic neurons and SLC1A2 (EAAT2) and SLC1A3 (EAAT1), both located on glia cells. In glial cells, GABA is metabolized to succinate that then enters the Kreps Cycle forming glutamine. The glutamine is transferred to the presynaptic neuron by System A-like and System N-like transporters, where it is decarboxylated to form glutamate.[35] This is only of direct significance in that one of the enzymes involved, GABA aminotransferase, is a target for the treatment of cocaine addiction (see Chapter 6).

GABA$_A$ Ion Channel   The GABA$_A$ chloride ion channel receptor exists as a pentamer of four TM subunits held together by intermolecular forces. In all, there are eight different main subunits with a total of 19 variations (e.g., there are six α sub-units) giving a possibility of 11,628 random combinations of pentamers! To date, three pentamers have been shown to be of particular importance in addiction. All three contain two α-, two β-, and one γ subunits. Benzodiazepines acting on a pen-tameric GABA$_A$ receptor containing the α1βγ subunit produce sedation while acting on a pentamer composed of the α2βγ subunits results in anxiolytic properties. The pentamer mediating addiction is the same as for sedation making the development of a nonaddictive benzodiazepine sedative very difficult.[68]

Depressant drugs such as alcohol, benzodiazepines, and barbiturates bind to GABA$_A$ at sites on the ion channel separate from where GABA itself binds. The abuse of alcohol and benzodiazepines then is of special significance with GABA. Binding of GABA to GABA$_A$ chloride ion channels present on a neuron causes an influx of negatively charged chloride ions into the neuron, the neuron becomes more negatively charged and the membrane potential is increased. The neuron becomes hyperpolar-ized; hyperpolarized neurons do not fire an action potential thus neurotransmitters will not be secreted at the axonal terminus. GABA can modulate dopamine in the following fashion. When GABAergic neurons innervate the VTA and synapse with dopaminergic neurons, the secreted GABA can interact with postsynaptic GABA$_A$ ion channels located on the surface of the dopamine neurons. This action causes the dopamine neurons not to fire, reducing the levels of synaptic dopamine released into the nucleus accumbens. If the levels of GABA are decreased, GABA can no longer function as an inhibitor of dopamine release so dopamine levels in the synapse will increase. This is what can occur during addiction as shown in Figure 2.13.

GABA$_B$ Receptor   GABA$_B$ receptors are Class C GPCRs and are coupled to G$_{i/o}$ G-proteins. They are similar in structure to the mGlu receptors and are formed from heterodimerization of the GABA$_{B1}$ and GABA$_{B2}$ seven TM subunits. GABA$_B$ recep-tors are located both presynaptic and postsynaptic on neurons throughout the brain.

The GABA$_{B1}$ subunit contains the GABA orthosteric binding site on the extracel-lular venus flytrap domain, whereas the GABA$_{B2}$ subunit mediates G-protein-coupled signaling. The two subunits interact by direct allosteric coupling such that GABA$_{B2}$ increases the affinity of GABA$_{B1}$ for agonists and GABA$_{B1}$ facilitates the coupling

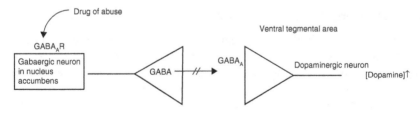

During addiction, binding of drugs to GABA$_A$ ion channels hyperpolarizes the cell, resulting in decreased levels of synaptic GABA. This disinhibits dopaminergic neurons, resulting in increased levels of dopamine

FIGURE 2.13   GABAergic synapse during addiction

of $GABA_{B2}$ to G-proteins.[42] Via the $\beta\gamma$ subunit of the $G_{i/o}$ G-protein, the $GABA_B$ receptors modulate a variety of signal transduction pathways such as attenuating the activity of Ca^{2+} channels and potentiating the activity of inwardly rectifying potassium channels ($K_{ir}3$).

Activation of $GABA_B$ receptors located on dopamine neurons also appears to exert an inhibitory effect on the neuron reducing the release of dopamine. By inhibiting the opening of Ca^{2+} channels, $GABA_B$ agonists will also hyperpolarize a neuron preventing an action potential.

GABAergic synaptic plasticity is well documented during both acute and chronic drug use.[69] GABAergic circuitry important for addiction originates in the nucleus accumbens sending efferent extensions to the VTA. There are also GABAergic interneurons in the VTA, these originate and end within the VTA. As mentioned above, GABA interacting on dopaminergic neurons in the VTA modulates the release of dopamine in the nucleus accumbens. Recent findings in fact suggest that GABA neurons in the VTA may potentiate dopamine neuron activity under conditions when reward expectation is high.[70] By some unknown mechanism, the GABA neurons somehow "determine" the value of a reward and control dopamine firing accordingly.

Benzodiazepines are the stereotypical drug that binds to $GABA_A$. When benzodiazepines bind to a $GABA_A$ receptor on a GABAergic neuron in the VTA they act as positive allosteric modulators, the neuron becomes hyperpolarized, and the amount of GABA that is released in the VTA is decreased. As GABA inhibits dopamine release, a decrease in synaptic GABA concentration results in an *increase* in dopamine release into the nucleus accumbens. We can say that the benzodiazepine *disinhibits* dopamine release. Opioids achieve the same effect by binding to μ-opioid receptors located on the GABA interneuron circuit in the VTA.

The importance of the $GABA_B$ receptor in addiction is validated by the use of the prototypic agonist baclofen in reducing the intake of alcohol, cocaine, and nicotine in humans.[71] Dysfunction of the GABAergic system also occurs in mood disorders. Lower levels of GABA have been measured in patients suffering from depression.[72] Hence, benzodiazepines have been used for the treatment of depression. Acting as a positive allosteric modulator, they can compensate for the lower levels of GABA by increasing the effect of GABA on the $GABA_A$ receptor.

2.5.3   Cholinergic System in Drug Addiction

Cholinergic neurons originate from clusters of cells in the brainstem, which have been labeled Ch1-Ch8 (Figure 2.14). Clusters Ch1-Ch4 project to the parts of the hippocampus, cortex, thalamus, and amygdala, where they modulate attention, learning, and memory. Ch5 and Ch6 neurons project to the thalamus, VTA, and the substantia nigra, where they modulate arousal, sleep, and control dopamine levels. These cholinergic cell clusters also contain GABAergic and glutaminergic neurons that project to the above-mentioned areas. There are also local circuits in the caudate nucleus and putamen components of the striatum.[73] The cholinergic neurotransmitter

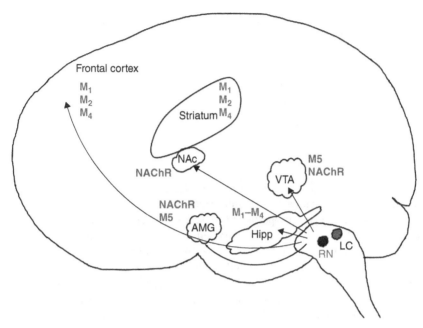

FIGURE 2.14   Cholinergic circuitry in addiction

acetylcholine binds to the muscarinic (mAChR) GPCRs and to the nicotinic (nAChR) ion channels. The physiological action of acetylcholine is different at central and peripheral endings. In the CNS the muscarinic receptors predominate, while in the peripheral nervous system nicotinic ion channels predominate.[31]

Acetylcholine is a simple molecule and was first identified as important in CNS function in 1907. Since the discovery of acetylcholine, a rich and extensive understanding of the medicinal chemistry and pharmacology of the cholinergic system was developed. Acetylcholine is synthesized in the presynaptic terminus by the enzyme choline acetyltransferase that catalyzes the transfer of the acyl group of acetyl CoA to choline. Once released into the synapse acetylcholine is removed by acetylcholine esterase. This extensively studied enzyme hydrolyzes acetylcholine back to choline that is then transported back into the neuron for resynthesis to acetylcholine. Inhibitors of acetylcholine esterase such as donepezil (Aricept) have found use in the treatment of Alzheimer's disease but not, so far, in addiction. A case study on the discovery and development of donepezil is available for the interested reader.[74]

There are five muscarinic receptors (M_1–M_5) that have been at times divided into two families based on function: M_2/M_4 and $M_1/M_3/M_5$. Modulation of the cholinergic system has been shown to be important in addiction. However, it is still equivocal if agonism or antagonism of the cholinergic system is required. The exact role of acetylcholine and the cholinergic system in addiction is an area of active research. The muscarinic receptors are expressed throughout the brain. With respect to muscarinic expression levels only and from the viewpoint of anatomy the distribution is as follows:[73]

- VTA and substantia nigra contain mostly M_5 expressed on dopaminergic neurons
- Hippocampus contains M_1–M_4
- Cortex is mostly M_1, M_2, and M_4 with $M_2 > M_4$
- Striatum is mostly M_1, M_2, and M_4 with $M_4 > M_2$
- Cerebellum is mostly M_2 and M_1

The physical and pharmacological properties of the muscarinic receptors have been extensively explored. Haga has written a comprehensive review using first-hand accounts on the molecular properties of muscarinic receptors.[75] The receptors have been solubilized and purified to homogeneity using affinity chromatography. The pharmacology has been studied using the nonselective but high-affinity ligands antagonists quinuclidinyl benzylate ([^{3}H]QNB) and N-methylscopolamine ([^{3}H]NMS) with K_d of 15–80 pM and 50–700 pM, respectively. After a 10-year period of extensive work, Haga in collaboration with Kruse was able to obtain a crystal structure of the M_2 receptor complexed to QNB with a resolution of 3 Å.[76] Different representations of the receptor structure are shown in Figure 2.15.

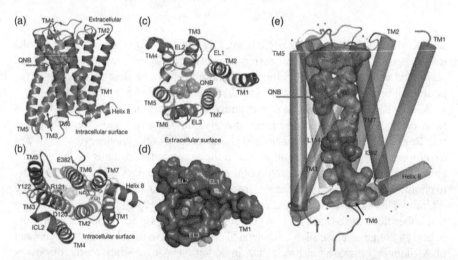

FIGURE 2.15 Structure of M_2 receptor. (a) M_2 receptor in profile bound to the antagonist QNB. (b) Cytoplasmic surface showing conserved DRY (aspartic acid:arginine:tyrosine) residues in TM3. (c) Extracellular view into QNB binding pocket. (d) Extracellular view with solvent-accessible surface rendering shows a funnel-shaped vestibule and a nearly buried QNB binding pocket. (e) An aqueous channel depicted as a solvent accessible surface model (green in color figure), extending from the extracellular surface into the transmembrane core, is interrupted by a layer of three hydrophobic residues (blue spheres in color figure) located at isoleucine I392 and leucine L114. Well-ordered water molecules are shown as red dots (Reprinted with permission from Haga et al.[76] Copyright © 2012 Macmillan Publishers Ltd.)

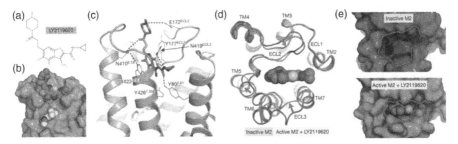

FIGURE 2.16 M_2 receptor complexed with allosteric and orthosteric ligands. (a) The M_2 receptor occupied by the orthosteric agonist iperoxo was crystallized in complex with the positive allosteric modulator LY2119620. (b) The allosteric ligand binds to the extracellular vestibule just above the orthosteric agonist. A cross section through the membrane plane shows the relative positions of the two ligands with the positive allosteric modulator LY2119620 binding above the orthosteric agonist iperoxo. (c) Several polar contacts are involved in LY2119620 binding, in addition to extensive aromatic stacking interactions with Trp422[7.35] and Tyr177[ECL2]. (d) Upon activation, the M_2 receptor undergoes substantial conformational changes in the extracellular surface, leading to a contraction of the extracellular vestibule. In particular, as depicted by the arrows, ECL3, TM6, and TM5 undergo a significant contraction. (e) This creates a binding site that fits tightly around the allosteric modulator, which would otherwise be unable to interact extensively with the extracellular vestibule in the inactive receptor conformation (Reprinted with permission from Kruse et al.[77] Copyright © 2013 Macmillan Publishers Ltd.)

In a most remarkable achievement, the crystal structure of the M_2 receptor complexed with both an allosteric ligand and orthosteric ligand was just published.[77] We can see in Figure 2.16 that both the allosteric and orthosteric ligands bind to the M_2 receptor in close proximity to each other.

The same group also succeeded in obtaining the crystal structure of the M_3 receptor crystallized in complex with tiotropium (Spiriva), a potent muscarinic inverse agonist used clinically for the treatment of chronic obstructive pulmonary disease.[78]

It has proven difficult to find truly selective orthosteric agonists at M_1–M_5. It is suggested that discovering allosteric agonists or positive allosteric modulators will prove more fruitful.[79] These crystal structures should be an invaluable asset for future drug discovery.

M_2 and M_4 Receptors   The muscarinic M_2 and M_4 receptors are coupled to $G_{i/o}$ proteins; agonist binding will inhibit adenylyl cyclase reducing cellular concentrations of the second messenger, cAMP. Agonism of M_2 in postsynaptic neurons inhibits neuron firing by the $\beta\gamma$ subunit of the G_i protein activating $K_{ir}3$. When M_2 or M_4 are located presynaptic, agonism can result in presynaptic inhibition due to the $\beta\gamma$ subunit of the G_o protein binding to voltage-gated Ca^{2+} channels and attenuating their activity.[80]

M_1, M_3, and M_5 Receptors   The M_1, M_3, and M_5 receptors are coupled to $G_{q/11}$ proteins. Agonist binding to these receptors will activate phosphoinositide-specific phospholipase C (PI-PLC), resulting in increased concentration of intracellular Ca^{2+} and neuron depolarization. The M_5 receptor is localized in the VTA and $M_5^{-/-}$ knockout mice had significant reductions in the self-administration of cocaine and morphine, which makes the M_5 receptor an attractive target for antagonist development.[81,82] Muscarinic M_1 and M_4 receptors are currently of most interest in stimulant addiction. Note that simultaneous agonism of M_1 and M_4 will activate different second messenger systems. This will be discussed more in Chapter 6.

nAChR Ion Channel   The nicotinic acetylcholine receptor is most associated with nicotine addiction so it will be discussed in more detail in Chapter 8. The nAChR ion channel consists of pentameric units, for example, the $\alpha 7$ nAChR consists of five $\alpha 7$ subunits with each subunit containing a binding site for acetylcholine. Modulation of nAChR is of great interest for the treatment of nicotine addiction and the study of this led to the discovery of varenicline (Chantix).

Drugs of abuse are known to modulate acetylcholine levels in the CNS. Remifentanil (μ-opioid agonist) at 0.032 mg/kg (IV) and cocaine at 0.33 mg/kg (IV) increased acetylcholine levels by 2.5-fold (cocaine) to fivefold (remifentanil) in the nucleus accumbens during the acquisition of drug reinforcement. Activation of both muscarinic and nicotinic receptors is required.[83]

Imaging studies have revealed that the cholinergic system differs in areas of the limbic system of cocaine- or nicotine-addicted individuals relative to drug abstinent individuals. The effects of the muscarinic and nicotinic cholinergic agonist physostigmine, the muscarinic antagonist scopolamine, and saline were investigated in 23 individuals. Regional cerebral blood flow after each infusion was determined using single photon emission computed tomography and a ^{99}Tc radiotracer.

Cocaine-addicted subjects had different blood flow patterns relative to the control group in discrete limbic regions such as the hippocampus, amygdala, and the insula that are important in the learning and memory of drug cues and associated craving. Disruptions in the posterior hippocampus were thought to be of particular importance. Changes in both nAChR and mAChR systems were observed.[84] The importance of these neuronal changes for the treatment of drug addiction remains to be seen but it does clearly show that chronic drug use causes changes to the cholinergic system.

2.5.4   Dopaminergic System in Drug Addiction

Dopamine has a profound effect on human behavior ranging from depression to euphoria. Two main dopaminergic circuits exist (Figure 2.17). The first originates from the substantia nigra and projects to the striatum (nigrostriatal tract) and is most closely associated with Parkinson's disease. The second pathway is the mesolimbic tract that originates in the VTA and projects neurons to many parts of the brain. This latter pathway is of most concern in addiction. Increased levels of synaptic dopamine are implicated in all drugs of addiction. Dopamine neurons are highly concentrated in the VTA with about 60-65% of the rat VTA being composed of dopamine neurons.[85]

As discussed in section 2.1, the presence of dopamine in the brain was reported in 1958. Dopamine is a catecholamine along with norepinephrine and epinephrine.

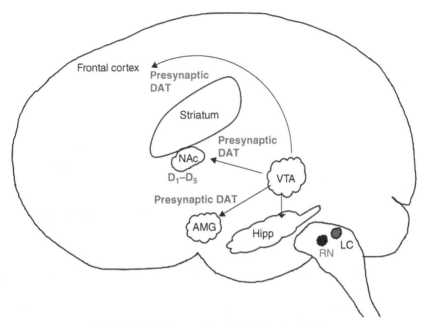

FIGURE 2.17   Dopaminergic circuitry in addiction

Dopamine and norepinephrine are considered neurotransmitters while epinephrine is a hormone and is located mainly outside the CNS. All are synthesized from tyrosine by the sequence

L-tyrosine → L-DOPA → Dopamine → Norepinephrine → Epinephrine

L-DOPA is of course significant as it is given in Parkinson's disease as an orally bioavailable prodrug of dopamine.

Dopamine is stored in presynaptic vesicles in dopaminergic neurons and upon release into the synapse is removed by the DAT. Within the presynaptic terminus levels of dopamine and the other catecholamines are controlled in part by monoamine oxidases. Phenethylamines are substrates for monoamine oxidases and are converted into aryl acetaldehydes by oxidative deamination. One medicinal chemical method of modulating dopamine levels has been by using monoamine oxidase inhibitors.

Dopamine when released into the synapse can bind to five different dopamine receptors, D_{1-5}. They are divided into D_1-like receptors that include D_1 and D_5 and D_2-like receptors that include D_2, D_3, and D_4. The literature can be very confusing with regard to this. One needs to be careful to recognize if an author is reporting on a D_2 receptor or is reporting on results from D_2-like receptors. In the latter case, any medicinal chemistry and pharmacological effects could be due to anywhere from one to three receptors. Dopamine displays modest binding to all five receptors with strongest binding to D_3 (K_i = 2300, 2000, 30, 450, and 230 nM for D_1, D_2, D_3, D_4, and D_5, respectively).[86] A detailed review by Beaulieu and Gainetdinov on the physiology, signaling, and pharmacology of dopamine receptors presents excellent descriptions of the signaling pathways for the dopamine receptors.[87]

Though dopamine signaling is of immense interest in addiction, direct modulation of the dopaminergic system has not proven fruitful for addiction medication development.

The D_1 and D_2 dopamine receptors seem to be critical for learning and memory mechanisms, such as working memory, that are mediated primarily by the prefrontal cortex. The D_3, D_4, and, potentially, D_5 dopamine receptors seem to have a minor modulatory influence on some specific aspects of cognitive functions that are mediated by the hippocampus. D_2 antagonism is critical for antipsychotic medication so modulation of D_2 clearly has psychopharmacological effects though full antagonism with slow dissociation leads to extrapyramidal effects. The D_3 receptor has received the most attention in addiction medication development.[88]

D_1 Receptor   The D_1 receptor is located postsynaptic and is coupled to a G_s protein. Agonist binding activates adenylyl cyclase that hydrolyzes ATP to cAMP with subsequent activation of PKA. D_1 activation by agonists plays an important role stimulating motor control.

D_3 Receptor   The D_3 receptor is localized in the reward system within the VTA, nucleus accumbens, amygdala, and prefrontal cortex. The D_3 receptor is located on both pre- and postsynaptic neurons and is coupled to a $G_{i/o}$ protein. Agonist binding to the receptor located postsynaptic results in inhibition of the enzyme adenylyl cyclase reducing cellular concentrations of the second messenger, cAMP. Presynaptic D_3 receptors on the terminus of the dopaminergic neuron function as an autoreceptor and agonist binding will cause a decrease in the tonic level of dopamine in the synapse.

Design of D_3 selective ligands has been aided by the determination of the crystal structure of the D_3 receptor complexed to the D_2/D_3 antagonist eticlopride at a resolution of 3.15 Å.[89] In the absence of a crystal structure of a receptor complexed with an agonist, different biochemical and pharmacological tools can be used. An example is that from the group of Kuzhikandathil where they were interested in locating sites on the D_3 receptor responsible for tolerance. Tolerance in this case is the decrease in receptor signaling function upon repeated stimulation by an agonist. To locate areas of importance on the receptor for agonist binding, chimeric receptors were prepared along with site-directed mutagenesis and the effects on functional activity were measured. Functional activity was determined using whole cell voltage clamp recording of K_{ir} (GIRK) activity.

It was determined that the D_3 receptor intracellular loop 2 (IL2), and in particular, the cysteine amino acid at position 147 (C147, Cys147) was important for the D_3 receptor tolerance property. In the studies using chimeric D_3-D_2 receptors, the substitution of D_3 receptor intracellular loops with D_2 receptor intracellular loops prevented the induction of the tolerance property. The opposite had no effect though,

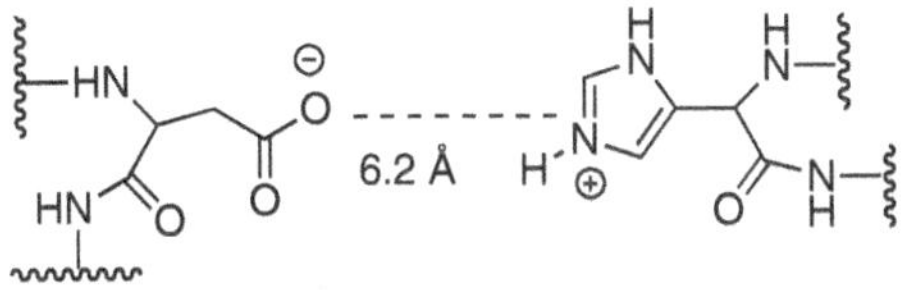

FIGURE 2.18   Molecular interactions in tolerance

reciprocal chimeras in which D_2 receptor intracellular loops were substituted with D_3 receptor intracellular loops failed to show tolerance. A further study showed that the extracellular loop 2 (EC2), and in that Asp187, was also important for tolerance. Molecular dynamic experiments suggested that a histidine at position 354 in EC3 interacted with the aspartic acid D187 via a salt bridge (Figure 2.18).

Indeed, for when histidine H354 was mutated to a nonpolar leucine, the agonist-induced GIRK response was absent and no tolerance was observed.[90] If correct, this and the IC2 interaction represent key molecular interactions required for the development of tolerance.

In studies using the selective D_3 antagonist SB-277011-A (D_3 K_i = 11 nM, D_2:D_3 = 120), it was shown that antagonism of D_3 can attenuate the effects of addictive drugs, in particular, cocaine and alcohol.[88] The mechanism of action is unclear but Heidbreder and Newman have proposed that based on animal models of addiction, selective D_3 receptor antagonists might directly modulate D_3 receptors located at critical junctures by three main pathways: (1) stress-induced reinstatement of drug-seeking behavior in the nucleus accumbens shell; (2) stimulus–reward associations mediating cue-triggered reinstatement of drug seeking in the basolateral amygdala, and (3) working memory and attentional processes in the mPFC.[91]

A positron emission tomography (PET) study using the D_2 radioligand [^{11}C]raclopride showed that even on the presentation of visual cues (people smoking cocaine) there is an increase in dopamine levels in the dorsal striatum of cocaine-addicted subjects.[92] Non-drug-induced cues then can increase dopamine levels in key parts of the brain implicated in relapse.

The use of PET imaging will be discussed later in Chapter 4, but another study can be mentioned that illustrates the importance of the dopamine system in addiction. PET imaging methods were used to examine the effects of cocaine abuse in non-human primates on D_2-like (D_2, D_3, and D_4) receptors.[93] There appeared to be an inverse relationship between D_2-like receptor availability and vulnerability to the reinforcing effects of cocaine. It was found that environmental variables such as the formation of social hierarchies could increase or decrease D_2 receptor binding. For example, dominant and subordinate monkeys were housed both individually and in a social setting. The distribution volume ratio (B_{max}/K_d) of [^{18}F]fluoroclebopride had a dramatic increase of 22% when the dominant monkeys were moved from individual housing to social housing. The distribution volume ratio in the subordinate monkeys remained the same. This was ascribed to either an increase in D_2-like receptors in the dominant monkeys and/or reduced levels of extracellular dopamine. Under acute cocaine self-administration protocols, the dominant monkeys had lower levels of cocaine self-administration than the subordinate monkeys. However, under chronic exposure, there was little difference. This suggests that changes in D_2-like receptor function appear to influence the vulnerability to abuse cocaine. Perhaps higher levels of D_2-like receptors could *reduce* the vulnerability of an individual to cocaine addiction during the social drug-taking phase of addiction. During maintenance, chronic cocaine exposure produced a decrease in D_2-like receptor binding, which may be a mechanism that contributes to continued drug use. Finally, during abstinence there are individual differences in rates of recovery of D_2 receptor availability.

DAT   The DAT is located on the presynaptic terminus of neurons and is responsible for removing dopamine from the synapse. The main site of action for cocaine is at DAT where it binds, blocking the transporter, and preventing the reuptake of dopamine. As a consequence, a large amount of synaptic dopamine is present. As one might expect an immense amount of work has been done on developing compounds that can modulate DAT activity, hence cocaine abuse. In short, it has not proven successful, at least in terms of medication development.

2.5.5   Adrenergic System in Drug Addiction

Adrenergic neurons originate in areas of the locus ceruleus and the midbrain and ascend to the amygdala, hypothalamus, thalamus, cortex, and hippocampus (Figure 2.19). The neurotransmitter norepinephrine (i.e., noradrenaline) acts upon the

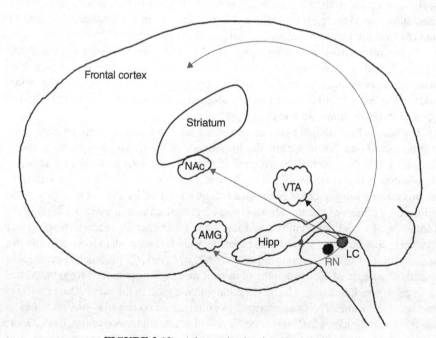

FIGURE 2.19   Adrenergic circuitry in addiction

GPCR α-adrenergic receptor (adrenoreceptor) and β-adrenergic receptor and on the NET to affect changes in the adrenergic system. Norepinephrine has been implicated as playing a role in selective attention, general arousal, and reaction to stress. The interaction of norepinephrine with drug addiction has been reviewed and the majority of comments below come from that review.[94]

The exact role of norepinephrine in addiction is still an open question but the interaction of norepinephrine with the α_1- and α_2-adrenoreceptors appears to be of importance in addiction. They are located presynaptic where they act as autoreceptors controlling the release of norepinephrine into the synapse. Agonists binding to the α_2-adrenergic autoreceptor will inhibit the release of norepinephrine while antagonist binding will increase concentrations of norepinephrine in the synapse. Norepinephrine does not appear to be a primary neurotransmitter in addiction but instead can modulate the levels of dopamine in different parts of the brain. It does, however, seem to be an important contributor in the self-administration and reinstatement of psychostimulants and for reinstatement of opioids and alcohol.

How norepinephrine modulates the synaptic concentration of dopamine is complex. One main pathway is through norepinephrine binding to α_1-adrenoreceptors imbedded in glutaminergic neurons in the prefrontal cortex. Binding appears to activate glutaminergic neurons that in turn innervate the VTA. Release of glutamate in the VTA causes in turn dopamine to be released into the synapse. The presence of norepinephrine acting on α_2-adrenoreceptors in the VTA, however, attenuates the firing of dopamine neurons reducing the levels of dopamine in the synapse.[95] Broadly speaking, antagonism of α_1-adrenoreceptors and agonism of α_2-adrenoreceptors will help decrease dopamine release in the VTA.

2.5.6 Serotonergic System in Drug Addiction

Though serotonin plays a critical role in depression and psychosis, its role in addiction appears to be less defined. As illustrated in Figure 2.20, serotonin neurons originate from the midline section of the brain stem, mainly from a group of cells called the raphe nuclei, and interact with reward system cells in the amygdala, striatum, VTA, hippocampus, hypothalamus, and neocortex. There are considered to be nine serotonin-containing cell bodies designated as B_1-B_9. The isolation, structural elucidation, and detection of serotonin in the brain occurred in the late 1940s and early 1950s. Serotonin is synthesized *in vivo* from tryptophan and is widely distributed throughout the body. The team of Nichols and Nichols has written a detailed review of the serotonin receptors.[96]

The serotonin system is quite diverse composed of the GPCRs $5\text{-HT}_{1A,B,D,F}$, $5\text{-HT}_{2A,B,C}$, 5-HT_4, 5-HT_5, 5-HT_6, and 5-HT_7, the ligand-gated ion channels $5\text{-HT}_3\text{A-E}$, and the serotonin transporter SERT. Several of the receptors have been implicated in addiction. The 5-HT_{2A} receptor may be the most infamous of the serotonin family as it is the target of the psychedelic drugs lysergic acid diethylamide (LSD), mescaline, and psilocybin.[97] Therefore, any drug possessing agonist activity at 5-HT_{2A} would be expected to have some level of psychosis as a potential side effect. The psychosis mimics the positive symptoms of schizophrenia such as hallucinations

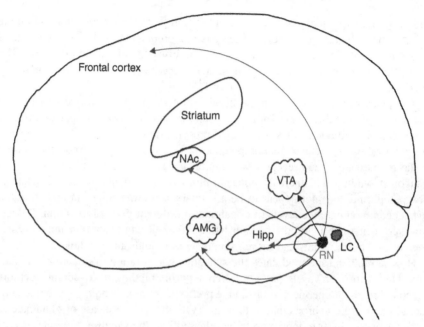

FIGURE 2.20   Serotonergic circuitry in addiction

and thought disorder. Likewise, antipsychotic drugs such as clozapine have potent 5-HT$_{2A}$ receptor antagonist activity.

The serotonergic system modulates feelings of emotion, cognition, and some motor functions. Its role in depression and anxiety is well documented with a growing interest in the role of serotonin in schizophrenia and Parkinson's disease. In these disorders, the 5-HT$_{1A}$R plays a major role.[98] A hypofunctional serotonergic system is also important in bipolar disorder. With respect to addiction, the 5-HT$_{1A}$, 5-HT$_{2A}$, 5-HT$_{2C}$ receptors, the 5-HT$_3$ ion channel, and SERT have been discovered to be important. The abuse of cocaine and alcohol has been most closely linked with the serotonin system; ethanol in fact is an agonist of the 5-HT$_3$ receptor.

5-HT$_{1A}$ Receptor   The 5-HT$_{1A}$ receptor is coupled to a G$_{i/o}$ protein and is located both presynaptic and postsynaptic. A large concentration of 5-HT$_{1A}$ receptors exists postsynaptic in areas of the limbic system while the presynaptic receptors are concentrated on neurons in the raphe nuclei. Agonists cause hyperpolarization of the postsynaptic neuron by opening G-protein-gated inwardly rectifying potassium channels causing an efflux of K$^+$ ions from the cytoplasm leading to reduced firing rates. Presumably, this would be achieved by the same mechanism as we discussed for the D$_3$ receptor. Cocaine and amphetamine indirectly attenuate cell firing by causing an increase in serotonin levels; serotonin then binds to the 5-HT$_{1A}$ receptor. If the 5-HT$_{1A}$ receptor is located presynaptic it acts as an autoreceptor; serotonin binding

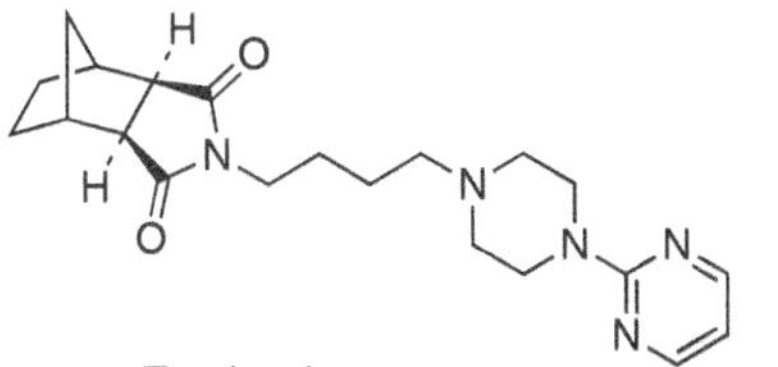

Tandospirone
5-HT$_{1A}$ K_i = 27 nM

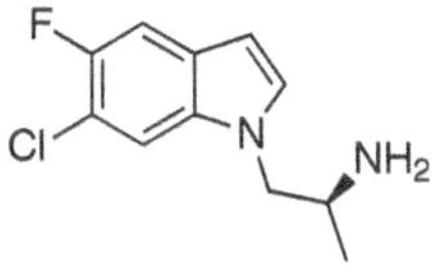

Ro 60-0175
5-HT$_{2C}$ K_i = 1 nM

FIGURE 2.21   Structures of 5-HT ligands

inhibits the release of serotonin into the synapse. Tandospirone (Figure 2.21) is a 5-HT$_{1A}$ receptor partial agonist that has been studied in Parkinson's disease models in rats. If we take this as a guide, 5-HT$_{1A}$ agonists may be exerting their effects through a nondopaminergic mechanism.[99] It is clear though that activation of 5-HT$_{1A}$ receptors by agonists increases dopamine levels and decreases serotonin levels in the frontal cortex of mice though the agonist effect on serotonin levels is more sensitive than on dopamine levels.[100]

5-HT$_{2A}$ Receptor   The 5-HT$_{2A}$ receptor is coupled to a G$_{q/11}$ protein and tends to be localized on postsynaptic dendritic termini though some are located presynaptic on axons. Antagonists of the 5-HT$_{2A}$ receptor can affect cocaine-seeking behavior. Infusion of a 5-HT$_{2A}$ antagonist directly into the medial prefrontal area of the cortex dose-dependently attenuated cocaine-induced reinstatement of cocaine-seeking behavior by rodents. The antagonist though had little effect on the self-administration of cocaine. This suggests that 5-HT$_{2A}$ antagonism is affecting cue-induced affective behaviors such as motivation and incentive.[101]

5-HT$_{2C}$ Receptor   The 5-HT$_{2C}$ receptor is also coupled to a G$_{q/11}$ protein and modulates dopamine levels in the mesolimbic system; the amygdala has a high concentration of 5-HT$_{2C}$ receptors. Agonist activation of the 5-HT$_{2C}$ receptor has an inhibitory effect on dopamine transmission and agonist activation of 5-HT$_{2C}$ can attenuate many of the subjective effects of cocaine. Administration of the 5-HT$_{2C}$ agonist Ro 60-0175 (Figure 2.21) to squirrel monkeys attenuated cocaine-induced reinstatement of cocaine use and also reduced the levels of dopamine in the nucleus accumbens when administered before a cocaine injection.[102]

As with the 5-HT$_{2A}$R study above, infusion of a 5-HT$_{2C}$ agonist directly into the medial prefrontal area of the cortex dose-dependently attenuated cocaine-induced reinstatement of cocaine-seeking behavior by rodents. The attenuation could be blocked by the administration of a 5-HT$_{2C}$ antagonist.[103]

5-HT$_3$ Ion Channel   The 5-HT$_3$ receptor is a cation-selective ion channel and is located both presynaptic and postsynaptic. It shares high sequence homology with nAChRs and like the nAChRs they are composed of five proteins that span the cell membrane. Binding of serotonin opens the ion channel allowing mono- and divalent cations to enter the cell causing desensitization. A key set of interactions consisting of a hydrogen bond and a cation-π interaction between serotonin and 5-HT$_{3A}$ along with hydrogen-bonding interactions between two loops in the binding site appears to be important for binding.[104]

A functional interaction between the 5HT$_3$ ion channel and the dopaminergic system was first reported in 1987. Use of the selective 5HT$_3$ antagonist ondansetron (Zofran) was found to attenuate locomotor hyperactivity in mice and primates induced by dopamine. Ondansetron has been investigated for the treatment of opioid addiction so it will be discussed further in Chapter 5. A mechanism of action was found that antagonism of 5HT$_3$ would inhibit the increase in dopamine levels in the nucleus accumbens. The exact mechanism on how an antagonist binding to a 5HT$_3$ ion channel can affect dopamine levels is still not clear. High levels of 5-HT$_3$ receptors are located in the prefrontal cortex, nucleus accumbens, and the amygdala. Presumably, then, the interaction of antagonists with 5HT$_3$ in these areas of the brain affects dopamine release in the nucleus accumbens. Studies using *in vivo* microdialysis procedures to measure dopamine release in terminal regions of the mesolimbic system in awake, freely moving animals found that 5-HT$_3$ ion channel antagonists have no effect in their own right on dopamine levels. However, antagonists reverse the increase in dopamine release resulting from stress or the administration of drugs such as morphine, nicotine, and ethanol.[105]

The 5-HT$_3$ ion channel appears to mediate ethanol-stimulated increases in dopamine as the addition of a 5-HT$_3$ antagonist to a perfusate during microdialysis experiments prevents ethanol from increasing both dopamine and serotonin levels.[106] In general, agonist binding to a 5-HT$_3$ ion channel in the nucleus accumbens leads to an increase in extracellular dopamine; the addition of an antagonist prevents this.

SERT   Serotonin is removed from the synapse by the serotonin transporter (SERT). SERT, like other transporters, is a large protein containing about 680 amino acids, 12 TM domains, and one extracellular loop. Both the amino and carboxyl termini are intracellular. SERT appears to be expressed exclusively in the serotonergic cells of the raphe nuclei. Binding of serotonin to SERT is quite strong with a $K_m = 0.2$-0.5 μM. The transporters SERT, NET, and DAT share about 50% absolute homology in the amino-acid sequences.[35]

The serotonin transporter SERT is most commonly associated with depression and the development of selective serotonin reuptake inhibitors (SSRIs). Modulation of SERT has also been associated with other mood disorders such as generalized anxiety disorder, panic disorder, social anxiety, and obsessive-compulsive disorders. As anxiety (stress) can be a trigger for relapse, it would seem natural to determine whether SSRIs could be beneficial for the treatment of drug addiction. The exact interplay between serotonin, dopamine, and stimulant or alcohol addiction is very complex and is still being investigated. Some facts though are that if serotonin levels are reduced the subjective "high" felt by cocaine users is reduced.[107]

2.5.7   Vesicular Monoamine Transporters

The vesicular monoamine transporter (VMAT) in of itself is not a neurocircuit but it is a fundamental component of the monoamine neurocircuitry. The role of the VMAT in addiction is an area of active investigation but it does appear to be especially important in methamphetamine addiction. A recent review by Eiden and Weihe discusses in detail the neuroanatomy, cell biology, and imaging of VMAT2 with respect to addiction and a review by Wimalasena focuses on the pharmacology and medicinal chemistry of the VMATs.[108,109] Unless noted otherwise, the following text is largely derived from these two reviews.

Vesicular monoamine transporters are located in storage vesicles in the presynaptic neurons, where they modulate the cytoplasmic concentrations of the monoamine neurotransmitters in the neuron. The vesicular amine transporters uptake singly positively charged amines (dopamine, serotonin, norepinephrine, epinephrine, and histidine) from the cytoplasm of the neuron into the secretory vesicles for storage. There are two transporters: VMAT1 and VMAT2; the former is located only in the peripheral, while latter is located in both the peripheral and CNS. Both are part of the solute carrier superfamily (Slc). Energy is required for the uptake of the monoamines and an ATP to ADP-driven proton gradient allows the transport of the monoamines into the vesicles. Upon depolarization of the neuron, the monoamine neurotransmitters are released from the vesicles into the synapse. All the monoamine neurotransmitters bind to VMAT2 with K_m's of approximately 2 µM.

The prototypical VMAT2 ligands are the natural product reserpine and the drug tetrabenazine (Xenazine) that is approved for use in Huntington's disease. As VMAT2 helps control monoamine levels, it is also of interest in other CNS disorders such as depression, schizophrenia, and Parkinson's disease.

The mechanism by which amphetamine-like compounds increase synaptic dopamine levels is still being explored. A working hypothesis though is that methamphetamine or amphetamine increases synaptic dopamine levels by entering the presynaptic neuron via DAT and proceeds to bind to VMAT2 that is located on the surface of synaptic vesicles. It is clear that methamphetamine is a substrate for VMAT2, not an inhibitor.[110] It is thought then that the increase in cytosolic dopamine concentration would be due to an exchange mechanism where methamphetamine is exchanged for dopamine. The increase in cytosolic dopamine would in turn potentiate the efflux of dopamine in the synapse.

Neurocrine Biosciences has, in the pipeline, a VMAT2 inhibitor (NBI-98854) in Phase II clinical trials for use in the neuromuscular disease tardive dyskinesia. The structure of the compound has not been released but based on Neurocrine Biosciences patent applications, the compound appears to be a derivative of the tetrabenazine metabolite dihydrotetrabenazine (Figure 2.22) where the –OH group is functionalized, perhaps as the amino-acid ester. Of note is inhibition or attenuation of VMAT2 does not appear to cause any serious adverse effects. It is not mentioned if Neurocrine Biosciences is interested in developing the drug as a therapeutic agent for drug abuse.

Of interest from a chemotype, structural point of view is the alkaloid stepholidine (Figure 2.22). The compound is isolated from the herb *Stephania*, is a D_1 partial

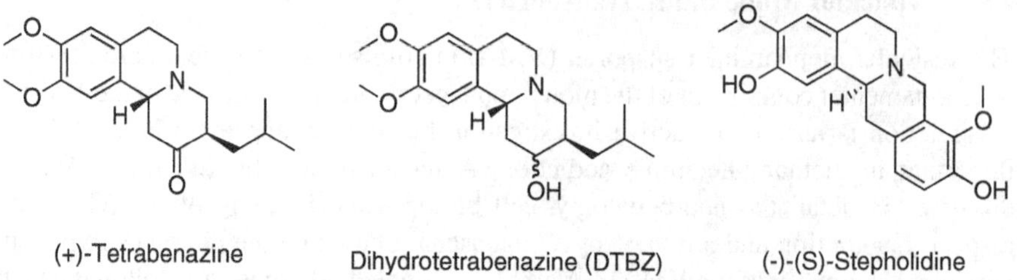

(+)-Tetrabenazine   Dihydrotetrabenazine (DTBZ)   (-)-(S)-Stepholidine

FIGURE 2.22   Structures of VMAT2 ligands

agonist and D_2 antagonist, and has been shown to inhibit the acquisition, maintenance, and reinstatement of morphine-induced conditioned place preference in rats.[111] It has not been reported if the compound binds to VMAT2.

2.5.8   Cannabinoid System in Drug Addiction

Endocannabinoids can be classified as neurotransmitters or neuromodulators with enough important interactions between the endocannabinoid system and the neurotransmitter systems described above to warrant discussion at this stage. Two cannabinoid receptors that bind to endocannabinoids are known: CB_1 and CB_2. Endogenous endocannabinoids (Figure 2.23) are derived from lipid fatty acids and are typified by anandamide (CB_1 $K_i = 32$ nM) and 2-arachidonylglycerol (CB_1 $K_i = 472$ nM).

The action of endocannabinoids on the CB_1 receptor has an interesting mechanism of action. Upon depolarization of postsynaptic neurons, endocannabinoids are released from these neurons into the synapse where they can bind to CB_1 receptors on the presynaptic neuron resulting in inhibition of the release of both excitatory and inhibitory neurotransmitters. The endogenous endocannabinoids are then removed from the synapse by a transporter located on neurons and glia cells. Once inside the neurons, the endocannabinoids are hydrolyzed by fatty acid amide hydrolase to fatty acids such as arachidonic acid.

With regard to addiction, the CB_1 receptor is the most important. Both receptors are located throughout the body but the CB_1 receptor is located mainly in the CNS with largest concentrations in the hippocampus, neocortex, basal ganglia, and the cerebellum. Moderate receptor levels are also present in the basolateral amygdala, hypothalamus, and the periaqueductal gray matter of the midbrain. As we have seen, these are areas involved in the control of pain, emotion, motivation, and cognition.[112]

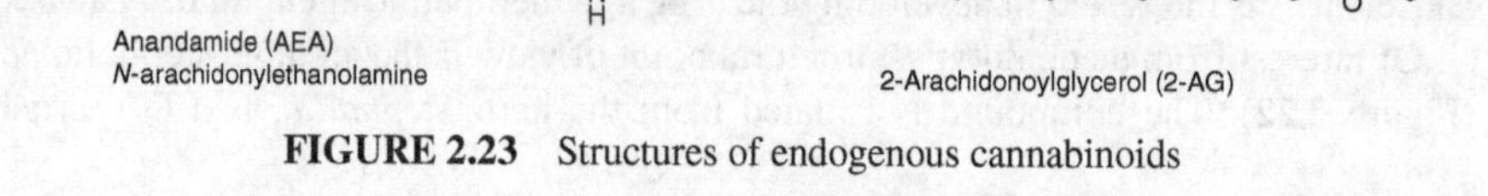

Anandamide (AEA)
N-arachidonylethanolamine

2-Arachidonoylglycerol (2-AG)

FIGURE 2.23   Structures of endogenous cannabinoids

The CB_1 receptor is located in many of the neurocircuitry systems we have discussed where it is able to modulate neurotransmitter release.

Both CB_1 and CB_2 are coupled to G_i proteins and G_o proteins and inhibit adenylyl cyclase. Agonism of CB_1 activates inwardly rectifying potassium channels and inhibits voltage-gated calcium channels via the $G_{i/o}$ protein. Inhibition of calcium channels located presynaptic reduces calcium influx into the neuron resulting in inhibition of Ca^{2+}-dependent neurotransmitter release from the presynaptic neuron. In general, activation of presynaptic CB_1 receptors can lead to inhibition of the evoked release of a number of different excitatory or inhibitory neurotransmitters both in the brain and in the peripheral nervous system. Although the primary effect of CB_1 receptor agonists on neurotransmitter release seems to be one of inhibition, this may sometimes result in enhanced neurotransmitter release at some point downstream of the initial inhibitory effect. For example, there is evidence that cannabinoids enhance release of the endogenous κ-opioid receptor agonist dynorphin within the spinal cord and that this effect depends on CB_1 receptor-mediated inhibition of tonically active neurons that exert an inhibitory influence on dynorphinergic neurons. There is also evidence from experiments both with whole animals and with brain slices that CB_1 receptor agonists can stimulate dopamine release in the nucleus accumbens, and it is likely that this effect stems from a cannabinoid receptor-mediated inhibition of glutamate release from extrinsic glutamatergic fibers. These are large fibers that form synapses in the nucleus accumbens with GABAergic neurons that project to the VTA to exert an inhibitory effect on dopaminergic mesoaccumbens neurons. It is possible that cannabinoid receptor-mediated disinhibition of dopamine release in the nucleus accumbens gives rise to increases in acetylcholine release in the prefrontal cortex. Thus, GABAergic neurons project from the nucleus accumbens to the prefrontal cortex, and it is thought that dopamine released in the nucleus accumbens may act on these neurons to disinhibit acetylcholine release in the cortex. Results from microdialysis experiments with rats have indicated that at low doses, intravenously administered cannabinoids can also act through CB_1 receptors to increase acetylcholine release in the hippocampus, whereas data from *in vivo* electrophysiological experiments suggest that systemically administered cannabinoids can enhance dopamine release from mesoprefrontal cortical neurons that project from the VTA to the prefrontal cortex. This stimulatory effect on cortical dopamine release may result from inhibition of GABA release mediated by CB_1 receptors that are presumed to be located on the terminals of prefrontal cortical GABAergic interneurons that modulate the activity of pyramidal neurons. These prefrontal cortical pyramidal neurons project to the VTA, where they form excitatory synapses on mesoprefrontal dopaminergic neurons that release GABA from the prefrontal cortical GABAergic interneurons that have been postulated to express CB_1 receptors.[113]

Agonism of CB_1 receptors in the VTA modulates dopamine release into the nucleus accumbens. Interaction of cannabinoids, or other agonist, on CB_1 receptors located in the VTA on presynaptic glutamatergic and GABAergic neurons acts as inhibiting modulators, influencing the release of neurotransmitters into the synapse where they can interact with their receptors on VTA dopaminergic neurons. For example, activation of CB_1 receptors on GABAergic neurons results in the attenuation of GABA release into the mesolimbic system. As GABA acts as a brake on the release of dopamine, the loss of GABA binding to GABAergic receptors on dopamine neurons in the VTA increases the levels of dopamine in the nucleus accumbens.[114]

The endocannabinoid system plays a regulatory role in controlling novelty-seeking behavior. Using a combination of pharmacological, genetic, and behavioral tools, it was found that CB_1 receptors located on glutaminergic neurons in the cortical regions of the brain favor novelty seeking.[115] However, CB_1 receptors located on GABAergic neurons inhibit novelty-induced behavior. Therefore, antagonism (or $CB_1^{-/-}$) of the CB_1 receptors located on glutaminergic neurons *inhibited* the approach behavior of mice to a novel object, whereas antagonism (or $CB_1^{-/-}$) of the CB_1 receptors located on GABAergic neurons *promoted* approach behavior; the mice lacked control of safety behavior.

Linking this study with the discussion of the GABAeric and glutaminergic systems, activation of the endocannabinoid system (e.g., Δ^9-THC) located on glutaminergic neurons could decrease behavioral inhibition resulting in impulsive behavior. This control mechanism could have profound effects not only in addiction, but also in impulse disorders such as attention deficit hyperactivity disorder (ADHD).

Acute use of cannabinoids in humans impairs response inhibition, increases risk taking, and dysregulates executive function and inhibitory motor control, and appears to increase certain types of impulsive behavior but not all.[116–118] Most interestingly, a genetic link between impulsivity and the cannabinoid system has been found. In a study of 251 individuals from Indian tribes, in southwest California, a linkage between four single nucleotide polymorphisms in or near the *CNR1* receptor gene and impulsivity was determined.[119]

2.5.9 Opioid System in Drug Addiction

The opioid system consists of three GPCRs classified as μ-, δ-, and κ-opioid receptors. Of the three, the μ-opioid receptor is most linked with addiction. All three opioid receptors are coupled to G_i and G_o proteins and inhibit adenylyl cyclase. Activation of the μ-opioid receptor activates inwardly rectifying potassium channels in a similar fashion as for the M_2, $5HT_{1A}$, and CB_1 receptors.[120] Coupling with G_o proteins will close Ca^{2+} ion channels.

The opioid system is modulated by endogenous opioid peptides, which bind to all the opioid receptors with endorphins and enkephalins having higher potency at the μ- and δ-opioid receptors and the dynorphins at the κ-opioid receptor. The opioid system controls a wide range of physiological functions. All three receptors modulate antinociception, the sensation of pain. The μ-opioid receptor also modulates reward, respiratory function, and gastrointestinal motility. The δ-opioid receptor modulates immune function and mood, and the agonists of the κ-opioid receptor induce dysphoria (a mood of general dissatisfaction, restlessness, depression, and anxiety) and reduce renal output (urine flow).

Analgesic drugs like morphine affect their primary response through agonism of the μ-opioid receptor. The three receptors share high homology and it can be difficult to develop selective ligands for each receptor. A huge amount of work has been

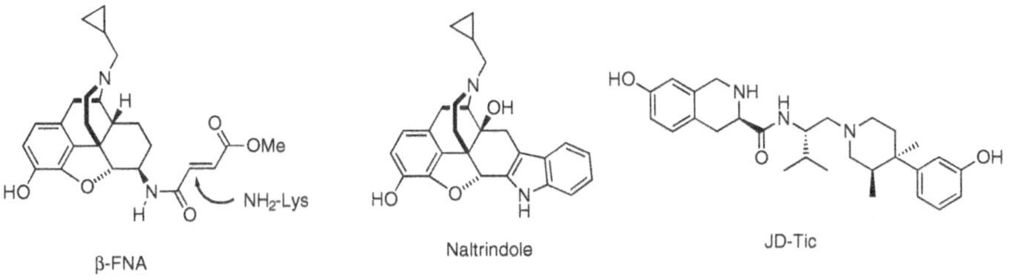

FIGURE 2.24   Structures of opioid ligands

done on exploring the SAR of opioid morphinan ligands derived from morphine. It has proven difficult though to find truly selective ligands, where the ratio of binding affinity or efficacy is >100. Selective ligands are available though by abandoning the morphinan structure and developing different chemotypes.

With the development of selective ligands, modulation of the receptors has been investigated with respect to drug abuse. Numerous studies have shown that selective agonism of the κ-opioid receptor attenuates cocaine, heroin, and amphetamine behavioral responses (e.g., self-administration, reinstatement) in rodents and non-human primates. The effect of selective agonism of the δ-opioid receptor is less clear.[121] A role in alcohol abuse appears to be possible.

Of huge interest is that X-ray crystallography structures of all three receptors bound to antagonists were all reported in an issue of *Nature* in 2012. A crystal structure of the mouse μ-opioid receptor was obtained by covalently binding it to the μ-opioid antagonist β-FNA (β-funaltrexamine).[122] The ligand β-FNA (Figure 2.24) is a well-studied compound that forms an irreversible adduct by conjugate addition of the amino group of Lys233$^{5.39}$ with an unsaturated fumarate group attached to position 6 of naltrexamine. The μ-opioid binding site appears to be largely exposed to the extracellular surface allowing rapid association and dissociation of agonists and antagonists. The δ-opioid receptor structure was obtained complexed with the reversible δ-selective antagonist naltrindole (Figure 2.24), while a crystal structure of the κ-opioid receptor was obtained complexed with the reversible κ-selective antagonist JDTic.[123,124] Naltrindole and β-FNA are both morphinan opioids while JDTic (Figure 2.24) is a longer, more linear totally synthetic compound. All three though showed certain binding similarities to conserved Tyr129$^{3.33}$, Asp128$^{3.32}$, Ile277$^{6.51}$, His278$^{6.52}$, and Val281$^{6.65}$ that are present in all three opioid receptors.

Though the work is certainly ground breaking for those in the opioid drug design field, the crystallographic structures are still those of the receptors with antagonists, hence in their inactive conformations. If similar crystal structures could be obtained with agonists then perhaps less addictive analgesic drugs could be obtained. However, as we will see with the arrestin study, the situation is more complex than following a simple structure-guided drug design protocol to enhance binding affinity and selectivity.

A substantial amount of work has established links between the opioid system and cocaine use. In general, elevated levels of μ-opioid binding in regions of the reward system have been correlated with cocaine use. In one study, patients were treated with the µ-opioid agonist [^{11}C]-carfentanil and levels of binding measured by PET. Those patients with higher levels of µ-opioid binding were found to have greater cocaine use during the trial and were also more prone to relapse after a 12-week treatment program.[125] The authors concluded that PET measurement of µ-opioid binding levels might serve as a predictive diagnostic tool toward the success of patients in cocaine treatment programs.

The interaction between δ-opioid receptors and cocaine use is more complicated.[121] While many preclinical rodent studies show that some interaction exists, it is not clear how to use it for medication development. It remains an area of active interest.

The role of β-arrestin in the addictive properties of opioids has become of interest. The normal course of GPCR activation is that once an agonist binds to a GPCR, the receptor is phosphorylated by protein kinases. After phosphorylation, the receptor can bind to intracellular proteins called arrestins. One result of arrestin binding to a phosphorylated receptor is internalization of the receptor making it unavailable for agonist binding. Studies in knockout mice where morphine was given to mice lacking β-arrestin-2 ($\beta arr2^{-/-}$) have shown that arrestin binding is necessary for the development of desensitization and tolerance. The mice lacking the gene producing β-arrestin were unable to develop tolerance to morphine, but the mice were still able to become physically dependent to morphine.[126] A recent study examined the effects of 20 opioids on the interactions of G-proteins and arrestin with µ-opioid and δ-opioid receptors using bioluminescence resonance energy transfer (BRET) experiments in cells.[127] Both β-arrestin-1 and β-arrestin-2 were labeled with green fluorescent protein for BRET (see Chapter 4 for a discussion of BRET) analysis but only β-arrestin-2 showed agonist-mediated resonance energy transfer signals. Addictive opiates like morphine and oxymorphone were found to be effective agonists for inducing receptor-G-protein coupling but acted as competitive antagonists for the binding of arrestin to δ-opioid receptors or weak partial agonists for arrestin binding to µ-opioid receptors. All the drugs had greater efficacy for promoting G-protein binding to the µ-opioid and δ-opioid receptors than promoting arrestin binding to the µ-opioid and δ-opioid receptors. For example, morphine has a potency of EC_{50} = 398 nM for promoting µ-opioid/β-arrestin binding but a potency of EC_{50} = 16 nM for promoting µ-opioid/G-protein coupling.

There are some intriguing results that point the way to a new approach. It is proposed that narcotics produce an analgesic effect when the drug binds to the µ-opioid receptor resulting in association of the (receptor)(drug) complex with a G_i protein. The normal sequence of receptor phosphorylation and subsequent binding of arrestin occurs. The sequence is represented as:

$$\text{drug} + \mu \xrightarrow{K_1} (\text{drug})(\mu) \xrightarrow{K_2} (\text{drug})(\mu)(G_i) \xrightarrow{K_3}$$
$$(\text{drug})[(\mu)(P)](G_i) \xrightarrow{K_4} (\text{drug})[(\mu)(P)(\text{arrestin})](G_i)$$

It is proposed that the addictive nature of narcotics is due to the initial binding of the (drug)(µ-opioid receptor) complex with G_i being stronger than subsequent binding of the (drug)(μ-opioid receptor)(G_i) complex with arrestin; $K_2 > K_4$ for addictive drugs.

If it could be found possible to weaken the binding affinity of the protein-drug complex with the G-protein and make step 2 equal or less than that of 4, the (drug)(μ-opioid receptor)(G_i) complex + arrestin step ($K_2 \leq K_4$) then a less addictive analgesic might be discovered.

This concept has become an area of active interest. A couple of recent studies on the κ-opioid follow. It has been known for some time that agonism of the κ-opioid receptor has an analgesic effect and does not lead to addiction. Unfortunately, as mentioned above, κ-opioid activation has several disagreeable side effects that prevent its clinical use. One of these, dysphoria, has been found to require arrestin recruitment, whereas the analgesic effects do not. The team of Javitch discovered that the well-known κ-opioid partial agonist $6'$-guanidinonaltrindole ($6'$-GNTI) is a potent partial agonist at the κ-opioid receptor for the G-protein activation pathway but it does not recruit arrestin. In fact, $6'$-GNTI functions as an antagonist to block the arrestin recruitment and kappa opioid receptor (KOR) internalization induced by other nonbiased agonists. They propose that $6'$-GNTI could be a promising lead compound in the search for nonaddictive opioid analgesic drugs.[128]

A compound that can selectively activate part of a signaling pathway is called a biased agonist. In a second multisite collaborative study, several κ-opioid agonists have recently been identified in an *in vitro* screening program that showed a range of signaling bias. The screening program analyzed, in parallel, the ability of the test ligands to initiate both β-arrestin recruitment and G-protein coupling upon ligand binding to the κ-opioid receptor. The ligand library used initially consisted of the National Institutes of Health (NIH) Clinical Collection (446 compounds) and NIH Clinical Collection 2 (281 compounds) of small molecules that have a history of use in human clinical trials. From this, only the known arylacetamide κ-opioid agonist GR-89696 was a potent biased ligand. With the arylacetamide compound as a template they conducted a similarity search of the ZINC database giving additional compounds. Finally, known κ-opioid peptide analogs of dynorphin A and Salvinorin A agonists were screened. From this effort, they determined that the arylacetamide ligands were either weakly G-protein biased or arrestin biased. The peptides tested were G-protein biased. They also determined that $6'$-GNTI was G-protein biased. Some benzomorphans with mixed μ-, κ-opioid agonist activity, for example, butorphanol showed slight preference for G-protein bias.[129]

A potential fourth opioid receptor was discovered in 1995. The current designation is N/OFQ receptor and is also known in the literature as ORL1 and OP_4. This receptor has high homology to the other three opioid receptors but is considered to be opioid-like. The endogenous peptide ligand is called nociceptin/orphanin FQ. It is widely present in cells of the reward system and also modulates antinociception and is a negative regulator of learning and memory. A recent review "The biology of Nociceptin/Orphanin FQ (N/OFQ) related to obesity, stress, anxiety, mood, and drug dependence" comprehensively details current knowledge on the receptor and suggests that N/OFQ agonists could be useful for the treatment of alcohol addiction. N/OFQ modulators are entering clinical trials, there are three agonists in Phase I-II for pain and an antagonist in Phase II for major depressive disorder.[130]

These results suggest, as have other unrelated studies, that the cellular processes of desensitization and tolerance are different than dependence and that the effects of opioid binding to their receptors result in complex cell signaling responses.

2.6   LOCATION OF RECEPTORS

An important component of neurobiology is to locate and quantify the concentration of receptors in the brain. There are three main methods that have been used: quantitative autoradiography, *in situ* hybridization, and immunocytochemistry. All these bear a resemblance to classic staining studies in that the brain must be removed, sliced into thin sections, and then processed by one of the above techniques. Autoradiography uses a radiolabeled ligand that will bind to its appropriate receptors in the brain slices. The radiation emitted from the radioisotope bound to the tissue is detected with radiation-sensitive film or photoemulsion layer. The sensitivity of this method will of course be reflected in the strength of the ligand binding (K_d). *In situ* hybridization measures the amount of a receptors mRNA in cells while immunocytochemistry relies on antibodies binding to the receptors.

Autoradiography visualization has the advantage of being amendable to *ex vivo* experiments. In *ex vivo* experiments, a drug is administered to an animal, after a period of time the animal is sacrificed and the concentration of the drug in the brain or the effects of the drug on the brain can be determined. Coupled with behavioral studies it can give valuable information at the amount of drug needed to affect certain behaviors. A recent example of this comes from Bristol-Myers Squibb where a tritiated dipyridyl triazolopyridazinone (BMS-725519) that is selective for the CB_1 receptor was investigated as part of a program in obesity.[131] The synthesis of the compound is shown in Figure 2.25.

The compound labeled parts of the brain rich in CB_1 receptors are cerebellum, hippocampus, striatum, and frontal cortex. By increasing the amount of drug given receptor occupancy could be determined (e.g., 12.7% at P.O. dosage 0.03 mg/kg to 74.1% occupancy at 30 mg/kg). If this information is coupled with behavioral information, then the effect of receptor occupancy, if any, on behavior can be determined.

Fluorescent probes that combine concepts of autoradiography and immunocytochemistry are being developed, which allow the avoidance of radioactive materials.

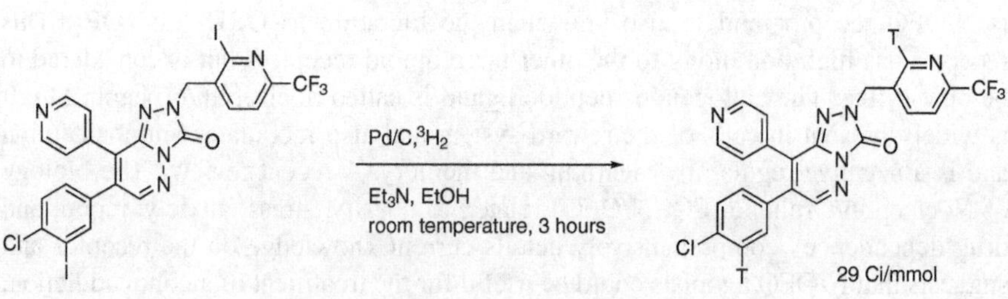

FIGURE 2.25   Synthesis of [^{3}H]-BMS-725519

A series of biotinylated endocannabinoid analogs are reported to be as sensitive as commercial immunocytochemistry assays in visualizing CB_1 receptors in transfected cells.[132]

A somewhat newer method is agonist-stimulated $[^{35}S]GTP\gamma S$ binding autoradiography.[133] This method can both identify the location of GPCRs and those that are functionally active. It relies on $[^{35}S]GTP\gamma S$ binding to active receptors in tissue slices and then the radioactive sulfur is detected by normal autoradiography means. When performed in the presence of different ligands, it can give information about the functional activity of the ligand such as it being an agonist or antagonist. Valuable information such as changes in receptor systems during the development of addiction and upon withdrawal can be obtained. For example, chronic morphine use (12 days) in rats reduces μ-opioid agonist (DAMGO) stimulated $[^{35}S]GTP\gamma S$ binding in regions of the brainstem (e.g., dorsal raphe nucleus and locus coeruleus), but not in the nucleus accumbens and amygdala.[134] This suggests that chronic morphine (or opioids in general) administration can affect the signaling processes controlled by activation of μ-opioid receptors in the brainstem. As the serotonergic (dorsal raphe nucleus) and adrenergic (locus coeruleus) neurocircuits originate from this area, levels of serotonin or norepinephrine could be altered by chronic morphine use. One advantage of this method is that different receptors can be located using only one radioactive compound ($[^{35}S]GTP\gamma S$). The majority of receptors of interest in addiction have been studied by this method.

From these types of studies, a generalized depiction of receptor locations in the reward system can be made as shown in Figure 2.26. Note that this is oversimplified, but the main components are present.

To illustrate the complexity of neuronal interaction, a depiction narrowing in on just the VTA is shown in Figure 2.26. Here, we can see GABA and glutamate neurons

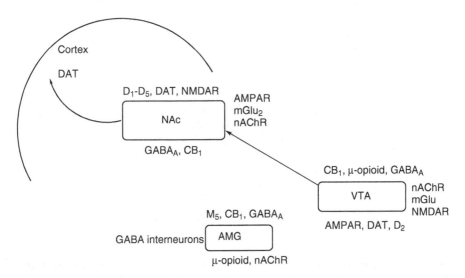

FIGURE 2.26   Location of receptors in the reward system

entering the cell area of the VTA and releasing GABA and glutamate, which then interact with the dopamine neuron whose cell body is in the VTA. Axons from the dopamine cell body travel to the nucleus accumbens. If the neurotransmitters hitting the dendrites of the dopamine cell body cause depolarization, an action potential will occur and dopamine will be released in the synapse in the nucleus accumbens where they will bind to D_1-like and D_2-like receptors located on the dendrites of other neurons. Note the clusters of receptors on all the presynaptic and postsynaptic regions of the neurons that drugs of abuse can bind to. Agonists and antagonists binding to these receptors will modulate the release of GABA and glutamate, ultimately controlling the release of dopamine in the nucleus accumbens.

2.7   AN EXAMPLE

Let us look at a concrete example using Figure 2.27 as a guide. As we have seen, the increase in synaptic concentration of dopamine is one of the characteristic effects of a

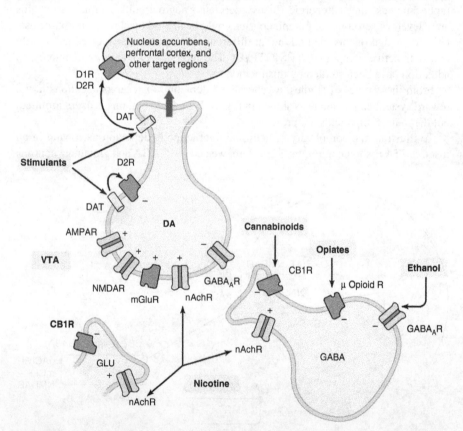

FIGURE 2.27   Receptor-level action of reinforcing drugs in the ventral tegmental area (Reprinted with permission from Zachariou et al.[38] Copyright © 2012 Elsevier)

drug of abuse. As discussed earlier, one function of GABA is the control of dopamine concentration. GABA will be released from the GABAergic neuron, diffuse through the synapse, and bind to $GABA_A$ ion channels on the surface of a dopaminergic neuron. Activation of the ion channel causes an influx of Cl^- ions increasing the membrane potential, resulting in hyperpolarization of the dopaminergic neuron. As hyperpolarized neurons do not fire, the end result will be *inhibition* of the release of dopamine. Binding of drugs of abuse to their receptors located on these neurons can result in a complex interaction between the GABA and glutamate neurotransmitter systems and those of other neurotransmitters, especially dopamine.

Opioids and the GABA system can be used as an example. Receptors for opioids are located on GABAergic neurons. Imagine morphine binding to a μ-opioid receptor located on a GABA neuron. Morphine acting as an agonist activates G_i proteins that activate K^+ channel opening and inhibits voltage-gated Ca^{2+} channel opening. Both actions cause a decrease in cellular + ion concentration, the GABA neuron is hyperpolarized resulting in *deactivation* of the GABAergic neuron and GABA will *not* be released. The binding of morphine then prevents the release of GABA from the GABAergic neuron. If the GABAergic neuron forms a synapse with a dopaminergic neuron (a neuron that releases dopamine), then the lack of GABA crossing the synapse and binding to $GABA_A$ or $GABA_B$ receptors and producing an inhibitory response results in an abnormal increase in dopamine concentrations. In other words, morphine prevents the normal braking or inhibitory response of GABA. If the increase in dopamine occurs in the VTA, a drug effect is observed.

2.8   USE OF BIOLOGICAL MARKERS

Biomarkers are usually taken to be any biological compound, for example, protein or DNA, whose levels can be quantified and correlated with a disease. The NIH Biomarkers Definitions Working Group has defined a biomarker strictly as "a characteristic that is objectively measured and evaluated as an indicator of normal biological processes, pathogenic processes, or pharmacologic responses to a therapeutic intervention." The new IUPAC Glossary of terms used in Medicinal Chemistry, Part II 2013, defines a biomarker as "Indicator signaling an event or condition in a biological system or sample and giving a measure of exposure, effect, or susceptibility." A key use of biomarkers is for them to be individualized. That is they allow classification of individuals into levels of risk. Ideal biomarkers could be obtained by noninvasive methods using biological samples from urine, hair, or blood.

We can take analysis of Fos as an example.

Acute and chronic drug use has been shown to produce a variety of neural adaptations.[135] Acute use of all the major drugs of abuse results in a transient increase in levels of transcription factors such as FOS (fos stands for Finkel & Osteogenic Sarcoma) and CREB. Transcription factors are proteins that control which genes are turned on or off in the genome. They do so by binding to DNA and other proteins. Once bound to DNA, these proteins can promote or block the enzymes that control the transcription of genes.

The *Fos* gene family consists of four members: FOS, FOSB, FOSL1, and FOSL2. These genes encode leucine zipper proteins that can dimerize with proteins of the JUN family, thereby forming the transcription factor complex AP-1. Alternatively, spliced transcript variants encoding different isoforms have been found for this gene (National Center for Biotechnology Information). One of these splice variants, ΔFosB, is a highly stable form of FosB (FBJ murine osteosarcoma viral oncogene homolog B) that has been shown to accumulate over longer periods of time (days) with repeated drug administration. Animals with activated ΔFosB have exaggerated sensitivity to the rewarding effects of drugs of abuse, and ΔFosB may help to initiate and maintain a state of addiction.

An increase in Fos protein expression is considered a histological marker for increased brain activity induced by a stimulus. The stimulus in this case is the activation of the second messenger cascade by the drugs of abuse binding to receptors. There is a lot of discussion if different parts or neurocircuits of the brain are responsible for distinct steps of the addiction process. To gain a fundamental understanding of physiological correlations with behavioral changes, changes in Fos protein expression have been examined in relation to identifying brain circuits associated with reinstatement.[136]

As an example, cocaine-induced conditioning effects and levels of Fos have been well studied. Rats were trained to self-administer cocaine in association with a cue, light, or a tone. After the cocaine was stopped (extinction), the ability of the same tone or a different tone to reinstate cocaine self-administration was studied along with the levels of Fos in different regions of the brain.[137] Fos levels were generally increased equally regardless of the nature of the cues. This suggests that similar neuronal circuits are responsible for cue-induced reinstatement regardless of the cue and the history of the reinforcement. On a molecular level, various theories for relapse involve long-term changes in ΔFosB.[138]

2.8.1   Alcohol Use Biomarkers

Some recent reviews have thoroughly discussed the current state of affairs with the discovery and clinical use of biomarkers for alcoholism.[139] Clinical endpoints can include quantification of alcohol consumption, analysis of drinking patterns such as binge drinking, and organ damage caused by chronic and excessive alcohol use. There are both direct and indirect measurements. Indirect markers measure the negative consequences of alcohol consumption on the body's anatomy and physiology. For example, elevated liver enzymes may reflect an alcohol-mediated consequence of heavy drinking. These indirect markers are due to long-term chronic use and may not accurately measure acute use. Direct markers though are compounds generated by metabolism of ethanol.

Two direct biomarkers used in the clinic are the measurement of the ethanol metabolites ethyl glucuronide (EtG) and ethyl sulfate (EtS). The Phase II nonoxidative hepatic glucuronidation of alcohol produces small amounts of EtG, while Phase II sulfonation of ethanol by 3′phosphoadenosine-5′phosphosulfate and a sulfotransferase produces small amounts of EtS, both metabolites can be detected in the urine and quantified. Lande and Marin of the Walter Reed National Military Medical Center have recently reported on the analysis of the above markers from 58 active duty service members.[140]

This study examined 374 combined EtG and EtS results obtained from 58 patients who were receiving care at the Walter Reed Department of Addiction Treatment Services from December 1, 2010, through June 30, 2011. The study group was principally comprised of men ($n = 53$, 91%). The majority ($n = 33$, 57%) of patients were 21–25 years old, unmarried ($n = 34$, 58%), and deployed to either Iraq or Afghanistan ($n = 40$, 69%). The patients' average Alcohol Use Disorders Identification Test scores ($n = 53$; mean $= 6.1$, SD 6.3) approached the standard screening level of eight, which suggests hazardous drinking, with 17 (32%) of these patients exceeding that threshold.

To illustrate some of the clinical nuisances involved in the analysis of biomarkers, it is worth examining their study.[*] First, ethanol is present in many consumer products as diverse as mouthwash and handy wipe sanitizers to gasoline. Low levels of ethanol metabolites would not be unexpected and must be distinguished from levels due to drinking. The investigators considered any EtG value equal to or exceeding 100 mg/mL and EtS values equal to or exceeding 50 mg/mL as positive test results. Compare this with levels expected from consumer products. The consumption of nonalcoholic beer, ripe bananas, apple juice, or grape juice can lead to EtG concentrations less than 100 ng/mL, which dissipated in less than 24 hours. Repeated daily use of alcohol-based hand sanitizers can also produce measurable amounts of EtG at values just slightly greater than 100 ng/mL. The researchers were interested in pairing analysis of both EtG and EtS biomarkers to determine whether false positives could be eliminated.

It seemed reasonable that there would be a match between paired analyses, this turned out not to be true. The paired tests were most often negative ($n = 295$, 78.9%). Only 10% of the paired tests gave a positive result, analysis of EtS gave a positive result in 8.6% of the patients while EtG gave a positive result in only 1.6% of the patients. The conclusion was that a fully validated, full-proof, laboratory method for detecting alcohol consumption does not exist and that analysis of EtS is more diagnostic than analyzing EtG. Of significant note for future consideration, a significant trend for testing positive EtS was noticed for service members with combat deployment.

As an example of indirect biomarkers, correlations of ethanol consumption and changes in protein concentration were detected in heavy drinking cynomolgus monkeys. Of the 40 or so proteins examined, the concentrations of 17 proteins were observed to be either increased or decreased relative to abstinent monkeys.[141]

2.8.2 Methamphetamine Use Biomarkers

Metabolomics is an interesting area of biomarker research that seeks to identify all endogenous small molecules within a sample in a nontargeted fashion. By not concentrating on analyzing specific candidate metabolites or metabolic reaction sets,

---

[*]Reprinted with permission from Lande and Marin.[140] Copyright © 2013 Taylor & Francis.

metabolomics attempts to quantitatively assay the entire range of metabolites present in a sample to characterize its overall biochemical state. This was recently applied toward analyzing behavioral sensitization in mice toward methamphetamine.[142] This could give valuable clues toward the identification of addiction biomarkers. The researchers used the standard analytical tools of liquid chromatography-mass spectrometry and gas chromatography-mass spectrometry to identify 301 metabolites with a >3 signal/noise measurement. The metabolites detected were then correlated with 10 different activity-based types of behavioral measurements indicative of sensitization. Correlations were detected with homocarnosine (a dipeptide of GABA) and histidine in total distance traveled sensitization, GABA metabolite 4-guanidinobutanoate and pantothenate in stereotypy (repetitive purposeless movements) sensitization, and *myo*-inositol in margin time (amount of time in seconds spent by the animal in close proximity, within 1 cm, to the walls of the cage) sensitization.

The first two metabolites, homocarnosine and 4-guanidinobutanoate, show the importance of the GABAergic system in sensitization. Both are known to produce CNS inhibitory effects. *Myo*-inositol is of interest as elevated levels of it produce anxiolytic effects. The margin time is a well-established index of anxiety in mice. As mice are inquisitive creatures, increased time spent near the walls of the chamber is indicative of anxiety. In this study as the *myo*-inositol levels detected increased, the margin time decreased.

2.9 MEMORIES AND ADDICTION

Memory remains a mysterious phenomenon. Memory maintains our experiences of the past that can be recalled in the future. Many questions exist, how are memories stored, formed, and recalled? It is a fascinating concept. Understanding the role of memory and learning can be central to the development of medications for addiction, especially for relapse. An important theory of memory research is that once a memory is consolidated, it can be reconsolidated at a future time and during this process of reconsolidation memories can be changed, that is, they can be changed or affected. This can have profound effects in the treatment of relapse. The operational process of changing consolidated memory is called exposure therapy.

Dudia has defined memory consolidation as "in the domain of memory research and theory, consolidation, or memory consolidation, refers to the progressive postacquisition stabilization of long-term memory, as well as to the memory phase(s) during, which such presumed stabilization takes place."[143] There are thought to be two types of consolidation. One occurs immediately in a time frame of minutes or hours and involves changes in synaptic connections. It is often referred to as synaptic consolidation, cellular consolidation, or local consolidation. The second type of memory consolidation occurs in the time frame of weeks to years. It is believed to involve reorganization over time of the brain circuits, or the systems, that encode the memory. This can referred to as system consolidation.

Reconsolidation of memory is now thought to be a process different from consolidation. Tronson and Taylor have defined reconsolidation as a process in which retrieval of a memory trace can induce an additional labile phase that requires an active process to stabilize memory after retrieval.[144] Another way of looking at it is that the consolidated memories upon retrieval are in a "fragile" state and to be used need to be reinforced or reconsolidated.

Somewhat coupled with consolidation and reconsolidation is the concept of extinction. According to Skinner, Pavlov coined the term "extinction" to explain a process by which a "conditioned stimulus loses its power to evoke the response when it (*the response*) is no longer reinforced."[6] Note that extinction does not refer to "forgetting." Modern behaviorists have reshaped extinction as "refers either to the learning process by which a cue (or action) previously associated with a reinforcer becomes newly associated with no outcome, leading to a decrease in the previously established conditioned response or to the procedure by which a cue or action previously paired with a reinforcer is now paired with no reinforcer."[144] An example can be modified from that of Skinner that may be appropriate to the audience. Imagine the behavior is one sending a grant proposal to NIH. The initial submission is successful and it gets funded. The action is the grant and the reinforcer is the grant money. Despite repeated attempts, further grant proposals however are not funded so eventually the researcher stops sending grant proposals. The behavior of submitting grants is extinguished.

It is known that neural circuits involved in *developing* addiction are also involved in extinction and reconsolidation of drug-associated memories. It is being discovered that learning and memory, that is, the processes of consolidation, reconsolidation, and extinction, are associated with physical and functional changes of neurons. These changes are referred to as plasticity. Synaptic plasticity has been well explored. Two major forms of long-lasting synaptic plasticity exist: long-term potentiation (LTP) and long-term depression (LTD). Long-term potentiation is a long-lasting (hours to days) increase in the synaptic response of neurons to stimulation of their afferents. Long-term depression is a long-lasting decrease in the synaptic response of neurons to stimulation of their afferents following a long patterned stimulus.[145] Long-term depression changes in synaptic plasticity are especially prevalent with regard to glutamate acting on NMDA, AMPA, and mGlu receptors. There are several forms of LTD and mechanisms of induction that we need not address. Of importance is that LTD is thought to be involved in information storage and learning. The majority of the neurotransmitters have been implicated in being involved in memory reconsolidation—glutamate, noradrenaline, dopamine acetylcholine, and GABA—as well as glucocorticoids and cannabinoids.[146] In particular, glutamate binding to NMDA ion channels and a complex interaction of the intracellular signaling molecule PKA and the transcription factor CREB, along with many others, appear especially to be important.

The amygdala is critical in detecting and responding to threats, in learning, and controlling the body's emotional and physiological response to danger. The prefrontal cortex links to the amygdala and can inhibit neural activity in the amygdala and suppress its reaction to a threatening experience. This circuitry from the prefrontal cortex is how exposure therapy suppresses anxiety and fear. Following is an example of a memory reconsolidation experiment in humans but before we proceed, let us look at how such experiments are conducted in animals.

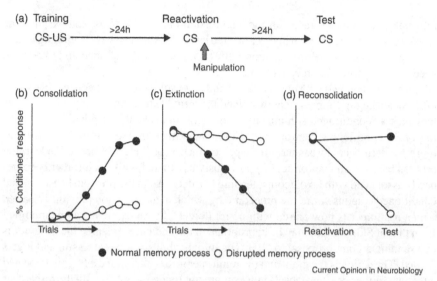

FIGURE 2.28 How reconsolidation is studied (Reprinted with permission from Tronson and Taylor.[683] Copyright © 2013 Elsevier Ltd.)

The protocol for studying memory reconsolidation typically involves three phases: a training phase, a reactivation phase after which manipulations occurs, and test. Memory reactivation is triggered by presentation of the conditioned stimulus (CS), without (and sometimes with) the occurrence of the unconditioned stimulus. Figure 2.28 is a schematic of typical reconsolidation paradigms. Animals are trained on a Pavlovian memory task with one or many trials. At least one day after the end of training, memory reactivation is induced by re-exposure to the CS, at which time manipulations are applied. At least 24 hours after reactivation, animals undergo a test of responding to the conditioned stimulus. During consolidation (Graph b in Figure 2.28) and extinction (Graph c in Figure 2.28) we infer an active process by observing a change in behavior (filled circles). In contrast, reconsolidation (Graph d in Figure 2.28) causes no observable change in behavior (filled circles). Evidence for an active process stems from changes in behavior as a consequence of manipulations (e.g., disruption, open circles).

Schiller, Kanen, Ledoux, Monfils, and Phelps investigated the involvement of the prefrontal cortex on inducing extinction during threat memory reconsolidation in humans using functional magnetic resonance imaging. They saw that under specific conditions of extinction, the prefrontal cortex was becoming active in addition to the amygdala, and the functional connections between it and the amygdala were growing stronger. The experiment, involved 19 participants, was conducted by having two colored squares as the conditioned stimulus (CS+) paired with mild electric shocks (unconditioned stimulus) to the wrist and a third colored square that was not paired with a shock (CS−). On Day 1, the acquisition stage, the two colored squares were shown to each participant accompanied at times with an electric shock (CS+), the third colored square was presented without an electric shock (CS−). Day 2 was the reactivation and extinction stage. On presentation of only one of colored squares was a shock delivered; this was called the reminded conditioned stimulus (reminded CS+). This will reactivate the threat response associated with a square. On presentation of the other two squares no shock was delivered. Thus, it was considered that the participants were not reminded that one of the squares presented a threat, this was called the nonreminded conditioned stimulus (nonreminded CS+). Then, they underwent an extinction process for all three conditioned stimuli (squares) to try to "erase" the threat stimuli associated with an electric shock. The extinction procedure was to first watch a video for 10 minutes, then to show 10 times the reminded CS+, 10 times the nonreminded CS+, then 11 trials of CS−. They believe that during this period the reminded CS+ undergoes extinction during the reconsolidation period while the nonreminded CS− undergoes standard extinction, or not during the reconsolidation period. Day 3 was reinstatement of threat memory followed by re-extinction. To reinstate the threat memory each participant was given four shocks without the colored squares being shown. Then, they underwent re-extinction as on Day 2. It was found that the prefrontal cortex was involved, activated, during extinction of the nonreminded CS+ but not during the reminded CS+. It appears that disrupting a threat memory during reconsolidation does not activate the prefrontal cortex thus breaking its linkage with the amygdala.[147] Physical manifestations of stress due to a threat memory would then presumably not occur. If the processes could be extended to addiction then relapse due to stress of perhaps cues might be reduced.

It is thought that cue-induced relapse may be due to an abnormally active memory process involved in encoding cue-drug associations. This would occur during reconsolidation. Manipulations of these drug-associated memories could be a useful strategy in preventing relapse. An interesting review by Torregrossa and Taylor explores this issue.[148] Though not directly related to drug abuse, use of the NMDA partial agonist D-cycloserine improved the extinction of fear in a trial involving 28 subjects.[149] The drug was used as an adjunct to exposure-based psychotherapy. Propranolol has been used in a study involving heroin addicts with evidence of reduced memory reconsolidation but if this reduced heroin consumption was not reported. There have been some beneficial uses of propranolol for the treatment of patients with post-traumatic stress syndrome (PTSD) in reducing anxiety when exposed to a stimulus that reminds them of the traumatic experience.

Drug use is associated with complex changes in regions of the mesocorticolimbic circuitry such as the VTA, the nucleus accumbens, and the medial prefrontal cortex. Long-term depression in the medial prefrontal cortex has also been linked to cue-induced relapse to drug-seeking behavior with heroin or cocaine. As a more specific example, physical changes in spine density and/or head diameter of medium spiny neurons located on the nucleus accumbens are correlated with memory. A larger head diameter is associated with LTP and a reduced head diameter is associated with LTD.[150]

Therapeutic targets involving changes occurring due to LTP/LTD may become relevant. Chronic severe pain such as neuropathic pain is treated with low doses of opioids. The opioids bind to μ-opioid receptors in the dorsal horn of the spinal cord where they decrease the pain stimulus signal from the peripheral nervous system to the CNS. Long-term potentiation of sensory neurons in the dorsal horn is thought to contribute to the development of chronic pain. Rodent studies have shown that a brief, high dose (0.45 mg/kg/h) of the μ-opioid agonist remifentanil actually reduced the LTP of the sensory neurons (up to 188% of control).[151] This suggests that the treatment of chronic pain could be achieved by acute use of high doses of opioids. It would seem that peripheral-restricted opioids would be ideal for this use.

In all it is a very interesting area filled with the potential of novel approaches to the treatment of addiction but could be controversial as it does involve, in principle, the "manipulation" of memory.

2.10   STRESS, THE HPA AXIS, AND ADDICTION

A dysfunctional endocrine system has been linked to mood disorders and addiction. The endocrine system responds to stress, which has been shown to be an important initiator of substance abuse and of relapse. The mediators of stress responses are a series of hormones and their cascade systems. In the brain, the hypothalamus lies at the center of the hormonal cascade systems. Of significant importance to addiction as while as mood disorders such as depression and anxiety is one of these cascade systems called the Hypothalmic-Pituitary-Adrenalcortical system (HPA axis).[152,153] In brief, the hormone CRF is secreted from the hypothalamus and binds to the GPCRs CRF1 and CRF2 that are located in the pituitary gland. This, in turn, results in the secretion of adrenocorticotropic hormone (ACTH) into the systemic circulation where it binds to ACTH receptors located in adrenal glands. The ACTH receptor is a GPCR coupled to G_s proteins. ACTH binding activates the synthesis and secretion of cortisol, a primary stress hormone in humans, into the systemic circulation. Glucocorticoids released during stress have a wide impact on peripheral tissues and the brain by binding to glucocorticoid receptor (GR) and mineralocorticoid receptor.

Glucocorticoid hormones refer to steroids that are synthesized in the adrenal gland, more specifically the adrenal cortex, and regulate glucose metabolism. Mineralocorticoid hormones are also steroids synthesized in the adrenal cortex but they regulate ions and water levels. The corticoid steroids can bind to both receptors, the main glucocorticoid hormone is cortisol (hydrocortisone) while aldosterone is the main mineralocorticoid hormone. Mineralocorticoid receptor has 57% amino-acid identity with glucocorticoid receptor in the ligand-binding domain and 94% in the DNA binding domain. The cloned and expressed glucocorticoid receptor has high affinity for the synthetic steroid dexamethasone, modest affinity ($K_d \cong 4$–5 nM) for the physiologic steroids cortisol and corticosterone, and low affinity for aldosterone, deoxycorticosterone, and the sex steroids. The mineralocorticoid receptor has high binding affinity and selectivity for corticosterone, aldosterone, cortisol, and progesterone.[154]

Both the glucocorticoid receptor and mineralocorticoid receptor are not GPCRs but are bound to chaperone proteins, upon ligand binding, for example, cortisol, both dissociate from their chaperone protein and bind to common nuclear hormone

Brain   Systemic circulation

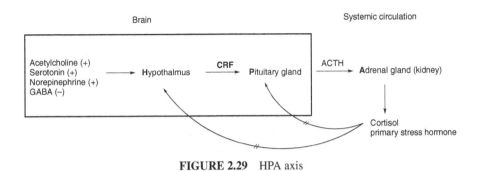

FIGURE 2.29   HPA axis

response elements that regulate transcription. Ultimately, then, perturbations in the HPA axis can result in changes in protein expression.

The HPA axis is controlled by several neurotransmitters that act on receptors in the hypothalamus and by negative feedback of cortisol as shown in Figure 2.29. Neurons containing CRF are located in many areas of the brain including the locus coeruleus, the origin of the adrenergic system, and the raphe nuclei, where the serotonergic system originates. Levels of both norepinephrine and serotonin are potentiated during periods of stress.

Studies show that opioids attenuate HPA axis activity, whereas other drugs of abuse potentiate HPA axis activity. The effects of *acute* drug use on the HPA axis have been reviewed.[152] The situation is complex, though. For example, in mouse gene inactivation studies, it was found that the glucocorticoid receptor on dopaminergic neurons affects behavioral responses to cocaine but not to morphine.[155] Inactivation of the glucocorticoid receptor gene reduced cocaine-induced conditioned place preference.

The effects of opioids on the HPA axis have received special attention. It has been argued that heroin-dependent patients suffer from a persistent hyperactive response to stress, even after detoxification. As an example, a recent small clinical trail involving 28 heroin-dependent patients showed that heroin administration reduces craving, withdrawal, and anxiety levels and leads to significant decreases in serum and saliva ACTH and cortisol concentrations.[156] Of special note is that the basal ACTH and cortisol levels were significantly higher in the heroin-dependent patients than in healthy controls. Therefore, in the heroin-dependent patients, the HPA axis appears to be abnormally active and the use of heroin restores hormone levels to their normal levels. The authors concluded that regular opioid administration protects addicts from stress and underscore the clinical significance of a μ-opioid agonist treatment for heroin-dependent patients.

One direction of developing drugs to act upon the HPA axis has been the investigation of CRF1 antagonists and CRF1 partial agonists. In addition to stress, the sensation of anxiety is controlled by the HPA axis. A hypothesis for the treatment of drug abuse during the withdrawal and abstinence stages is to reduce the hyperactive HPA axis found in substance-addicted individuals. A CRF antagonist would block binding of CRF to CRF1 and CRF2 receptors thus reducing the secretion of corti-sol. As will be discussed in Chapter 7, two clinical trials sponsored by the National Institute on Alcohol Abuse and Alcoholism are planned for investigating the use of CRF1 antagonists in the treatment of alcoholism.

The HPA axis is also of importance in PTSD. As addiction and PTSD can be comorbid, it is worth examining PTSD at this point.

Bessel A. van der Kolk of Boston University has explained PTSD as follows: "the biology of routine stress responses and the biology of trauma are fundamentally different: Stress causes a cascade of biological and physiological changes that return to normal after the stress is gone or after the organism has established a new homeostasis. In contrast, in PTSD, the biological alterations persist well after the stressor itself has disappeared. The fundamental problem in PTSD is a "fixation of the trauma". Trauma during childhood can be particularly troublesome. Compared with normals, people with histories of severe child maltreatment were of 4 to 12 times greater risk to develop alcoholism, depression, drug abuse, and suicide attempts, and a 2 to 4 times greater risk for smoking. A review of the neuroendocrine findings in PTSD shows very specific abnormalities in this disorder, compared with other psychiatric problems. The most prominent of these abnormalities appear to be in the HPA axis. PTSD patients show evidence of an enhanced negative feedback inhibition characterized by an exaggerated cortisol response to dexamethasone, an increased number of glucocorticoid receptors, and lower basal cortisol levels. These findings contrast with the reduced cortisol response to dexamethasone, the decreased number of glucocorticoid receptors, and the increased basal cortisol levels described in major depression. In a normal stress response cortisol acts on the regions of the brain that initiate cortisol release. Once the acute stress is over the HPA axis activates negative feedback inhibition, leading to the restoration of basal hormone levels. It is thought that in individuals suffering from PTSD that upon receiving a stressor, the levels of cortisol produced are lower than those without PTSD. This could result in a tonic level of the sensation of stress."[157]

Before we continue, let us look briefly at anxiety. The anatomical focal center for anxiety is considered to be the amygdala. The amygdala receives input from the sensory thalamus, the cortex (higher function, reasoning), and the hippocampus (memory). Outputs from the amygdala innervate the hypothalamus (part of the HPA axis in Figure 2.29), locus ceruleus (main site of noradrenergic neurons), and the periaqueductal gray region. The periaqueductal gray is located in the brain stem and is the origin of a descending pain-control pathway. It also coordinates complex behavioral responses to threatening stimuli by sending neuronal projections to other parts of the brain stem. As the amygdala is also part of the addiction neurocircuitry pathway, one can certainly imagine how perturbations of different components could modulate feelings of anxiety and stress leading to an increased risk of relapse. In fact, we will see examples where the principle mode of action of a drug used for the treatment of relapse or maintenance is as an anxiolytic.

Mice overexpressing CRF have a "high-anxiety" phenotype and the startle response is increased. Mice lacking the *CRF1* gene display behavioral phenotypes of low anxiety.

The startle response is an automatic, reflexive response to a loud stimulus. Measurement of the startle response is recognized as the premier animal model to study schizophrenia. We are always being bombarded with sensory information of taste, smell, sight, hearing, and touch. Our brains have adapted mechanisms to filter out what is unnecessary and concentrate on what is important. A predominant hypothesis for schizophrenia is that patients have a lower threshold of filtering unnecessary information, that is, sensory and cognitive gating. In effect they receive too much information and cannot correctly process it. In the startle response, model animals are trained to respond to signals, usually a loud noise. Before the noise though, the animals are given a low frequency sound that "alerts" them that a loud noise will shortly follow. Once the mice have been trained, they associate the low frequency sound with the startle signal and the resultant startle signal has a much lower effect; perhaps they do not jump as high. This process is known as prepulse inhibition.

Once established, drugs can be tested to discover their effects on prepulse inhibition and the startle response. The development of CRF antagonists is an active area of research due to the general trend of a hyperactive HPA axis in addiction and mood disorders.[158]

3

BEHAVIORAL PHARMACOLOGY AND ADDICTION

Behavioral pharmacology is a branch of science which studies the behavioral effects of drugs. It incorporates approaches and techniques from neuropharmacology, animal behavior, and behavioral neuroscience, and is interested in the behavioral and neurobiological mechanisms of action of drugs, normally with the emphasis on the whole organism. Another goal of behavioral pharmacology is to develop animal behavioral models to screen chemical compounds with therapeutic potentials. It is a relatively new area of science that started in the 1950s. Robbins and Murphy have written an interesting article on the development of behavioral pharmacology in the United Kingdom.[159]

They state that there can be considered three goals for the discipline. The first goal was to develop and refine behavioral procedures that would be effective in helping to screen pharmacological agents for potential clinical effectiveness. The second goal was to employ sophisticated behavioral techniques to analyze the mechanisms of action of behaviorally active drugs. Work towards this goal has remained the core of behavioral pharmacology. The third goal is the use of drugs as a means or tool for the analysis of complex behavioral processes.[160]

3.1 ANIMAL MODELS OF ADDICTION

The use of animal models that mimic stages of drug addiction in humans plays a very important role in the nonclinical analysis of the potential medications. The animal models are used to help identify mechanisms that are responsible for the transitions

Drug Discovery for the Treatment of Addiction: Medicinal Chemistry Strategies, First Edition. Brian S. Fulton.
© 2014 John Wiley & Sons, Inc. Published 2014 by John Wiley & Sons, Inc.

between stages of the addiction cycle. The use of animal models and a close working relationship between a medicinal chemist and a behavioral pharmacologist are indispensible in the field of medicinal chemistry and drug addiction. It is impossible to develop medication for the treatment of addiction simply on the basis of *in vitro* ligand binding and efficacy measurements. In fact, arguments could be made for going directly from the flask into rodents, foregoing traditional ligand-binding studies till later.

While animal models can never reproduce the complex social and often personal reasons why people abuse drugs, they nevertheless provide a rigorous means to precisely control environmental context, drug exposure, as well as assessing behavioral and cognitive performance prior to drug administration. They also allow neural manipulations (e.g., intracerebroventricular injection of selective ligands, microdialysis) and so establish the causal influences of putative neural loci and, in turn, the cellular and molecular substrates of drug addiction. Thus, to date, animal models provide a valuable means to investigate the different stages of the drug addiction cycle, including especially the initiation of drug taking, the maintenance phase (which is often accompanied by binges and an escalation of drug intake), and finally, the switch to compulsive drug intake defined operationally by an increased motivation to take the drug, an inability to inhibit drug seeking, and continued drug use despite negative or adverse consequences.[161]

As we discussed in Chapter 2, the different stages of addiction to be modeled are:

(1) Acute reinforcement/social drug taking/impulsive use
(2) Escalating/compulsive use (e.g., binge drinking)
(3) Dependence
(4) Withdrawal
(5) Relapse to the compulsive stage (craving)

For each of these stages, animal models have been developed. Rodents (mice and rats) are the most common animals used in behavioral studies. An immense amount of knowledge is available on the neuroanatomy of mice and their behavior. By studying their normal behavior, one can develop models that might mimic human behavior. Drugs could then be given to adjust this behavior.

The concept of reinforcement has provided the cornerstone for current theories and animal models of drug addiction. A reinforcer is defined as "any event that will increase the probability of a response" and, though not strictly correct is sometimes used interchangeably with "reward." In general, drugs function as positive or conditioned reinforcers by virtue of their rewarding effects, and reward often connotes additional attributes of a drug (e.g., pleasure) that cannot be easily defined operationally.[162]

The development of valid animal models is a complex area that we will not address in any detail. The general principles originate from the behavioral sciences.[163] In brief, the model can focus on specific features or symptoms of drug addiction or it can address more clinical dimensions of the disorder.

The validation of animal models of addiction is based upon the same principles that have been established for models in general, namely fulfilling standard criteria among which reliability and predictive validity are the most important. However, there are other criteria that have been used widely in validating animal models of drug addiction, including face validity and construct validity. Briefly, reliability refers to the consistency and stability with which the independent and the dependent variables are measured. Thus, a reliable model of drug addiction must allow for a precise and reproducible manipulation of the independent variable (e.g., the amount of drug) and an objective and reproducible measure of the dependent variable (behavior) in standard conditions. A further key criterion for the validation of an animal model is its predictive validity. A valid animal model should predict either the therapeutic potential of a compound in humans or a variable that may influence both the dependent variable of the model and the process under investigation in humans.[161,164]

We have looked at margin time as a behavioral assay for anxiety in Chapter 2. As an example, let us explore this field in more detail. A simple behavioral model in mice is called the Anti-Anxiety or Light-Dark Box test.[165] Mice have a natural tendency to explore a novel environment but also to avoid brightly lit spaces. Mice are placed in a box consisting of two chambers that are connected, one is exposed to light and the other is dark. They are allowed to explore this and the number of times they cross from one chamber to the other is recorded. The mice are then given a drug and the number of crossings for a given amount of time is again observed. The natural tendency of the mice will be to stay in the dark chamber. It has been found that anxiolytic drugs such as benzodiazepines increase the number of crossing between chambers. Most important, the rate of crossing changes with the dose, there is a dose-dependent increase in crossings. We can interpret this observation as the mice are "less anxious." There are many ways this simple behavioral assay can be used. Drugs that increase anxiety should have the opposite effect. Mutant mice lacking certain receptors could be used to ascertain what receptor systems may be important in anxiety. Is the drug just acting as a stimulant? Behavioral studies specific for motor activity can be added to examine this or the time spent in the lit part of the box could be used as another measure of general stimulant activity.

A fundamental behavioral assay in addiction is measuring the self-administration of a drug. Humans seem to have little difficulty self-administrating drugs and while other animals might not naturally self-administer a drug, they can be trained to do so. This process can require several weeks of work during which the animal can be trained to distinguish between a drug and another rewarding choices such as food. The entire process relies on the concept of a drug acting as a positive reinforcer. Positive reinforcers can be defined as any stimulus that increases the tendency of a behavior it is contingent on. Note that it is the behavior of the animal that is the dependent variable. The models can examine the "set"—physiological/mental state and the "setting"—circumstances of drug taking. An aspect of drug addiction that is often explored is the difference between wanting to take a drug and liking to take

TABLE 3.1   Human versus Non-human Laboratory Models of the Stages of the Addiction Cycle

Stage of addiction cycle	Non-human laboratory models	Human laboratory models
Binge–Intoxication	Drug or alcohol self-administration Conditioned place preference Brain stimulation reward thresholds Increased motivation for self-administration in dependent animals	Self-administration in dependent subjects Impulsivity
Withdrawal–negative affect	Anxiety-like responses Conditioned place aversion Elevated reward thresholds Withdrawal-induced increases in drug self-administration	Acute withdrawal Self-medication Mood induction
Preoccupation–anticipation	Drug-induced reinstatement Cue-induced reinstatement Stress-induced reinstatement	Drug reinstatement Cue reactivity Emotional reactivity Stress-induced craving Resistance to relapse Cue-induced brain imaging responses

Source: Reprinted with permission from Koob et al.[166] Copyright © 2009 Nature Publishing Group

a drug. Assuming we know what a rodent likes, thinks, or wants, wanting can be described as salience behavior and liking as hedonic behavior.

Table 3.1 summarizes correlations between human and non-human (mainly rodent) models of drug addiction.[166]

In general, the following three steps can be considered common to all preclinical studies investigating drugs for medication.[167]

(1) The animals are trained to self-administer the drug of abuse.
(2) The animals are then treated with a range of test medication doses to evaluate effects on drug self-administration.
(3) Determine that the medication-induced decreases in drug self-administration are due to the diminished reinforcing effects of the abused drug and not due to nonselective impairment of the subject's behavior such as sedation.

The studies can be done under acute or chronic use of the drug but results due to chronic use are most applicable to deciding if a new chemical entity should be used in a clinical trial. The magnitude of behavioral results is usually quantified by the expression of an ED_{50} (effective dose that gives a 50% response) in units of g drug/kg body weight.

3.2   SELF-ADMINISTRATION

The positive reinforcing properties (rewarding) of a drug are examined using the self-administration procedure. The majority of drugs of abuse are self-administered by animals such as rodents and non-human primates. It can be viewed as determining if the animal is willing to "work" to obtain a dose of the drug. The normal method of evaluating "work" is through the animal pressing a lever or button to obtain a fixed dose of the drug. The potential abuse liability of any drug is studied by self-administration. The study of self-administration is very important in behavioral pharmacology of drugs of abuse so it is worth our time to study it in detail and learn its benefits and potential pitfalls.

The series *Current Protocols in Neuroscience* gives detailed procedures for experimental techniques in the neurosciences, just as *Organic Synthesis* gives detailed experimental procedures for synthetic chemists. Thomsen and Caine have detailed procedures for Chronic Intravenous Drug Self-Administration in Rats and Mice and Platt, Carey, and Spealman have done the same for Intravenous Self-Administration Techniques in Monkeys.[168,169]

Medicinal chemists routinely analyze binding affinity and *in vitro* functional activity as part of the process of optimizing the activity of a drug. For those chemists in CNS research the added burden of being able to interpret behavioral pharmacology experiments is required. Keeping with this spirit we will examine the procedures of self-administration in rodents in some detail using as a guide protocols developed by Morgane Thomsen and S. Barak Caine.[1]

3.2.1   Chronic Intravenous Drug Self-Administration in Rats and Mice[168]

Intravenous drug self-administration in animals is a procedure for measuring the reinforcing effects of drugs, and is frequently used as an "animal model" of human drug abuse and dependence for scientific studies. Some drugs of abuse are commonly administered by intravenous injection and/or inhalation in humans, including heroin, cocaine, and nicotine. For a variety of reasons, including pharmacokinetic factors, intravenous drug self-administration provides a more direct model of human drug abuse than would oral self-administration of these substances. The observation that over 20 psychoactive drugs that are abused by humans also produce reinforcing effects in animals suggests that the procedure is useful as a predictor of abuse liability in humans. Drug self-administration studies in animals are also widely used for preclinical evaluation of candidate medications for the treatment of drug abuse and dependence. In addition, basic research studies suggest that the underlying pharmacological mechanisms and neural circuitries mediating the reinforcing effects of drugs in animals overlap considerably with those mediating the abuse related effects of drugs in humans. Thus, studies of intravenous drug self-administration in experimental animals may provide useful information regarding the abuse liability of novel drugs in humans, the efficacy of candidate treatment medications for drug dependence,

---

[1] Adapted with permission from Thomsen and Caine.[168] Copyright © 2005 John Wiley & Sons, Inc.

the biological basis of drug addiction in humans and the function of reward-related brain systems.

Intravenous drug self-administration may be studied in a variety of experimental animals, and rodent species offer several advantages. Relative to larger animals, rodents are economical, suitable for invasive and novel procedures (e.g., brain manipulations and drug treatments having limited safety information) and amenable to genetic studies (e.g., inbred strains and transgenic and knockout mice). One limitation is that both the life span and the technical feasibility of maintaining chronic intravenous drug administration are limited in rodents relative to larger animals, hampering long-term studies of drug self-administration behavior. Nevertheless, technical innovations made in the last 10 years have vastly improved the feasibility of long-term studies in rodents.

Operant Chambers   Operant chambers can be designed either as permanent housing for the test animal or as a temporary test apparatus that can accommodate successive sessions with different animals throughout a day. The former approach is advantageous for studies that require 24-hour access to drug self-administration and the latter approach allows for a much higher throughput. The rat operant chamber shown in Figure 3.1 was designed for the temporary approach.

Essential components of the chamber are as follows. A catheter through which a drug solution of saline can be administered has been surgically implanted in the

FIGURE 3.1   Operant chamber for self-administration (Reprinted with permission from Thomsen and Caine.[168] Copyright © 2005 John Wiley & Sons, Inc.)

jugular vein. The catheter is attached to a swivel allowing the rat freedom of movement. In this chamber the rodent will choose between food (delivered in cup c), drug, or saline by pressing on a lever (d in the figure) with its paws. In other operant chambers nose-poke holes are available instead of levers. Rats and mice readily acquire drug self-administration with either levers or nose-poke holes, although acquisition of operant behavior may be more rapid with nose-poke holes. Discriminative cues such as light or sound can be added that are associated with either lever or nose-poke hole.

There are many approaches to training animals to self-administer a drug in an operant task. The three most common methods are (1) initial training of food-restricted animals with food reinforcement, and then replacing food with intravenous drug as the reinforcer, keeping animals in a food-restricted state; (2) initial training of food-deprived animals with food reinforcement, then replacing food with intravenous drug as the reinforcer, with food freely available; and (3) direct training with intravenous drug reinforcement with no prior operant training.

Following training a crucial component of the design of the self-administration experiment is the selection of a schedule of reinforcement and the unit doses of self-administered drug (the dose delivered per injection). Many types of schedules have been developed but some common schedules of reinforcement worth discussing are fixed-ratio schedules, multiple schedules, second-order schedules, and progressive-ratio schedules.

Fixed-ratio schedules require a certain number of lever presses before the reinforcer is delivered. It can be used as a measure of the reinforcing effect, for example, a rodent may press the lever many times to receive a drug if the drug is strongly reinforcing. If the drug is not the rodent may stop after several lever presses. Behavior that is maintained by drug injections or by presentation of another reinforcer in alternate components within the same test session may be referred to as a multiple schedule. In a second-order schedule experiment responding is maintained by a cue associated with drug delivery prior to reinforcement with the drug itself. A schedule in which the response requirement is increased for successive reinforcer presentations is termed a progressive-ratio schedule. Implicit in the use of this schedule is the view that over a broad range of ratio values, the point at which behavior is extinguished (termed the "breaking point") should be a good measure of the relative effectiveness of the reinforcer. The time required for a session depends on the type of schedule; for direct assessment of self-administration behavior (e.g., on a fixed-ratio schedule), sessions of 3-4 hours may model human drug intake better than briefer sessions (such as 1 hour).

Data and Interpretation   Under most schedules of reinforcement, including fixed-ratio schedules, unit doses of drug below a certain threshold fail to maintain responding in self-administration procedures, presumably because they fail to be adequately reinforcing, and unit doses above a certain threshold produce erratic patterns of self-administration or cessation of responding, perhaps as a result of convulsant, sedative, or other undesirable effects that are inconsistent with self-administration behavior. Importantly, within the range of doses that engender stable responding, as the unit dose

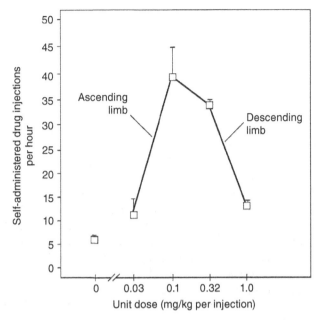

FIGURE 3.2 Dose-effect curve for cocaine self-administration under a fixed ratio 5. Abscissa: dose of self-administered cocaine per injection, log scale except for zero dose. Ordinate: rate of self-administration expressed as injections per hour (Reprinted with permission from Thomsen and Caine.[168] Copyright © 2005 John Wiley & Sons, Inc.)

is increased, self-administration rates decrease, and, conversely, as the unit dose is decreased, self-administration rates increase. When self-administration data are plotted with ascending unit dose on the abscissa and rates of self-administration on the ordinate, this inverse relationship between unit dose and rates of self-administration comprises what is referred to as the descending limb of the dose-effect function for drug self-administration (Figure 3.2).

Lower unit doses than those on the descending limb of the dose-effect function, that is, doses ranging from those, which fail to maintain responding up to doses that produce intermediate rates of responding, comprise what is referred to as the ascending limb of the dose-effect function. Peak rates of responding are observed at the nexus (*this will tell the pharmacologist what dose of a drug should be given in subsequent studies*) between the descending and ascending limbs of the dose-effect function. There are critical reasons for determining complete dose-effect functions for self-administration in any study. For example, if an experimental manipulation increases self-administration rates of a single unit dose of drug that maintains stable self-administration behavior and lies on the descending limb of the dose-effect function under control conditions, this effect may be interpreted as being comparable with decreasing the unit dose of drug. Conversely, if an experimental manipulation increases self-administration rates of a single unit dose of drug that did not maintain stable rates of self-administration in all subjects and lies on the ascending limb of the dose-effect function under control conditions, this effect may be interpreted as being comparable with increasing the unit dose of drug. In summary, increased rates of self-administration may be the result either of functionally increasing or decreasing the effects of the self-administered drug, depending on the shape and position of the dose-effect function for drug self-administration, and the position of the unit dose on that curve. Decreased rates of self-administration are even more difficult to interpret, and evaluation of complete dose-effect functions may aid in drawing firm conclusions regarding meaningful pharmacological interactions between a self-administered drug and an experimental manipulation. There are also pragmatic reasons for determining complete dose-effect functions for drug self-administration in every experiment (rather than relying on data from prior studies). First, a broad range of experimental variables influences the shape and position of inverted U-shaped dose-effect functions for drug self-administration. Accordingly, interpretation of self-administration data depends critically upon knowledge of the shape and position of the complete dose-effect function for self-administration under the experimental conditions particular to each experiment. Second, some manipulations alter self-administration only of lower unit doses and others alter self-administration only of higher unit doses. Thus, analysis of individual doses may lead to false-negative results regarding experimental manipulations that alter drug self-administration. Third, a hallmark of any drug effect, including a reinforcing effect, is a dose-dependent relationship between drug concentration and the observed effect. Accordingly, reinforcing effects of drugs are most clearly demonstrated when they are dose-dependent. For all these reasons, complete dose-effect functions for drug self-administration should be determined for comprehensive interpretation of self-administration data. Completion of an entire self-administration study can take as little as a few weeks to never if complications arise.

3.2.2   Intravenous Self-Administration Techniques in Monkeys

The same basic principles used for rodents apply also to self-administration studies in monkeys. Some unique considerations will be mentioned from Platt, Carey, and Spealman.[169]

Self-administration studies in primates require a supply of healthy monkeys and the facilities to house them. The use of primates, particularly large primates such as macaques, in biomedical research is highly regulated at the institutional, local, and federal levels. The facilities needed to house primates are expensive to build and maintain. Staff needs to be familiar with the care of large and powerful animals for several reasons, not the least of which is the safety of both humans and animals. Monkeys can harbor a variety of diseases that can readily be transmitted to humans (e.g., herpes B virus in macaques). Other needs include access to a well-equipped veterinary operating room, familiarity with sterile procedure, and expertise in basic vascular surgery.

Self-administration in monkeys is a time-consuming undertaking. Monkeys initially need to be habituated to the handling procedures associated with the remote-chamber method or to the tether in the home-cage method. Depending on the age, size, and disposition of the individual monkey, this can take from one to several weeks. At this point, the monkey is shaped to respond in the operant chamber, either using food reinforcement first, or beginning directly with drug infusions. Again, this process is highly individualistic and can take from 2 to 10 days before monkeys respond consistently. Once animals are at this stage, catheter implantation is next. Following recovery (up to 4 days), monkeys are shaped to respond to relatively high doses of drug. Again this process is highly individualistic, but animals usually initially acquire self-administration of an efficacious reinforcer such as cocaine within 2-4 days. In an average self-administration study, monkeys are given access to vehicle or a particular dose of drug for several consecutive days, or until there are no increasing or decreasing trends in response rate or number of injections per session. Therefore, if the effects of vehicle and four doses of drug are to be determined, several weeks may be required before results for the complete dose range are available. The time frame given above is meant to serve as a guide, as there are many factors that can dramatically lengthen the amount of time needed to complete self-administration experiments.

3.3 CONDITIONED PLACE PREFERENCE

Conditioned place preference (CPP) is also used to study the rewarding effect of a drug. This method is a nonoperant procedure using classical Pavlovian conditioning. In a simple form, animals are exposed to two different environments where one environment is paired with certain cues and a drug is administered. The other environment is paired with other cues and no drug is given. The animals are allowed to explore each environment and the time spent in the drug-paired environment is considered to be indicative of the reinforcing effect of the drug.

As an example, we can examine a study about the effect of morphine injections on rhesus monkeys under a CPP paradigm.[170] The infrastructure of the experiment is shown in Figure 3.3.

Room 2 was the start room with two inner doors while in Room 1 was a window and in Room 3 there was a refrigerator. The monkeys could move freely into the other two rooms through these doors. The upper part of the figure shows that the three rooms were monitored by a camera. Room 3 shows a monkey walking in the room.

Five rhesus monkeys were used for this experiment. For the training portion, the monkeys were placed in Room 2 and then allowed to freely explore the other two rooms for 50 minutes on three consecutive days. Fruits and vegetables were placed in each room during the training period. Each monkey tended to form a preferred attachment to either room one or three during the training period (habituation of pre-CPP). The next phase was the CPP conditioning period that lasted for 12 days. During this, each monkey was injected with a dose of morphine ranging from 1.5 mg/kg to 4.5 mg/kg in an ordered manner on the odd-numbered days. They were led to their nonpreferred room where they remained for 50 minutes. During this process, the monkey is learning to associate that room with the drug, this room is called the drug-paired room. On the even-numbered days, each monkey received an injection

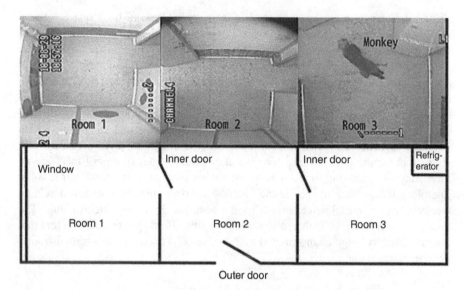

FIGURE 3.3 Structure of the conditioned preference place paradigm (Reprinted with permission from Wang et al.[170] Copyright © 2012 John Wiley & Sons, Inc.)

of saline and was placed in their preferred room. This room then is the saline-paired room.

At the end of the 12 days, each monkey was tested for CPP. They were each placed in the middle room (Room 2) and allowed to enter either Room 1 or Room 3. The same test was conducted about 15 months later to see whether long-term CPP would develop. They found that 24 hours after the 12-day CPP conditioning, all of the monkeys preferred the drug-paired room, which was not their favorite room during the training period. In addition, after 15 months, three of the monkeys still preferred the drug-paired room. The study went into more detail such as time spent in each room, distances traveled, etc.

It is clear that rhesus monkeys could establish CPP that lasted at least 15 months. They interpret this finding to indicate that rhesus monkeys can form long-term morphine-associated memory. The authors conclude "this demonstrates that a long-lasting morphine-induced CPP in a primate species is a good basis for further investigations on the neurobiological and pharmacological mechanisms involved in the acquisition, maintenance, and potential decay of such conditioned rewarding effects and their relevance to human drug addiction and relapse."

3.4 TOLERANCE

All drugs of abuse produce some level of tolerance. This means that larger quantities of the drug must be taken to achieve the same level of a "high" or other effects. The development of tolerance to a drug can be ascertained by observing rightward or downward shifts in dose-effect curves. If tolerance is developing, there will be a rightward shift in the dose-effect curve. This reflects that a larger amount of drug is required to achieve the same effect.

The physiological mechanisms responsible for tolerance are not yet clear. It has been widely shown that repeated stimulation of a receptor by an agonist leads to a reduced functional response over time. This process by which the receptors become less efficient is called desensitization. Molecularly, this is thought to occur by intracellular loops of the 7TM becoming phosphorylated, preventing binding of the G-protein to the receptor. The effect of agonist desensitization is not uniform; each agonist causes its own unique level of desensitization. When receptors become desensitized, they do not respond to the same concentration of the drug as they had before. A second response that can occur on receptor overstimulation is called downregulation. Here, the amount of receptor present in the cell membrane is decreased by the receptor being internalized inside the cell by β-arrestin. Once the receptor is internalized, it is either stored for translocation back to the cell membrane or is destroyed in the endosome.

3.5 EXTINCTION/WITHDRAWAL

Withdrawal can be modeled by studying behavioral and physiological changes or by the use of operant procedures. Changes in operant behavior can be a change in lever pressing. It has been observed that animals trained to press a lever for food (operant behavior) will exhibit decreases in response rates during withdrawal. One advantage of studying operant behavior is quantification; the severity of withdrawal might be ascertained by the magnitude of decreased response rates. Withdrawal is generally induced by drug abstinence or by precipitated withdrawal when an antagonist is administrated. The main difference is that in the latter the drug of abuse is still "on board." The antagonist will be competing with the drug for binding to the receptor.

3.6 REINSTATEMENT (ANIMAL MODELS OF RELAPSE)

In general, reinstatement can be defined as the resumption (after extinction has occurred) of a previously learned response (behavior) that occurs upon exposure to certain sensory stimuli. Several methods have been devised to stimulate an animal to retake a drug following extinction. In the pharmacology literature, this is referred to as drug-seeking behavior. The methods used may involve some form of stress, drug priming, or the use of drug-associated cues. There are different approaches to study this depending on whether one uses the self-administration or CPP model.

The CPP reinstatement model works on the basis of classical conditioning and is considered to model contextual cue-elicited drug-seeking behavior. In our morphine CPP example with the monkeys, each room also contained a colored piece of paper. The colored paper can be considered a contextual cue if the monkey was able to associate the color with the drug-paired room. Self-administration meanwhile is considered to model drug-taking behavior in humans and evaluates the primary reinforcing properties of drugs.

To study reinstatement, the animal is first trained to self-administer the drug, usually in association with a sensory cue. Extinction is then induced by *not* administering the drug or the cue. Following a period of time, the animal is placed back in a testing chamber, the sensory cue is reintroduced and it is observed if the animal will resume pressing the lever that previously administered the drug. However, instead of a drug being self-administered the animal will be self-administering saline. If the CPP model is used, a drug-priming injection (the injection of a small amount of the drug) or stressor (foot shock) is applied to determine whether acute exposure to the drug causes reinstatement.

Let us explore the reinstatement model in more detail since relapse is such an important component of drug addiction and is a component that should be treatable by pharmacological intervention. Our monkeys did not need a drug-priming injection or have a stressor applied, as they had not undergone extinction.

The general experimental design for an extinction/self-administration procedure is depicted in Figure 3.4.[171]

The animal is allowed to self-administer a drug (acquisition) until a stable level of responding occurs (maintenance). The time for this can vary depending on the type of animal used and the drug being tested. When a steady state of drug taking is reached, the drug is then withheld causing extinction. If the animal was addicted to the drug, then it could model withdrawal in humans. In behavioral pharmacology terminology the "drug-taking behavior is extinguished by withholding the drug reinforcer." In some cases, a saline solution instead of the drug solution would be administered when the animal presses a lever. Extinction is considered to have occurred when the number of responses (lever presses, nose pokes, etc.) is about 20% of that of the last session. The exact number of responses is a function of the experiment though and can depend on the species of the animal. At this point, a triggering reinstatement cue can be given. Different tests can be drug priming, stress, drug-paired environmental

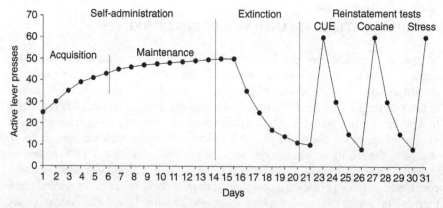

FIGURE 3.4   Experimental paradigm for reinstatement of drug-seeking behavior (Reprinted with permission from Yahyavi-Firouz-Abadi and See.[171] Copyright © 2009 Elsevier Inc.)

cues (e.g., light, noise), etc. Reinstatement is said to occur if the animal responds at a rate above that observed before the extinction process. There are many variations of this that have been developed but overall this is the basic process. It should be mentioned that the predictive ability of the reinstatement model as used for animals for the clinical endpoint of relapse in humans has been questioned and needs to be further investigated.[172]

Are there molecular explanations for reinstatement or relapse? This is an area of active research. A recent example of this research is as follows.[173,174] Rats were trained to self-administer cocaine in a design in which pressing one of two levers (the active lever) triggered an intravenous infusion of the drug along with a light and a tone (cue); pressing the other (inactive) lever had no consequence. The rats were trained for 10 days and then underwent extinction where upon pressing the active lever neither an intravenous infusion of cocaine nor the cue was present. When it appeared the rats had undergone complete extinction, the cue was again presented with the lever pressing. Rats proceeded to press the lever upon presentation of the cue in an attempt to obtain cocaine. What is interesting is that at different times during the experiment some of the rats were sacrificed and the researchers examined tissue slices from the nucleus accumbens and measured the size of medium spiny neurons and the amount of currents induced by α-amino-3-hydroxy-5-methyl-4-isoxazolepropionic acid (AMPA) and N-methyl-D-aspartate (NMDA) ion channels.

Medium spiny neurons are the major neuronal cell type in the striatum and have long been implicated as playing an important role in cocaine addiction. These medium spiny neurons are located postsynaptic and receive glutamate inputs from the cortex and other limbic neurons. It is proposed that the medium spiny neurons process information being delivered from the cortex and limbic neurons and transmit it to areas of the brain dealing with motor control and goal-directed behavior. It is proposed that activation of AMPA ion channels located on medium spiny neurons by glutamate mediates cocaine-seeking behavior.[175]

They discovered that after the training and extinction process the diameter of the medium spiny neurons increased in size relative to control rats from 0.327 to 0.415 μm. Immediately after cue-induced reinstatement, the diameter of the spine heads increased further to 0.491 μm and then decreased to a prereinstatement diameter of 0.427 μm. Note though that these head diameters are still larger than the non-cocaine exposed control rats. There was also a higher level of AMPA currents than NMDA currents indicating increased activity at the synapse. These changes are correlated with the number of lever presses during reinstatement. Most interesting is that these effects were *not* seen with sucrose indicating that the rewarding mechanisms of sucrose are different than that of cocaine.

3.7   DRUG DISCRIMINATION

The drug discrimination test evaluates the stimulus similarity between a test compound and a reference agent. Animals can be trained so that when they are administered a particular drug at a specific dose, they respond on one lever for a reinforcer, such as a food pellet. When administered saline, they respond on the opposite lever for the same reinforcer. A large number of compounds, many of them are drugs of abuse, have been established as a discriminative stimulus in several species, including mice, rats, pigeons, monkeys, and humans. A number of features make drug discrimination a useful test for assessing the abuse potential of novel compounds. First, the effect demonstrates that the agent acts within the CNS. Compounds that fail to cross the blood-brain barrier (as determined by pharmacokinetic studies) do not function as discriminative stimuli or antagonize the discriminative stimulus effects of other agonists. Second, the discriminative stimulus properties of a compound are pharmacologically selective. That is, agents with a mechanism of action that is similar to the mechanism of the training compound will substitute for it. Replacement strongly suggests the new drug elicits the subjective properties similar to the reference drug. In other words, if an animal cannot discriminate the effects of a new drug from, say, cocaine, there would be reason to believe the new drug can be used as a substitute for cocaine to achieve the same end effects as cocaine. Third, the drug discrimination assay serves as a reliable model of the subjective effects of drugs that are known to play a major role in their abuse. In addition, drug discrimination studies can be used to establish duration and onset of action.[176] It needs to be emphasized that discrimination tells you if the test drug "feels the same as the training drug," but not whether the feeling is good, bad, neutral, high, etc.

Drug discrimination in humans can be as simple as giving subjects two pills on different days and asking them if they can tell which pill contains a drug that gave the same effects as a previously given drug.

A classic drug discrimination study is shown in Figure 3.5. In this study, Swiss-Webster mice were trained to discriminate 10 mg/kg cocaine administered intraperitoneally 10 minutes before the session from saline, also administered intraperitoneally 10 minutes before the session. Liquid food (25 μL of Ensure protein drink, vanilla) was used as the reinforcer, with a maximum of 30 reinforcers available per 20-minute session. Here, nose pokes were used so if the mice were injected with cocaine they would receive a drink of Ensure if they poked the appropriate hole. The average time required for the mice to meet criteria for cocaine discrimination was 15.1 ± 1.3 weeks (range, 6–31 weeks). Amphetamine and the κ-opioid agonist U-50488 were then tested to see whether they would substitute for cocaine.[177]

Responding in the drug discrimination assay was quantified as the percentage of drug-appropriate responding. Drug-appropriate responding is obtained as described by Glennon and Young. An animal's degree of progress to learn a two-lever drug discrimination task was determined by an evaluation of its distribution of presses on the levers either prior to, or up to, the delivery of the first reinforcement. Thus, when the FR 10 schedule of reinforcement (for *Fixed Ratio* = 10 lever presses must be performed before the reinforcer—food—is given) was in effect, discrimination learning was assessed for each subject by dividing the number of responses that occurred on the drug-designated lever by the total number of responses that occurred on both levers up to the delivery of the first reinforcement; percent drug-appropriate lever responding was then obtained by multiplication of that value by 100. For instance, imagine the efficacy of a diazepam for the treatment of anxiety is being

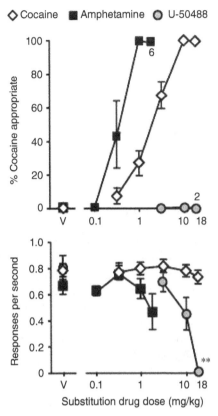

◇ Cocaine ■ Amphetamine ○ U-50488

FIGURE 3.5 Effects of cocaine, D-amphetamine, and the κ-agonist U-50488 in mice trained to discriminate cocaine from saline. Abscissa: drug dose in mg/kg; "V" indicates vehicle. Ordinates: percentage cocaine-appropriate responses (top), response rate in responses per second (bottom) (Reprinted with permission from Thomsen et al.[177] Copyright © 2010 American Society for Pharmacology and Experimental Therapeutics)

examined; the right-side lever will be designated as the diazepam-appropriate lever. On Monday, the animal is injected with 3.0 mg/kg of diazepam (15-minute pre-session injection), placed in its assigned operant chamber, and proceeds to press on the left-lever 9 times and the right-lever 10 times; food reinforcement (in this example) would be presented after the 10th right-side lever press. For this day, discriminative control would be assessed at 53% diazepam-appropriate responding (i.e., 10 right-side lever presses divided by 19 total lever presses × 100). On Tuesday, that same rat is injected with vehicle (15-minute pre-session injection), placed into its designated chamber, and presses the right lever 4 times and the left lever 10 times; food is presented after the 10th left-side press. On this day, discriminative control would be assessed at 29% diazepam-appropriate responding (i.e., 4 divided by 14 × 100).[178]

Back to our experiment, as shown in Figure 3.5, cocaine produced drug-appropriate responding in a dose-dependent manner, reaching 100% in all mice. Amphetamine reached 100% drug-appropriate responding in all mice, whereas no appreciable drug-appropriate responding was observed after U-50488 treatment up to a dose that almost eliminated responding.

This shows that amphetamine, as expected, is discriminated by the mice as "like cocaine" but the κ-opioid agonist is discriminated as "not like cocaine." From the leftward shift, we can also conclude that amphetamine is a more potent drug than cocaine.

For more details on drug discrimination, the excellent book *Drug Discrimination: Applications to Medicinal Chemistry and Drug Studies*, edited by Glennon and Young, is highly recommended.[178]

3.8   OPERANT SENSATION SEEKING MODEL

A new operant model termed "operant sensation seeking" was designed to analyze the effects of novelty and sensation seeking on drug use. In this model, the mice were presented a visual (light flash) and auditory stimulus (infusion pump noise) on pressing a lever. Note that no normal reinforcer such as food is given in response to the lever press. It was found the mice would do this without prior training. The validity of the model was supported in that like other operant studies using a reinforcer, mice responded on fixed- and progressive-ratio schedules, were resistant to extinction, and responded to dopamine antagonists.[179]

3.9   USE OF ANIMAL BEHAVIORAL MODELS

Do the results of preclinical animal models translate to human studies? There has been an immense amount of work published on the investigation of drugs of abuse in behavioral pharmacology models of drug addiction. How does it translate? Certainly, it is clear that all major drugs of abuse (except for some hallucinogens) that are taken by humans are self-administered by rodents and non-human primates. The abuse liability of drugs determined in animals translates well to humans. Unfortunately, drugs that appear to attenuate relapse in rodent models have not generally translated well in human clinical trials. As mentioned earlier, there is though some controversy in behavioral pharmacology on the validity of reinstatement models. Also this lack of translation may simply reflect the depth and complexity of addiction.

In a literature meta-analysis study, the preclinical versus clinical effects of 14 drugs acting on the monoamine transporter systems were analyzed under an agonist substitution paradigm. It was concluded that though the reinforcing and rewarding animal models were predictive of abuse liability in humans the animal models were not predictive of effective treatment outcomes.[180]

In an extensive and thorough study, Haney and Spealman investigated how medications based on an antagonist, agonist, or partial agonist approach influenced heroin and cocaine self-administration by rodents, non-human primates, and humans and to compare these results with clinical outcomes. They concluded that if the self-administration model was used it can reliably identify medications to treat opioid and cocaine addiction.[181]

ACKNOWLEDGMENTS

I would like to thank Morgane M. Thomsen, Ph.D., Assistant Professor of Psychiatry at McLean Hospital, Harvard Medical School, for a critical review and many helpful comments during the writing of this chapter. Errors if any are mine.

4

MEDICATION DEVELOPMENT FOR THE TREATMENT OF DRUG ADDICTION

Despite intensive efforts by scientists over the last 60 years, the development of safe and effective medications for the treatment of psychiatric disorders remains a great challenge.

We will examine pharmacological approaches used to treat addiction to opioids, stimulants, depressants, nicotine, and marijuana. Of these, pharmacological treatments of addiction to opioids, alcohol, and nicotine are most advanced. This could be due in part to the common occurrence of relapse in these disorders. As no drugs are yet approved for stimulant addiction this area is receiving extensive attention. In the following chapters, human clinical data will be emphasized as much as possible. There have been many studies that have investigated different drugs in animal models of self-administration, withdrawal, and relapse, with many promising results. Unfortunately, when these drugs and results have been transferred to the clinic, with very few exceptions, they have all failed.

There are many review articles available related to preclinical and clinical aspects of medication development approaches for substance abuse; however, there is a noticeable lack of review articles on the medicinal chemistry of addiction that have been written by medicinal chemists with their fellow medicinal chemists in mind. Recent reviews oriented toward the medicinal chemistry of addiction by Carroll, Newman, and Prisinzano contain structure–activity relationship (SAR) information that will be of interest to chemists.[182–184] This book will not discuss the chemistry of drugs of abuse. That subject has been amply addressed in many books and review articles. We will concentrate instead on the chemistry of compounds used to treat addiction.

The intricacies of the drug discovery process have been exhaustively discussed, so there is no need to add to that. Suffice it to say that drug discovery for CNS disorders has to address the added burden of drugs crossing the blood-brain barrier (BBB). For oral drugs that means there are three compartments that must be considered: passage through the intestinal membranes into systemic circulation, transport of the drug in blood to the brain, and then, transfer through another hydrophobic environment into the brain. I have hoped to show that many of the theories for drug addiction ascribe certain areas of the brain to distinct aspects of the addiction cycle. Then, in theory, targeted drug delivery to these anatomical distinct areas of the brain might be able to selectively alleviate aspects such as relapse. For now, this is beyond our reach so the medicinal chemical approach must still be the entry of a drug though the BBB and widespread distribution throughout the CNS.

Drug discovery for the treatment of addiction (or mood disorders in general) can be approached by two basic strategies. One is the standard method of understanding the therapeutically relevant biochemical target(s) of existing drugs and then using that knowledge to design new drugs against the known, and validated, target. A second would be to develop behavioral models of the illness (e.g., relapse) and then discover drugs to attenuate or prevent these processes. The first approach starts with *in vitro* models while the second uses *in vivo* models. In practice, no one method is superior and both are used in tandem to varying degrees depending on the information at hand.

It should be noted that the bulk of medication development for addiction is conducted by academic laboratories and by intramural programs at National Institute of Drug Abuse (NIDA). With perhaps the exception of the commercialization of varenicline (Chantix) for nicotine addiction, there does not appear to have been a serious effort by the major pharmaceutical companies to discover and develop drugs to treat addiction beyond nicotine. Several of the drugs we will examine have been approved to other indications. As their mechanism of action predicts they may have a therapeutic effect in addiction, they have been used in small-scale clinical trials. Although this off-label approach is important, it has not proven fruitful in discovering a drug. It does, however, add to our information about the disease state and can allow refinement of models and theories of addiction.

4.1   LEAD DISCOVERY

The accepted methodology of drug discovery and development is to discover a compound or set of compounds (hit), which have a level of accepted pharmacological activity against a validated target, optimize the pharmacological activity of the compound to generate a lead, and then develop the lead into a new chemical entity (NCE). The NCE will be subjected to a series of clinical trials to hopefully become a drug.

Lead discovery is concerned with finding the hit. There are many approaches ranging from testing thousands of compounds for binding or functional activity in high-throughput screening (HTS) to simply fortuitously discovering that something works.

Then, an obvious question is where to start. To do medicinal chemistry, the chemist needs a target, normally a receptor or enzyme linked with the disease. Addiction and other CNS disorders are difficult therapeutic areas as no single target has been identified, that if acted upon by an agonist or antagonist, could completely attenuate the disorder. As a starting point for the identification of targets, several areas of interest have been identified by NIDA and have been summarized in the 2010 Addiction Treatment Discovery Program (ATDP) Guide for Compound Submitters. From this compounds that are of special interest to NIDA are listed below along with their rationale for interest.

4.1.1   NIDA Addiction Treatment Discovery Program

CRF-1 Receptor Antagonists   The role of corticotropin-releasing factor (CRF) in drug addiction and the rationale for the development of CRF-1 receptor antagonists as treatments for drug dependence have been extensively reviewed.[153] In rat models of stress-induced relapse to drug use, CRF-1 antagonists have been shown to block footshock-induced reinstatement of responding for cocaine, heroin, and alcohol. These data suggest efficacy of CRF-1 antagonists in counteracting the widely acknowledged ability of stress to trigger relapse to multiple drugs of abuse. Such efficacy in multiple drug addiction disorders would be beneficial because abuse and addiction to a single compound is less common than polydrug abuse and addiction.

Multiple pharmaceutical companies have been working toward the development of CRF-1 antagonists for the treatment of depression and/or anxiety, but safety issues have greatly hampered the development of CRF-1 antagonists by the pharmaceutical industry thus far. The NIDA Medications Development Program nonetheless remains committed to facilitating the development of a CRF-1 antagonist for the treatment of drug addiction and would welcome private sector partnership when a developable compound is identified. Evaluation of such a compound in stress-induced relapse models within the ATDP would be a logical first step.

CB$_1$ Receptor Antagonists   Evidence that cannabinoid-1 (CB$_1$) receptor antagonists may prove useful in treating drug addiction disorders has been the subject of two recent reviews.[185,186] Particularly notable is the ability of CB$_1$ receptor antagonists to modulate the pharmacology of tetrahydrocannabinol (THC), nicotine, cocaine, methamphetamine, opiates, and ethanol, which has generated a high level of interest in this class of compounds. Unlike compounds that block the ability of stress to trigger drug-seeking behavior in animal models of relapse, CB$_1$ antagonists have been reported to act either by blocking the subjective/rewarding effects of drugs like THC, or by blocking the ability of conditioned cues to promote reinstatement of drug-seeking behavior in animals extinguished from drug self-administration.[187] A possible goal for medicinal chemists working toward the discovery of "second generation" compounds is the discovery of neutral CB$_1$ antagonists. Because it is possible that such compounds would demonstrate improved tolerability in drug-addicted populations, the NIDA ATDP would be interested in evaluating neutral CB_1 antagonists in relevant animal models of addiction.

Orexin Receptor Antagonists   The potential significance of orexin receptors (hypocretin receptors) as targets for the discovery of addiction treatment medications is best considered in the context of the rationale for pursuing CRF_1 and CB_1 antagonists (see above). It is noteworthy that the orexins appear to participate in the hypothalamo-pituitary-adrenal axis, with orexins mediating "hyperarousal" and "overactivation of emotional systems" following stress.[188] Further, it has been demonstrated that orexin-containing neurons in the lateral hypothalamus are innervated by CRF-containing nerve terminals and that stress-induced activation of c-fos in orexin-containing neurons is impaired in mice that are deficient in CRF_1 receptors. It follows that orexin receptor blockade may produce effects that are similar to CRF_1 receptor antagonism. In addition, CB_1 receptors and orexin-1 receptors appear capable of heterodimerization, with CB_1 receptor activation sensitizing orexin-1 receptors to activation by orexin A. If some CB_1 antagonist effects are mediated through the blockade of endocannabinoid-facilitated orexin receptor activation, then direct antagonism of orexin receptors may produce similar effects. Thus, the NIDA ATDP would welcome opportunities to evaluate orexin antagonists in animal models of stress-induced relapse to drug use and in models of conditioned cue- and drug priming-induced relapse. The orexin-1 receptor antagonist SB 334867 blocks the expression of morphine-conditioned place preference, suggesting an attenuated response to conditioned cues.[189] In addition, SB 334867 has been shown to block footshock-induced reinstatement of cocaine self-administration behavior. Further studies of this class of compounds in animal models of addiction are warranted.

Kappa-Opioid Receptor Antagonists   The concept of a protracted abstinence syndrome following withdrawal from chronic μ-opioid use is a well-known phenomenon characterized by dysphoric mood state. This dysphoric state may contribute to relapse, in that one possible reason for the resumption of μ-opioid use is the desire to ameliorate dysphoria with a euphorigenic drug. The hypothesis that the dysphoria of the protracted abstinence syndrome results from an up-regulation of the endogenous κ-opioid system is consistent with studies suggesting that chronic μ-opioid agonist treatment up-regulates κ-opioid receptors in mice and enhances behavioral responsivity to κ-opioid agonists in primates.[190–192] Administration of κ-opioid agonists in man is associated with dysphoria and psychotomimetic effect, which is also consistent with the proposed role of dynorphin, the endogenous κ-opioid agonist, in mood states associated with protracted abstinence.[193] Taken together, these findings suggest that κ-opioid receptor antagonists may block dysphoria experienced during protracted abstinence and, in turn, this may decrease the likelihood of relapse.

While the above rationale is specific for the potential usefulness of κ-opioid antagonists in treating addiction to heroin and other μ-opioids, recent studies have demonstrated a role for the κ-opioid system in the response of animals to stress, expanding the potential application of κ-opioid antagonists to the treatment of cocaine and other drug addiction disorders. Most notably, the κ-opioid antagonists nor-BNI (binaltorphimine) and JDTic have been shown to block stress-induced potentiation of cocaine-conditioned place preference and to block footshock-induced reinstatement of cocaine self-administration behavior, respectively.[194,195] The ability of a drug to prevent stress from triggering relapse has important implications for the treatment of polydrug abuse and addiction.

Glutamate Modulators   Reported interactions of virtually all drugs of abuse with glutamatergic systems in brain provide strong rationale for the pursuit of several related biochemical targets in NIDA's medications discovery and development efforts. Glutamatergic mechanisms in addiction include (1) a role for glutamate in stimulating dopamine systems related to reward and (2) a dopamine-independent role for glutamate in altering the effects of conditioned stimuli on behavior.[196] It has been proposed that the hallmark of addiction—an unmanageable motivation to take drugs—results from pathological changes in prefrontal-accumbens glutamate transmission.[197]

Data supporting a role for both group I and group II metabotropic glutamate receptors in addiction have been extensively reviewed.[198] A rationale for pursuing $mGluR_1$-selective agonists is supported by *ex vivo* results suggesting the reversal of cocaine-induced plastic changes in AMPA receptor redistribution by $mGluR_1$-activated long-term depression.[199] Support for pursuing $mGluR_1$-selective antagonists for the treatment of addiction comes from studies demonstrating that the synthetic antagonist EMQMCM inhibits cue-induced and drug priming-induced reinstatement of nicotine-seeking behavior in rats as well as the expression of sensitization to the locomotor effect of morphine in mice.[200,201] A rationale for pursing $mGluR_5$ antagonists as addiction treatments is supported by the results of $mGluR_5$ knockout studies and by reported effects of the $mGluR_5$ antagonist MPEP on self-administration of cocaine, nicotine, and alcohol.[202,203] Additionally, a rationale for pursuing $mGluR_{2/3}$ agonists is supported by the efficacy of LY379268 in rat models of cue-induced relapse to cocaine and heroin.[204,205] Three other potentially promising mechanisms of glutamate modulation for addiction treatment are AMPA receptor antagonism, N-acetylated-α-linked-acidic dipeptidase inhibition, and selective antagonism of NMDA receptors containing the 2B subunit.[206,207]

GABA Mimetics   Evidence suggests a role for $GABA_B$ receptor agonists and indirect GABA agonists such as vigabatrin (a GABA-transaminase inhibitor) and NNC-711 (a GABA uptake inhibitor) in the pharmacological treatment of addiction to cocaine and other drugs of abuse.

The role of the $GABA_B$ receptor in addiction and the development of agonists for the treatment of drug addiction have been reviewed.[71,208] For example, preclinical studies have found that the $GABA_B$ agonists baclofen and CGP 44532 reduce cocaine self-administration in the rat at doses that have minimal or no effect on food-maintained responding. This finding with baclofen and cocaine has been extended to include other drugs of abuse, including heroin, nicotine, and ethanol. Vigabatrin has been shown to block cocaine- and nicotine-induced conditioned place preference as well as cocaine self-administration in the rat. A rationale for how $GABA_B$

agonists produce the above effects has been proposed based on the finding that GABA$_B$ receptor agonists inhibit mesolimbic dopamine neurotransmission. To the extent that the mesolimbic dopamine system has been implicated in mediating the rewarding effects of drugs of abuse, this inhibitory effect of GABA$_B$ receptor agonists on dopamine activity could explain the attenuation in the reinforcing effects of cocaine and other drugs of abuse produced by direct and indirect GABA$_B$ agonists. Taken together, these findings suggest a role for GABA$_B$ receptors in attenuating the reinforcing effects of a number of different drugs of abuse and GABA-mimetic drugs might therefore be considered as potential broad-spectrum antagonist therapies for drug addiction. In addition, it is noteworthy that vigabatrin blocks an increase in nucleus accumbens dopamine caused by cocaine-associated cues. This suggests that GABA mimetics may have an important second mechanism of action in treating drug addiction; they may also be effective against cue triggers of relapse.[209] Compounds with GABA-mimetic activity have been or are in clinical trials for the treatment of addiction.

Dopamine D$_1$ Receptor Agonists There is converging evidence that dopamine D$_1$ receptors are important targets for therapeutic intervention in cocaine dependence. Dopamine D$_1$ receptors are down-regulated by cocaine self-administration in primates, and dopamine D$_1$ receptor agonists have been shown to block both cocaine priming and initiation of cocaine self-administration in rodents.[210,211] In squirrel monkeys trained to self-administer cocaine, dopamine D$_1$ agonists blocked or attenuated the effects of different priming doses of cocaine on reinstatement.[212] Finally, a dopamine D$_1$ agonist administered intravenously to humans in a proof of concept study was reported to blunt the subjective effects of smoked cocaine and to decrease cocaine craving.[213]

Dopamine D$_1$ receptors, like dopamine itself, are implicated in the effects of a number of drugs of abuse, including alcohol, morphine, nicotine, and methamphetamine, however, only a few studies have evaluated the effects of dopamine D$_1$ agonists on reinstatement or drug-taking behavior. Those that have been published have shown that dopamine D$_1$ agonists can dose-dependently decrease self-administration of ethanol in mice.[214]

The effects of dopamine D$_1$ agonist on cognition are also of interest and are suggestive of other mechanisms by which dopamine D$_1$ agonists may be useful treatments. Deficits in working memory are associated with dopamine dysregulation in the prefrontal cortex and have been extensively studied in experimental paradigms relevant to the cognitive deficits of schizophrenia. In a number of studies, it has been shown that dopamine D$_1$ agonists produce improvements in working memory function in primate and rodent models.[215] In light of the identification of working memory deficits that interfere with treatment in both methamphetamine and cocaine abusers, it is possible that dopamine D$_1$ agonists may be beneficial for (1) reducing drug use, (2) preventing relapse, and (3) for improving drug-induced cognitive deficits.[216,217] These effects could also be found for multiple drugs of abuse, suggesting that a dopamine D$_1$ agonist treatment would be highly desirable.

Dopamine D_3 Receptor Agonists and Antagonists   The role of the dopamine D_3 receptor in addiction and the development of ligands for the treatment of drug addiction have been well reviewed and will be discussed in more detail in Chapter 6. Preclinical results are summarized below. The dopamine D_3 receptors were cloned in 1990 and have been of particular interest to drug abuse researchers in part because they are selectively located in brain regions that are affected by drug abuse, and are up-regulated in the brains of cocaine overdose fatalities.[218] Agonists of these receptors produce behavioral effects in rodents that do not resemble stimulants, but are perceived as cocaine-like by rodents and primates. The potency of compounds that activate dopamine D_3 receptors is related to their ability to decrease cocaine self-administration in rats, suggesting the involvement of these receptor types in cocaine drug taking. In addition, dopamine D_3 partial agonists have been shown to block the behaviorally activating effects of cues that have been paired with cocaine in rats, suggesting potential usefulness in blocking relapse following contact with environmental cues associated with drug use. Dopamine D_3 antagonists have been reported to block nicotine-primed reinstatement of nicotine self-administration in rats as well as cocaine-primed cocaine seeking in rats. A dopamine D_3 antagonist has also been reported to dose-dependently block footshock-induced reinstatement of cocaine self-administration in rats, overall suggesting a potential role for dopamine D_3 antagonists in preventing two triggers of relapse. A dopamine D_3 antagonist has also been shown to block enhancement of electrical brain stimulation reward by cocaine, and D_3 antagonists have been reported to block both the acquisition and expression of nicotine, cocaine, and heroin-conditioned place preference in rats. Taken together, results from different laboratories using different behavioral endpoints and different compounds suggest that both dopamine D_3 partial agonists and D_3 antagonists may be useful treatments, and may be effective for more than one drug of abuse.

Compounds Inhibiting Reuptake or Stimulating Release of Biogenic Amines   Cocaine and methamphetamine withdrawal, which are characterized by hypoactive dopaminergic, noradrenergic, and/or serotonergic systems, purportedly can motivate continued drug use.[219] It has been suggested that indirectly acting "agonist therapies" (e.g., reuptake inhibitors, releasers) used to stimulate biogenic amine receptors may normalize brain function and break the cycle of drug use. This hypothesis has driven medication discovery and development efforts for more than a decade and has received substantial support from NIDA. Clinically available compounds, including marketed antidepressants and appetite suppressants, have been evaluated in clinical trials with primarily negative results, and numerous chemistry grants have been awarded to support the synthesis of novel compounds that act indirectly to stimulate biogenic amine receptors. In support of the hypothesis, a promising medication for the treatment of cocaine addiction at this time appears to be modafinil, and one of its pharmacological actions is modulation of biogenic amines.

In terms of funding for future research NIDA believes that given the substantial efforts that have already been devoted to compounds targeting biogenic amine reuptake and release, future drug development efforts involving this mechanism will require substantial justification or unique compound attributes. Chemists seeking support for continued work in this area may wish to focus on the discovery of uptake inhibitors with novel, theoretically desirable transporter selectivity profiles or on the discovery of compounds with novel NE (norepinephrine), DA (dopamine), and 5-HT (serotonin) releasing profiles. Notably, compounds lacking effects on NE reuptake or release but having equivalent effects on DA and 5-HT reuptake or release would be unique and of interest to the NIDA ATDP for profiling in preclinical contracts.

VMAT2 Inhibitors   In contrast to cocaine, which increases the levels of extra-synaptic neurotransmitters by inhibiting biogenic amine reuptake at membrane trans-porters, the actions of amphetamines are considerably more complex, suggesting additional medication targets such as the CNS-specific vesicular monoamine trans-porter (VMAT2) that was discussed in Chapter 2.5.7. Amphetamines, like cocaine, inhibit the reuptake of biogenic amines at their cell surface transporters; however, unlike cocaine, amphetamines also function as competitive substrates that are trans-ported into nerve terminals. Amphetamines have the additional ability to enter the cell by passive diffusion across the membrane. Once inside the cell, amphetamines promote exocytotic, calcium-dependent release of transmitter into the extracellular space. In addition to promoting exocytosis from vesicles near the cell membrane, amphetamines also cause vesicular leakage of biogenic amines into the cytosol through both alkalinization and interactions with the VMAT2. At the VMAT2, amphetamines bind to the tetrabenazine site, block reuptake of biogenic amines into the vesicle, and promote release of biogenic amines into the cytosol, where the neurotransmitters are available for extracellular release through reversal of the cell membrane transporters.

There is experimental evidence that compounds interacting with the tetrabenazine-binding site on the VMAT2, such as lobeline, have the ability to functionally antagonize the neurochemical and behavioral effects of amphetamine and metham-phetamine. Lobeline not only blocks the discriminative stimulus effects of metham-phetamine in rodents, but also it reduces methamphetamine self-administration.

Muscarinic M_5 ACh Receptor Agonists and Antagonists   The unique distribution of muscarinic M_5 ACh receptors (M_5 receptors) in brain and their apparent ability to modulate dopaminergic neurotransmission in brain areas relevant to the reinforcing effects of drugs are noteworthy. Pioneering studies in the early 1990s in rodents showed while muscarinic M_5 receptors represent only 2% of all muscarinic recep-tors in brain, they are the only muscarinic receptor subtype found on dopaminergic neurons of the substantia nigra and the ventral tegmental area.[220–222] Dopaminergic transmission from these midbrain neurons via projections to the nucleus accumbens has been hypothesized to mediate the reinforcing effects of most drugs of abuse, including cocaine, amphetamines, opiates, ethanol, and nicotine.

Lacking selective ligands for muscarinic acetylcholine receptor subtypes, researchers investigating the role of these receptors in normal physiology as well as their possible significance to drug abuse and addiction have relied on receptor knockout studies. Muscarinic M_5 deficient mice also exhibit attenuated withdrawal symptoms after prolonged morphine treatment and subsequent naloxone administration and show attenuated cocaine withdrawal-associated anxiety using the elevated plus maze. Mice deficient in M_5 receptors shows a reduced responsiveness to morphine and cocaine in conditioned place preference studies, suggesting a decrease in the reinforcing effects of these drugs.[82,223] Decreased sensitivity to cocaine in muscarinic M_5 deficient mice has been supported by recent cocaine self-administration studies in which decreased sensitivity was apparent at low to moderate unit doses of cocaine.[81]

Although knockout studies must be interpreted with caution, the results are intriguing and argue for evaluation of selective M_5 receptor agonist and antagonists in animal models relevant to drug addiction. Such studies will be critical for understanding the role of M_5 receptors in drug addiction disorders and for determining the merits of related medication development efforts.

Nociceptin/orphanin FQ (N/OFQ, NOP, ORL-1) Receptor Agonists　The opiate receptor-like nociceptin/orphanin FQ receptor shares a high degree of sequence homology with the other opioid receptors. Identified in 1994 by molecular cloning, the nociceptin/orphanin FQ receptor was initially a receptor with no known endogenous ligand or function (i.e., an "orphan receptor"). In 1995, the 17-amino-acid peptide that serves as the endogenous agonist for the nociceptin/orphanin FQ receptor was independently discovered by two laboratories and was named "nociceptin" by one laboratory and "orphanin FQ" by the other.[224,225] Although it is structurally related to the opioid peptide dynorphin A, nociceptin/orphanin FQ does not bind to any of the traditional opiate receptors. Likewise, opioid peptides do not activate the nociceptin/orphanin FQ receptor and most small drug molecules (including naloxone and naltrexone) that bind to μ-, δ-, κ-opioid receptors do not show appreciable affinity at the nociceptin/orphanin FQ receptor. While the nociceptin/orphanin FQ receptor was initially thought to have pronociceptive activity, the results of subsequent studies suggest that blockade of stress-induced analgesia was responsible for the related findings.[226] Consonant with an ability to oppose stress, nociceptin has demonstrated positive effects in preclinical screens for anxiolytic activity, and knockout mice lacking the nociceptin/orphanin FQ gene exhibit increased susceptibility and impaired adaptation to stress.

Given the established efficacy of naltrexone in treating alcohol dependence and the apparent ability of the nociceptin/orphanin FQ receptor system to functionally oppose many effects of traditional opioid agonists, several laboratories have evaluated nociceptin/orphanin FQ receptor agonists in animal models of alcoholism, with promising results. A recent review details investigations into the role of the nociceptin/orphanin FQ receptor in addiction.[130] Nociceptin and/or the synthetic nociceptin/orphanin FQ receptor agonist Ro 64-6198 have been shown to: (1) block the reinforcing effects of ethanol in both conditioned place preference and self-administration studies; (2) block the ability of ethanol to reinstate ethanol-conditioned place preference; (3) block the ability of conditioned cues to reinstate ethanol self-administration behavior; and (4) block the ability of a footshock stressor to reinstate ethanol self-administration behavior. While many of the observed effects are similar to those seen with naltrexone, the ability of nociceptin to block stress-induced reinstatement suggests a potential advantage of nociceptin/orphanin FQ agonists over naltrexone, which is inactive in this model. NIDA is very interested in profiling the effects of synthetic, long-acting N/OFQ agonists. Compounds under consideration for development as treatments for alcohol dependence would be of particular interest to determine their potential in treating polydrug abuse.

For interested academic and industrial researchers, NIDA conducts testing of exploratory compounds that address the above receptor systems, as while as other, under the ATDP.[227] The stated mission of ATDP is to discover potential pharmacological treatments for substance abuse, with an emphasis on relapse prevention, in humans through preclinical testing and evaluation of compounds. The ATDP accepts compounds from specific pharmacological classes such as above and for which there is preliminary data. Compounds of known pharmacology are profiled in relevant animal models, which will vary depending upon the compound's mechanism of action. Because of the focus on relapse prevention as a clinical endpoint, the program has a number of reinstatement models using different drugs of abuse. The ATDP evaluates compounds in models of relapse to cocaine, heroin, or methamphetamine, using stress, conditioned cues, or drug primes to produce reinstatement in rats whose self-administration behavior has been extinguished. Compound testing is shaped by existing data in rodents and the sequence of testing is determined in collaboration with the compound submitter. Other tests such as *in vitro* receptor-binding assays, toxicology, and mutagenicity are also available.

4.2 PHARMACOLOGICAL ASSAYS

Pharmacological assays to test the potency of drugs have ranged from using live animals, to using tissues that contain the receptors, and to the modern binding and functional assays that use whole cells or cell membrane fragments in which the cloned receptors of interest have been expressed. The latter method allows one to obtain a true binding or inhibition constant without the complication of excessive errant binding to off-target proteins. There are pros and cons to all methods and in practice combinations of all three are often used. Using live animals allows one to help determine a drugs bioavailability and if it achieves its purpose, however, one may lose information as to how the drug works. Using cloned receptors, one obtains very specific and detailed information on how a drug binds to a receptor, however, you lose information on how a drug may act in a living body.

Modern medicinal chemistry tends toward, first, obtaining binding affinity and functional information, and then, progresses to animal studies. It is worth taking some time then to review certain fundamental aspects of pharmacology. The following discussion is not meant to be an exhaustive survey of assay methods and theory but more toward how one can use the data. For detailed definitions of terms used in pharmacology, the International Union of Pharmacology Committee on Receptor Nomenclature and Drug Classification has published recommended symbols and terms for quantitative pharmacology.[228]

4.2.1 *In vitro* Binding Assays

In drug discovery, the most common assays are competition-binding assays where one examines the ability of the test compound to displace radiolabeled ligands from the receptor. The magnitude of binding will be expressed as a K_i or IC_{50} concentration value. To be able to interpret these values, it is worth some time and effort to explore how they are obtained. The following discussion on *in vitro* binding assays is taken from *Analyzing Binding Data* by Motulsky and Neubig.[1]

The first step is to obtain an equilibrium binding constant for the ligand that will be used as the radiolabeled standard. The equilibrium binding constant, K_d, is obtained by performing a saturation binding experiment. Binding of a ligand to a receptor is assumed to proceed by a model called the law of mass action. The binding is depicted as

$$\text{Ligand (L)} + \text{receptor (R)} \rightleftarrows \text{ligand} - \text{receptor complex (LR)}$$

or

$$L + R \underset{k_{off}}{\overset{k_{on}}{\rightleftarrows}} LR$$

and is based on the following ideas:

- Noncovalent binding occurs when ligand and receptor collide (due to diffusion) with the correct orientation and sufficient energy. The rate of association (number of binding events per unit of time) equals $[L][R]k_{on}$, where k_{on} is the association rate constant in units of M^{-1} min^{-1}.

- Once binding has occurred, the ligand and receptor remain bound together for a random amount of time. The rate of dissociation (number of dissociation events per unit time) equals $[LR]k_{off}$, where k_{off} is the dissociation rate constant expressed in units of min^{-1}.

- After dissociation, the ligand and receptor are the same as they were before binding.

and on the following assumptions:

- All receptors are equally accessible to the ligands
- All receptors are either free or are bound to the ligand (ignores partial binding).
- Neither ligand nor receptor are altered by binding (not true for agonists, which cause conformational changes in the receptor)
- Binding is reversible

With these ideas in hand, a saturation binding experiment is done where the receptor is treated with a wide concentration of ligand covering a 6-12 concentration range,

---

[1] Adapted with permission from Motulsky and Neubig.[684] Copyright © 2010 John Wiley & Sons, Inc.

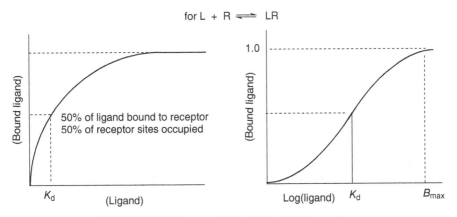

FIGURE 4.1 Prototypical saturation binding curves

and the amount of ligand that binds is determined. The ligand will also bind to other proteins and surfaces in the assay so this nonspecific binding is taken into account. When the amount of binding is plotted versus ligand concentration, graphs such as Figure 4.1 are typically obtained.

Several interesting pieces of information can be obtained from these graphs. The K_d, in units of mol/L, represents the concentration of ligand that occupies 50% of the receptors at equilibrium. When all the receptor sites are filled, we have maximum binding, B_{max}. We can get an idea of the affinity of the ligand for the receptor; if K_d is small the affinity is high. What is high affinity? The answer is subjective but for our discussions binding measured in pM to about 200 nM range will be considered high affinity, low to moderate affinity is in the 200 nM-µM range, and mM binding affinity will be considered as poor affinity. Poor affinity at the lead-finding stage is not necessarily bad. Fragment-based screening discovery tools, for example, SAR by NMR, often produce hits with K_d's of 100 µM to 10 mM. Typical values of K_d for useful ligands are 10 pM to 100 nM and typical values of B_{max} are 10–1000 fmol binding sites per mg of membrane protein. We can obtain values of K_d and B_{max} from the graphs above but in practice a Scatchard plot or a nonlinear regression method is used to obtain these values.

We now know the affinity of a ligand for its receptor, in practice the ligand used would have been radiolabeled with tritium (^{3}H) to allow binding to be detected and quantified. In theory, we can do this for any ligand or drug and measure its binding affinity for any receptor. Radiolabeling every compound you make in order to measure its binding affinity is not however a very efficient process of finding a drug; especially if you discover the compound you invested time and energy into introduce a radiolabel has poor binding affinity. To make this process more efficient, competition-binding experiments are done instead. Here, we will measure the ability of a nonradiolabeled compound to displace a tightly bound radiolabeled compound from the binding site.

Several pieces of information can be obtained:

- Pharmacologically identify a binding site. Perform competitive binding experiments with a series of drugs whose potencies at potential receptors of interest are known from functional experiments. Demonstrating that these drugs bind with the expected potencies, or at least the expected order of potency, helps prove that the radioligand has identified the correct receptor. This kind of experiment is crucial, because there is usually no point studying a binding site unless it has physiological significance.
- Determine whether a drug binds to the receptor. Thousands of compounds can be screened to find drugs that bind to the receptor. This can be faster and easier than other screening methods.
- Investigate the interaction of low-affinity drugs with receptors. Binding assays are usually only useful when the radioligand has a high affinity (K_d < 100 nM or so). A radioligand with low affinity generally has a fast dissociation rate constant, and so will not stay bound to the receptor while washing the filters. To study the binding of a low-affinity drug, use it as an unlabeled competitor.
- Determine receptor number and affinity by using the same compound as the labeled and unlabeled ligand.

When the experiment is performed, we will use the ligand from the equilibrium binding experiment as our standard. When the binding study is carried out, several concentrations of the unlabeled ligand or drug will be used and only one concentration of the radioligand standard is required. A typical result is shown in Figure 4.2.

The concentration of unlabeled drug that results in radioligand binding halfway between the upper and lower plateaus is called the IC_{50} (inhibitory concentration 50%), also called the EC_{50} (effective concentration 50%). The IC_{50} is the

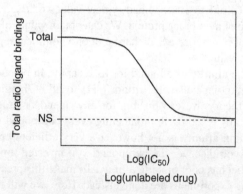

FIGURE 4.2 Plot of competition-binding experiment. NS, nonspecific binding (Reprinted with permission from Motulsky and Neubig.[684] Copyright © 2010 John Wiley & Sons, Inc.)

concentration of unlabeled drug that blocks half the specific binding, and it is determined by three factors:

- The K_i of the receptor for the competing drug. This is what is to be determined. It is the equilibrium dissociation constant for binding of the unlabeled drug, the concentration of the unlabeled drug that will bind to half the binding sites at equilibrium in the *absence* of radioligand or other competitors. The K_i is proportional to the IC_{50}. If the K_i is low (i.e., the affinity is high), the IC_{50} will also be low.
- The concentration of the radioligand. If a higher concentration of radioligand is used, it will take a larger concentration of unlabeled drug to compete for the binding. Therefore, increasing the concentration of radioligand will increase the IC_{50} without changing the K_i.
- The affinity of the radioligand for the receptor (K_d). It takes more unlabeled drug to compete for a tightly bound radioligand (small K_d) than for a loosely bound radioligand (high K_d). Using a radioligand with a smaller K_d (higher affinity) will increase the IC_{50}.

The final value of interest, K_i, is calculated from the IC_{50}, using the Cheng and Prusoff equation.

$$K_i = \frac{IC_{50}}{1 + \dfrac{[\text{radioligand}]}{K_d}}$$

Remember that K_i is a property of the receptor and unlabeled drug, while IC_{50} is a property of the experiment. By changing experimental conditions (changing the radioligand used or changing its concentration), the IC_{50} will change without affecting the K_i.

This equation is based on the following assumptions:

- Only a small fraction of either the labeled or unlabeled ligand has bound. This means that the free concentration is virtually the same as the added concentration.
- The receptors are homogeneous and all have the same affinity for the ligands.
- There is no cooperativity, binding to one binding site does not alter affinity at another site.
- The experiment has reached equilibrium.
- Binding is reversible and follows the law of mass action.
- The K_d of the radioligand is known from an experiment performed under similar conditions.

The analysis we have done assumes there is only one binding site and the radioligand and the drug being tested compete to bind to that one site. This can be determined

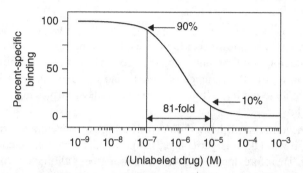

FIGURE 4.3   Competitive binding curve (Reprinted with permission from Motulsky and Neubig.[684] Copyright © 2010 John Wiley & Sons, Inc.)

by analyzing the shape of the curve in the semilog plot of [ligand] versus [bound ligand]. If the two ligands are competing for the same binding site, the slope of the curve is determined by the law of mass action. In Figure 4.3, the curve descends from 90% specific binding to 10% specific binding over almost 100-fold change in concentration.

If the slope of the line between 90% and 10% binding is −1.0, then it is said competitive binding is occurring. This is known as the Hill Slope. There can be several reasons explaining deviations from a slope of −1.0 such as different receptors in the assay, which compete with the ligands or more than one binding site (allosteric site) on the receptor where binding to one site influences the affinity of binding to another site.

This discussion was made to distinguish the importance between an IC_{50} and a K_i. They are not the same and need to be interpreted differently. An IC_{50} of 100 nM for compound A in a binding assay does not have the same binding affinity of compound B that has a $K_i = 100$ nM in the same assay. A frustrating component of medicinal chemistry is the wide range of binding constants that are sometimes reported for a receptor. We see how this can be a function of many variables. Philip Strange has written an interesting discussion on this topic focusing on measuring the interaction of antipsychotic drugs with the dopamine receptors and is well worth reading.[229]

For most common receptor-binding assays, radiolabel ligands are commercially available for use as a standard. If the receptor is new, then the researcher must develop an appropriate ligand and radiolabel it for use in the binding assays.

4.2.2   *In vitro* Functional Activity

After a binding constant is obtained, the next crucial piece of information is to measure the compound's functional activity to determine whether the compound is acting as an agonist, antagonist, or something in-between. As we discussed in

Chapter 2, agonists produce a response and antagonists do not. In principle, then any type of assay can be used as long as a response can be measured. At first this is done using animal tissue; for example, a detailed protocol describes the use of a classical model to study opioids using muscle from the guinea pig ileum, which has a large number of μ-opioid receptors.[230]

It is now common practice to express the receptor of interest in a cell line and then prepare a membrane homogenate that contains the receptor. As an example of this detailed procedures for the preparation of opioid and nociceptin/orphanin FQ receptor membrane homogenates are available.[231] For GPCRs, modern methods of determining agonist activity is by measuring the increase in $[^{35}S]GTP\gamma S$ binding or by the resultant effects of G-proteins on the second messenger components. Remember that upon the agonist binding to the receptor, the G-protein associates with the receptor and GTP displaces GDP. Use of radiolabeled GTP allows one to quantify this initial interaction. For second messenger analysis, an example is, agonist binding to a receptor coupled to $G_{i/o}$ proteins such as an opioid receptor will cause a reduction in cAMP production. The change in cAMP levels is determined, which is taken as a measure of agonist potency. The type and number of functional assays are quite varied and can be highly dependent on the physiological role of the receptor whether it is a GPCR, ion channel, or transporter.

Agonist activity is usually expressed in terms of EC_{50}, where EC_{50} is the concentration of an agonist that produces 50% of the maximal possible effect of the agonist. The maximal response is designated as E_{max}. In screening assays, the EC_{50} will be relative to a standard that is considered to give the maximal effect. Note that, then, it is possible to have >100% agonist potency, it just depends on the standard. Antagonist functional activity can be expressed as IC_{50}, which is usually interpreted as the concentration of an antagonist that reduces the response to an agonist by 50%. The term "pA_2" is sometimes used for an antagonist. It is the concentration of an antagonist that makes it necessary to double the concentration of an agonist that gave a maximal response in the absence of the antagonist. Antagonists can be further characterized as a competitive antagonist or a noncompetitive antagonist. A competitive antagonist will compete with the endogenous ligand for binding at the same location on the receptor, whereas a noncompetitive antagonist will not. A noncompetitive antagonist will bind elsewhere on the receptor and reduce or prevent the *action* of the agonist with or without any effect on the binding affinity of the agonist.

Complicating this picture is when the receptor is constitutively active. With a GPCR it means there will be certain basal level of activity even without an agonist being present. The theory behind this is that there are two different receptor states: an active form and an inactive form, which are in equilibrium with each other. If the ligand binds to the inactive form, it decreases the proportion of the active form and the amount of basal activity is reduced.

In these cases, we might find that a ligand affects functional activity by *reducing* the basal activity of the receptor. The ligands that do so have been called inverse agonist or negative antagonist. Some compounds that were originally classified as antagonists are now recognized as being inverse agonists. The antipsychotic drugs olanzapine and risperidone are now considered to be inverse agonists.[232] In fact,

Kenakin has calculated that of 380 antagonists studied in 73 GPCR targets, 322 were actually inverse agonists.[233]

It needs to be stressed that this phenomenon is observed under *in vitro* assay conditions, if the distinction between inverse agonist and antagonist is relevant to the clinic is open to discussion.[234] At issue for the chemist is whether it is relevant to make a "true" silent or neutral antagonist, or design an inverse agonist.

4.3   PARTIAL AGONIST APPROACH

What is clearly relevant to the clinic though is the concept of partial agonism. The drugs that have been approved by the FDA for the clinical treatment of drug addiction tend to share a common pharmacological mechanism of action by acting as partial agonists. In essence, this method "titrates" the patient to induce the feeling of a "high" to help reduce the probability of relapse. Partial agonists bind to a receptor at an orthosteric binding site. The orthosteric binding site is the region of the receptor where the endogenous neurotransmitter binds. The drug then competes with the neurotransmitter with binding to the receptor. A partial agonist will produce less of a functional effect than a full agonist. One should not conclude that a partial agonist automatically has a weaker binding affinity than a full agonist. There is not necessarily a proportional correlation between binding affinity and agonism.

What is the cause of partial agonism? It has been suggested that partial agonism found with the μ-opioid receptor is due to the number of receptors that are activated relative to the number activated by a full agonist. In other words, an agonist or a partial agonist will stabilize the same active conformation or conformations of the receptor, but the number of active conformations the partial agonist binds to is less than a full agonist.[235]

Some excellent reviews are available that expertly summarize both preclinical and clinical studies with respect to addiction using a partial agonist approach.[236,237]

4.4   ALLOSTERIC MODULATORS

Alternatively, drugs can bind to an allosteric site on the receptor. Allosteric binding sites are defined as being distant and separate from orthosteric sites. It is possible then that the drug and endogenous neurotransmitter can be simultaneously interacting with the receptor. Allosteric drugs can potentiate the effects of the natural ligand in which case the drug is called a positive allosteric modulator (PAM) or they can attenuate the effects of the natural ligand in which case the drug is called a negative allosteric modulator (NAM). Another way of thinking about negative allosteric modulators is that they are noncompetitive antagonists. The drug and the endogenous ligand do not compete with each other for the same binding site on the receptor. It is being discovered that some drugs originally thought to be agonists are in fact PAMs. One impetus to creating allosteric modulators is when high homology exists in a receptor family. In these situations, the amino acids lining the binding site within a receptor family may be very similar making it difficult to discover a highly selective drug. Perhaps then amino-acid differences outside the binding site can be used to differentiate between similar receptors.

On the basis of sequence homology and functional roles, 7TM receptors are commonly divided into three classes: A (e.g., M_1 mAChR), B (e.g., CRF_1), and C (e.g., $mGlu_5$). Most Class A (rhodopsin-like) GPCRs possess a distinctive orthosteric binding site (i.e., domain involved in docking of the endogenous ligand with the receptor), either deep within the helical bundle for small ligands or superficially across the extracellular loops and surface helical regions for larger ligands (e.g., small neuropeptides). The binding of an allosteric modulator may cause a conformational change in the receptor protein that is transmitted to the orthosteric site (and vice versa), in essence creating a "new" GPCR with its own set of binding and functional properties. In addition, allosteric modulators may engender collateral efficacy by biasing the stimulus, thus leading to signaling-pathway-selective allosteric modulation (either enhancement or blockade).[238] An example of the three classes of 7TM receptors and their orthosteric and allosteric binding sites and ligands is shown in Figure 4.4.

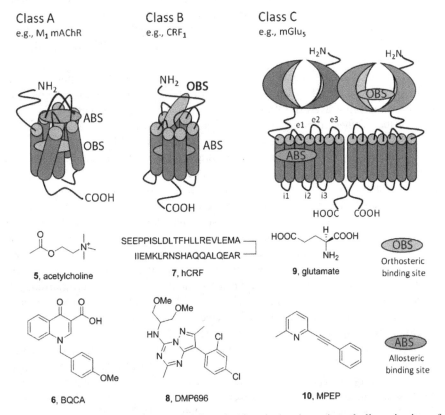

FIGURE 4.4 Structural topology and ligands of typical orthosteric and allosteric sites of 7TM receptors (Adapted with permission from Melancon et al.[239] Copyright © 2012 American Chemical Society)

It is important to remember that most allosteric modulators are pharmacologically inactive in the absence of the endogenous (orthosteric) ligand. This can be attractive for the design of therapeutic drugs. Allosteric modulators, with little inherent intrinsic activity that acts by enhancing or attenuating the response elicited by the endogenous ligand, offer several potential advantages over conventional agonists and antagonists in treating complex conditions and syndromes. Allosteric modulator binding effects are often saturable and therefore less likely to elicit adverse effects from overdose. As their functional effects are often exerted primarily in the presence of the endogenous ligand, modulator activity is tied to the homeostatic patterns of synaptic signaling or endocrine hormone release. Therefore, they only amplify or reduce the receptor signal when the hormone or neurotransmitter is released. Hence, an allosteric modulator will preserve the physiological receptor signaling hierarchy in complex neuroendocrine axes and neurotransmitter events, while at the same time, boosting the efficiency of the endogenous neurotransmitter/hormone. Furthermore, a lack of chronic receptor activation by the allosteric modulator may cause less receptor desensitization or internalization over time, overcoming the problem of diminishing therapeutic efficacy that is seen with many chronically administered orthosteric agonists. Allosteric modulators can also bias signal output in favor of only part of the receptor response profile. This results from conformational constraints placed on the receptor that limit its ability to engage effector/accessory proteins, for example, GRKs or arrestins, as well as G-proteins.[239]

This property is well known for orthosteric ligands, such as morphine, which promote G-protein coupling without causing arrestin-dependent desensitization. The physiological responsiveness to psychostimulants and morphine suggests the involvement of these other GPCR regulators in neurological/pathological states, such as addiction, Parkinson's disease, aging, mood disorders, and schizophrenia.[238]

To help identify allosteric binding sites, binding models have been developed for allosteric modulation. The simplest model is called the allosteric ternary complex model and assumes mass action. From this, one can derive a cooperativity factor α. The cooperativity factor is a thermodynamic measure of the strength and direction of the allosteric change in affinity for one site when the other is occupied. Positive allosteric modulators can be defined as $\alpha > 1$ and negative allosteric modulators as $\alpha < 1$. We will not have a detailed discussion on the measurement of allosteric modulation. A simplistic view is one determines if a compound has functional activity on (1) its own and in the (2) presence of the endogenous ligand (e.g., neurotransmitter). If the answer is "no" to (1) and "yes" to (2), then you have an allosteric modulator. Leach, Sexton, and Christopoulos have written detailed protocols for assay development and data interpretation for allosteric modulation.[240]

The Vanderbilt University team of Lindsley and Conn has pioneered the development of allosteric modulators for the acetylcholine muscarinic and metabotropic glutamate receptors. They have recently written an excellent review on the use of allosteric modulation for CNS drug discovery.[239] Some examples of allosteric modulation functional assays from this review are shown in Figure 4.5.

Plot A shows an example of a "triple add" screen that allows for, in a single assay, the identification of PAMs, negative allosteric modulators, agonists, and antagonists.

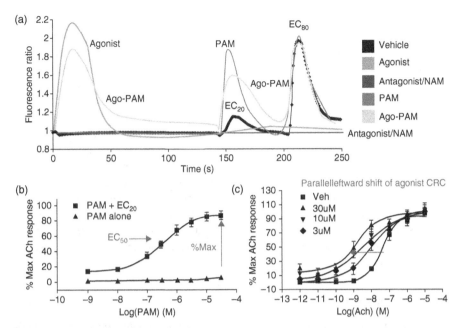

FIGURE 4.5   Functional assays, measuring calcium fluorescence as a surrogate for 7TMR receptor activation, employed to identify and profile 7TMR allosteric modulators (Adapted with permission from Melancon et al.[239] Copyright © 2012 American Chemical Society)

The "triple add" paradigm can be used for both HTS campaigns and primary assay for lead optimization. The vehicle is the black trace and an EC_{20} of orthosteric agonist is added 150 seconds into the kinetic run, followed by an EC_{80} of orthosteric agonist at about 220 seconds. Compounds are added at $T = 0$, and an agonist (green in ebook edition) elicits calcium fluorescence immediately upon addition. Secondary assays with orthosteric radioligands and/or mutant receptors will determine whether the compound is an orthosteric or allosteric agonist. An antagonist (red) will block both the EC_{20} and EC_{80}; once again, secondary assays will distinguish competitive from noncompetitive antagonists. A pure PAM (blue) will not elicit receptor activation alone but will potentiate the EC_{20} to varying degrees of efficacy, while an agonist-PAM (orange) will activate the receptor alone, plus potentiate the EC_{20}. Hence, a single assay protocol will identify agonists, allosteric agonists, PAMs, ago-PAMs, antagonists, and negative allosteric modulators.

Plot B shows the *in vitro* pharmacology of a mAChR PAM, once again with calcium fluorescence readout. The PAM has no effect alone on receptor activation, but in the presence of an EC_{20} (or subthreshold concentration of orthosteric agonist, acetylcholine in this case), a classical concentration–response curve results, from which an EC_{50} for potentiation can be calculated. Also, the %Max, the degree of potentiation above the EC_{20}, can be measured, and both the EC_{50} and %Max must be optimized.

Plot C examines a "fold shift" assay. Here, the concentration of the orthosteric agonist (acetylcholine) is held constant, and increasing concentrations of the PAM causes a parallel leftward shift of the acetylcholine concentration–response curve, in effect making acetylcholine a more potent agonist. If the fold shift of the agonist curve is to the left then the allosteric modulator potentiates the activity of the acetylcholine, while if the shift is right/down, it antagonizes the activity of the acetylcholine. In this case, we can see that the compound is a PAM.

The pharmaceutical industry has developed an interest in the development of allosteric modulators. In fact, the Swiss company Addex Therapeutics has built a pipeline of CNS-active compounds based on allosteric modulation.

In summary, a compound could be an orthosteric ligand where it competes with the endogenous ligand at the binding site and shows functional activity as a full, partial, or inverse agonist, or be an antagonist. Alternatively, the compound could be an allosteric ligand where it binds at a site distinct from the endogenous ligand and acts as a positive allosteric modulator or a negative allosteric modulator. These are our main choices.

4.4.1 Biased Ligands

Related to the concept of allosteric modulation is that of biased ligands. Many ligands that behave as inverse agonists or classical "neutral" antagonists for one effector pathway have been found to exert opposing effects, for example, agonist or partial agonist activity, for receptor coupling to an alternative G-protein or G-protein-independent pathway. The phenomenon of ligand coupling a receptor to only a subset of its potential effectors is known by several different terms, including functional selectivity, agonist-directed trafficking of receptor stimulus, biased agonism, differential engagement, and stimulus trafficking.[238] It is considered that the receptor can exist in different conformational states, where each state has a different level of activity. Binding to one state may produce a particular functional response, whereas binding to a different conformational state will produce a different functional response. It is proposed that allosteric modulator binding may also affect receptor conformation to favor certain active states or change the interaction of the receptor with other trans-membrane proteins, biasing the signal output generated by endogenous orthosteric ligands.[241] Kenakin and Christopoulos have reviewed different binding models of biased agonism and potential impacts on drug discovery.[242]

4.5 FUNCTIONAL INTERACTIONS BETWEEN RECEPTORS

There is accumulating pharmacological evidence that functional interactions exist between different neurotransmitter systems. More specifically, different receptors in a synapse appear to interact in some fashion to influence their response to agonist binding. The molecular mechanism by which this occurs is unknown. The concept of proteins forming dimers or oligomers is not new. We have seen in fact how the ion channels are composed of pentameric protein subunits that are inactive on their own.

Only upon combination of the five protein subunits is a fully functional ion channel formed. What is new here is the thought that receptors may allosterically interact to form "new" functional signaling complexes. It may be possible to have "indirect" allosteric modulation involving the regulation of GPCR heterodimers, wherein an orthosteric or allosteric ligand for one receptor modulates the signaling of the dimer partner through conformational changes transmitted by contact between receptor transmembrane domains.[238]

The functional interactions or dimerization of opioid receptors have received extensive attention so we can use this as an example. The existence of opioid dimers was proposed to explain pharmacological results that were originally ascribed to subtypes of the μ-, δ-, and κ-opioid receptors.[243,244] The possibility of G-protein-coupled receptor opioid heterodimers as drug targets has been reviewed and several excellent reviews have been written about GPCRs and on approaches to calculating affinity constants of ligands with dimeric receptors.[245–251]

There is substantial *in vitro* and *in vivo* pharmacological evidence for the existence of opioid heterodimers or at least a functional interaction of some nature between receptors. It is worth examining some of these pharmacological studies in more detail.

In vitro studies showed that the pharmacological profile of cells coexpressing μ-opioid and δ-opioid receptors was different from cells singly expressing the receptors.[252] The μ-opioid receptor was tagged with a c-Myc epitope (10 amino acids) and the δ-opioid receptor was tagged with a FLAG epitope (eight amino acids) and the presence of dimers and higher order oligomers was determined using gel electrophoresis. When the receptors were expressed separately no distinct heterodimer or hetero-oligomer bands could be detected, whereas the monomers, homodimers, and tetramers of each were easily discerned. When the receptors were coexpressed in the same cell the presence of an interaction was determined by differential immunoprecipitation, where c-Myc-tagged μ-opioid receptors were coprecipitated by the anti-FLAG antibody and FLAG-tagged δ-opioid receptors were coprecipitated by the anti-Myc antibody.

There was a reduction in affinity of selective μ-opioid agonists and δ-opioid agonists toward cells coexpressing the receptors. For example, morphine bound to the μ-opioid receptor with a $K_i = 0.22$ nM, but to the μ-δ coexpressed receptor mixture with $K_i = 1.6$ nM. More striking was the binding affinity change for the selective μ-opioid peptide agonist DAMGO. DAMGO bound to μ-opioid receptor with a $K_i = 3.3$ nM and to the δ-opioid receptor with a $K_i = 5296$ nM; however, with a $K_i = 31$ nm to the μ-δ coexpressed receptor mixture. The same trend was observed for δ-opioid selective ligands.

Effects on desensitization and internalization were observed. In cells coexpressing μ-opioid and δ-opioid receptors DAMGO-induced internalization of receptors was enhanced while the δ-opioid peptide agonist DPDPE did not induce internalization.

The functional activity was also affected. A common cell biology technique to determine whether the G_i-protein is involved in signal transduction is to add pertussis toxin. This toxin catalyzes the addition of ribose to the α_I-subunit of the G_i-protein, the resultant glycoprotein conjugate (ribose-G_i protein) binds poorly to receptors, in this case the μ- and δ-opioid receptors. In the presence of pertussis toxin, agonism of the μ- and δ-opioid receptors will result in no functional effects being observed. Agonism of the $G_{i/o}$-coupled μ- and δ-opioid receptors will inhibit adenylyl cyclase reducing levels of cAMP. Membranes expressing only the μ-opioid receptor revealed inhibition of cAMP production by DAMGO in a dose-responsive manner with an $EC_{50} = 6$ nM. When pertussis toxin was added, the activity was completely abolished and there was no effect on cAMP production. Similarly, in membranes expressing δ-opioid receptors, DPDPE inhibited cAMP production with an $EC_{50} = 10$ nM; again pertussis toxin prevented DPDPE from inhibiting cAMP production. However, in membranes coexpressing the μ- and δ-receptors, both DAMGO and DPDPE were able to inhibit cAMP production even in the presence of pertussis toxin. The authors suggest this may be due to the recruitment of another G-protein that is not pertussis toxin sensitive.[252]

In a study suggesting a functional interaction between opioid receptors, both agonists and antagonists of δ-opioid receptors have been shown to potentiate the actions of morphine on the μ-opioid receptor. In particular, the δ-opioid peptide agonist, deltorphin II, increased the effect of morphine-mediated $[^{35}S]$GTPγS binding in mouse spinal cord membranes from $EC_{50} = 198$-83 nM. In CHO cells expressing only the μ-opioid receptor, coexpressing μ- and δ-opioid receptors, and in K-N-SH cells that endogenously express μ- and δ-opioid receptors, deltorphin II approximately doubled the amount of bound DAMGO and morphine.[253]

Other evidence of opioid receptor interactions comes from animal studies. *In vivo*, the δ-opioid peptide agonist DPDPE has been shown to potentiate the antinociceptive activity of the μ-opioid agonist DAMGO when administered to the spinal cord of mice.[254] This potentiating affect was lost though in morphine tolerant mice. In another spinal study in mice, it was shown via isobolographic analysis that DPDPE had a multiplicative effect on the dose-dependent increase in hot plate response latency of the μ-opioid agonists DAMGO and morphine.[255] Interestingly, administration of DPDPE (intracerebroventricular; i.c.v.) to mice potentiated the antinociception of the μ-agonist morphine and normorphine but not that of other μ-agonists such as etorphine and sufentanil.[256]

Behavioral pharmacological studies in rhesus monkeys have further demonstrated functional interactions between opioid receptors. The mixed μ/κ-opioid agonist MCL-101 could reduce cocaine self-administration without the side effects seen with a pure κ-opioid agonists.[257]

A polypharmacy dose containing the μ-opioid agonist heroin and the selective δ-opioid agonist SNC80 ($IC_{50} = 0.94$ nM)[258] given to rhesus monkeys produced additive effects in a responding assay and superadditive effects in a thermal nociception assay. The presence of SNC80 did not, however, affect the reinforcing effects of heroin in the self-administration assay.[259]

The effect of μ/δ-opioid and δ/κ-opioid drug combinations on sedation and analgesia in rhesus monkeys was investigated via a dose-addition analysis. It was found that SNC80 had no measurable analgesic effect but did produce dose-dependent sedation. As expected, the μ-opioid agonists methadone, fentanyl, and morphine produced dose-dependent analgesia and sedation. Interestingly, a combination of SNC80 and an μ-opioid agonist (e.g., fentanyl, methadone) induced a leftward shift (superadditivity) on a thermal analgesic response. This was not seen with an δ-opioid agonist/κ-opioid agonist combination. The μ/δ-opioid combination effect on sedation was either additive or subadditive and δ-opioid agonists also did not enhance the abuse-related effects of the μ-opioid agonists.[260] This demonstrates that an agonist acting on both the μ-opioid and δ-opioid receptors may possibly produce a strong analgesic effect without a concomitant increase in side effects such as sedation and abuse liability. The authors conclude that the pharmacological profile of a μ/δ-opioid agonist combination may be clinically favorable as an analgesic compared with that of the component drugs separately.

With regard to drug abuse treatment, μ-opioid agonist/partial agonist therapy, for example, methadone, has proven useful for the treatment of opioid addiction. It has been shown that chronic treatment of rhesus monkeys with drugs that are κ-opioid agonists and μ-opioid partial agonists can dose-dependently reduce cocaine self-administration.[257]

The μ/δ-opioid connection has been further clarified by Negus where they found that SNC80/methadone mixtures at proportions previously shown to enhance methadone-induced antinociception did not enhance the abuse-related effects of the μ-opioid agonist methadone in rhesus monkeys. These results support the proposition that δ-opioid agonists may selectively enhance μ-opioid agonist analgesic effects without enhancing μ-opioid agonist abuse liability. It was also found that SNC80 alone did not affect cocaine self-administration.[261]

4.5.1 CB$_1$/μ-opioid Synergism

There is growing evidence of a (μ-opioid)(CB$_1$) functional interaction, especially with regard to pain. The psychoactive compound Δ^9-THC has for some time been known to reduce opioid withdrawal symptoms. It appears clear that at a fundamental neurological level a functional interaction between the CB$_1$ and μ-opioid receptor occurs.[262] Recent studies in rodents have indicated that when an opioid agonist is co-administered with a cannabinoid agonist, at sub-analgesic doses of both drugs, an analgesic effect of normal magnitude is achieved. For example, a sub-active oral dosage of 20 mg/kg morphine in combination with a sub-active oral dosage of 20 mg/kg Δ^9-THC was as effective as a high dose of morphine at 80 mg/kg p.o. (ED$_{80}$ morphine p.o. is 80 mg/kg).[263]

In another study by Welch who explored potential mechanisms of synergism, rats received morphine and/or Δ^9-THC and the effect on a paw withdrawal test was determined. The paw pressure test consisted of gently holding the body of the rat while the hind paw was exposed to increasing mechanical pressure. The force (measured in grams) at which the rat removed its paw was defined as the paw pressure threshold. Antinociception was then quantified by calculating % Maximum Possible Effect (% MPE).

The effects of chronic morphine treatment were determined by subcutaneous (s.c.) implantation of six 75-mg morphine pellets (Figure 4.6A) or by injections of 100–200 mg/kg morphine twice daily (Figure 4.6D). Chronic Δ^9-THC was administered

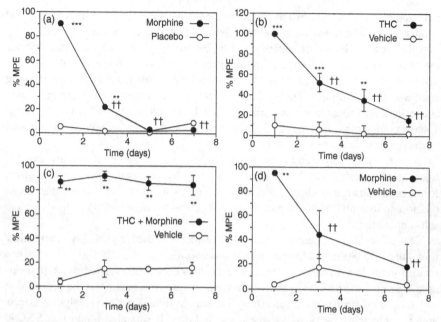

FIGURE 4.6 Morphine THC synergism (Reprinted with permission from Smith et al.[685] Copyright © 2007 Elsevier B.V.)

by injection at 4 mg/kg i.p. twice daily (Figure 4.6B). The low dose combination was 0.75 mg/kg each of morphine (s.c.) and THC (i.p.) twice daily (Figure 4.6C). The study was conducted for 6.5 days.

The low dose combination produces the same antinociceptive effect (%MPE) after 7 days as either morphine of THC produced at day 1. Interestingly, we can see whereas chronic administration of morphine or THC produced antinociceptive tolerance to the respective drugs, combination treatment did not produce tolerance. They also determined in tissue functional assays that combination treatment did not alter CB_1 receptor- or μ-opioid receptor-stimulated G-protein activity. Both CB_1 and μ-opioid receptors are G_i/G_o coupled and activation of both by agonists (e.g., Δ^9-THC and morphine) attenuates receptor-stimulated G-protein functional activity such as $[^{35}S]GTP\gamma S$ binding.[264]

This indicates that the two drugs, activating different neuronal circuits, act in a synergistic fashion. This synergism of antinociception can occur without the development of tolerance to either agonist and without attenuation of receptor activity. How this occurs is unknown though it is clear that there exists either an indirect or direct interaction between the cannabinoid and opioid neuronal systems.

Of the three opioid receptors, this interaction appears to predominately involve the μ-opioid receptor.[264] The antinociceptive studies tend to involve the use of Δ^9-THC or other highly potent cannabinoid ligands that bind to both the CB_1 and CB_2 receptors. CB_1 tends to be distributed in the central nervous system while CB_2 is distributed

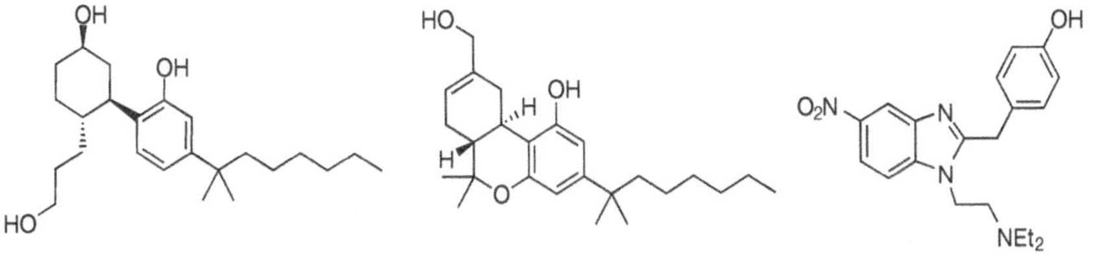

CP55940
K_i CB$_1$ = 0.58 nM
K_i CB$_2$ = 0.69 nM

HU 210
K_i CB$_1$ = 0.0.061 nM
K_i CB$_2$ = 0.52 nM

Etonitazene
K_i μ-opioid = 0.2 nM
K_i δ-opioid = 176 nM
K_i κ-opioid = 233 nM

FIGURE 4.7   CB$_1$ and μ-opioid agonists

mainly in the peripheral nervous system.[113] Presumably, the neural analgesic effects and interaction with the opioid system involves the CB$_1$ receptor but the CB$_2$ cannot be necessarily ruled-out.

Simultaneous modulation of the cannabinoid and opioid systems has been shown to reduce drug self-administration. In particular, the combined self-administration of the nonselective CB agonist CP55940 (Figure 4.7) and heroin in rats resulted in a significant reduction in the drug combination-associated lever pressings compared with the drugs alone.[265] A likewise dramatic effect was found with CP55940 and the selective μ-opioid receptor agonist etonitazene (Figure 4.7). Furthermore, the endogenous CB$_1$ agonist 2-arachidonoylglycerol and the synthetic CB$_1$ agonist HU-210 (Figure 4.7) have been shown to attenuate the effects of naloxone-precipitated withdrawal in morphine-dependent mice.[266] This indicates that compounds with dual agonist pharmacological properties at the CB$_1$ and μ-opioid receptors may be of use in the treatment of addiction and require further investigation.

4.5.2   Bivalent Ligands and Receptor Dimers

A question now arises as to: Is this interaction due to receptors physically interacting with each other in the membrane to form a complex that behaves in a functionally different manner? Or, does simultaneous binding to the "free" receptors by different ligands cause some change in the resultant second messenger cascade? One method to explore the former hypothesis has been the development of bivalent ligands. These ligands have been synthesized to explore potential interactions with homo- or heterodimers and what functional effects might result. In theory if two receptors are interacting in a membrane and their respective ligand binding sites are not blocked or perturbed by this interaction, a compound containing agonists or antagonists for each binding site should bind to both receptors. One way to achieve this is to simply take the two ligands and connect them together by a long enough linking unit that can span across the receptors and allow each ligand to bind. Any combination of ligands is feasible. One can link identical agonists to form a homobivalent agonist ligand or link the agonist with a different agonist to form a heterobivalent ligand. One can readily

α-oxymorphamine   7'-aminonaltindole

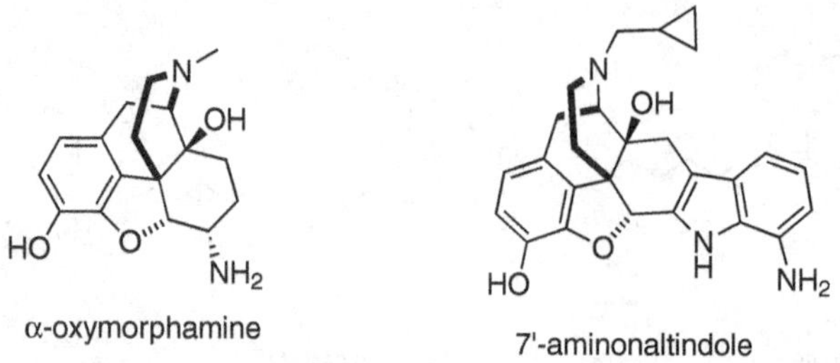

FIGURE 4.8   Ligands used for MDAN series

imagine then different combinations of agonist-antagonist heterobivalent ligands, antagonist-antagonist homo- and heterobivalent ligands, etc. As most drugs contain functional groups such as amines or alcohols, the synthesis of the bivalent ligands can be straightforward. One can take, for example, an α, ω-dicarboxylic acid (e.g., adipic acid $HOOC(CH_2)_4COOH$) and couple it to two different amine-containing drugs to form a bis-amide.

It should be noted that this is at times a controversial subject. The bivalent ligands by nature are large hydrophobic compounds whose bioavailability to the brain is questionable. One problem is determining whether the binding affinity and functional activity of the bivalent ligand are dramatically different than the two ligands administered together. To help address this, some binding models have been developed to examine binding data to determine whether bivalent ligand cooperativity exists.[249,267,268]

A number of groups, especially those of Portoghese and Neumeyer, have investigated the SAR of bivalent ligands in opioid dimerization. From the studies of Portoghese, the MDAN series of bivalent ligands have been examined in some behavioral pharmacology assays. We can examine this series as an example.

MDAN stands for μ-opioid receptor (M)-δ-opioid receptor (D)-agonist (A)-antagonists (N), where the μ-opioid agonist α-oxymorphamine was connected to the δ-opioid antagonist 7'-aminonaltrindole (Figure 4.8) by a linker consisting of a central diamine flanked by diglycolic acids on each end. The pharmacophores are connected to the linker or spacer by amide bonds. The total space between the pharmacophores is 14–21 atoms.

The pharmacological properties of the compounds were examined using a radiant heat tail-flick assay as a measurement of their analgesic properties. The bivalent ligands functioned as agonists in acute intracerebroventricular studies in mice, with potencies ranging from 1.6- to 45-fold greater than morphine. As a somewhat confusing issue that is often observed, the monovalent μ-opioid agonist α-oxymorphamine analogues (7'-aminonaltrindole not attached) were substantially more potent than the MDAN bivalent ligands and were essentially equipotent with one another and oxymorphone. Chronic intracerebroventricular studies revealed that MDAN ligands whose spacer was 16 atoms or longer produced less dependence than either morphine or a monovalent μ-opioid control, in fact no dependence or tolerance was observed

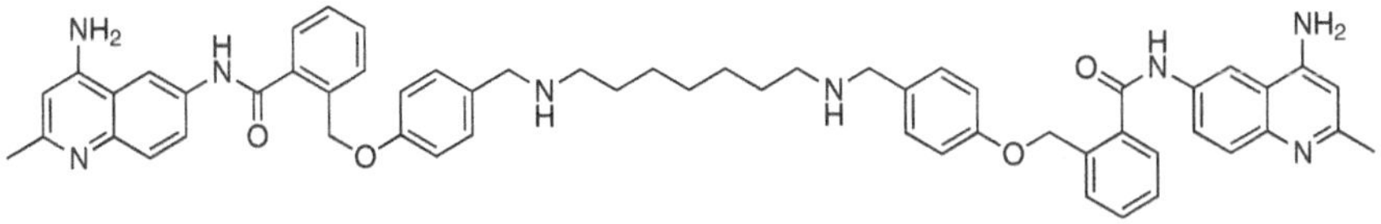

NOP antagonist
IC$_{50}$ = 31 nM

FIGURE 4.9   Structure of NOP bivalent antagonist

with an atom spacer of $\geq$ 19 atoms though analgesic activity was maintained. Most interesting was that one of the bivalent ligands with a spacer of 21 atoms was 50-fold more potent than morphine regardless of whether administration was intracerebroventricular or intravenous, strongly suggesting the compound can cross the BBB, which seems quite remarkable.[269]

A second study compared the rewarding properties of three bivalent MDAN ligands and a μ-opioid receptor agonist monovalent ligand with those of morphine in the conditioned place preference assay in mice after intravenous administration. Place preference developed to morphine and to the monovalent agonist, but not to the MDANs. Reinstatement was also evaluated after extinguishing morphine-conditioned place preference; morphine and the monovalent agonist reinstated morphine-conditioned place preference but the bivalent ligands did not. The authors suggest that the bivalent ligands are less rewarding compared with morphine in opioid-naïve mice and do not induce reinstatement in previously morphine preferring mice.[270]

How these bivalent ligands may be interacting with the receptors is an interesting question. Beyond pharmacological and cell biology procedures such as fluorescence-detected coimmunoprecipitation and bioluminescence resonance energy transfer (BRET), little is known. BRET is often used to determine whether receptor dimerization is occurring. An item of interest to the author that may relate to this question is the recent study by the research group of Borioni, where they used 700 and 400 MHz NMR spectroscopy to investigate the conformation of a nociceptin opioid peptide (NOP) bivalent antagonist (Figure 4.9) in liposomes.

Using 2D NOESY data and molecular modeling calculations as conformational restraints, it was determined that the bivalent ligand in the presence of liposomes strongly interacted with the bilayer and assumed a preferential folded conformation with the quinoline arms superimposing on each other as shown in Figure 4.10.[271]

The reader interested in more information on receptor dimerization in addiction and on bivalent ligands is directed to some recent reviews.[273,274]

4.6   MULTI-TARGET DRUGS

Drug discovery programs typically take the one-compound-one-target approach, where the goal is to identify and validate a target unique to the disorder. Then,

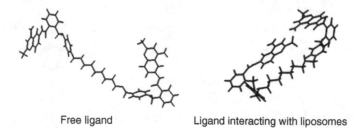

Free ligand   Ligand interacting with liposomes

FIGURE 4.10   Solution versus liposome conformations of NOP bivalent ligand (Reprinted with permission from Borioni et al.[271] Copyright © 2010 Elsevier Inc.)

design, synthesize, and develop a compound with outstanding selectivity and binding affinity for that target. In theory, side effects due to extraneous off-target binding or due to high drug dosage would be eliminated or reduced. Most modern drugs are designed along this principle. Though great advances have been made in, especially in terms of safety, the pharmacological treatment of psychiatric illness is still difficult. The first drugs developed in the 1950s for the treatment of schizophrenia and depression proved to be compounds with complex pharmacological profiles in which the drug acted upon various receptors. The drugs were a breakthrough in the treatment of mental illness as before them, electroshock therapy or other methods were the only "treatments" available. However, the initial drugs, for example, imipramine, chlorpromazine, had multiple side effects that made them unattractive.

The PDSP database lists submicromolar affinities (K_i) for chlorpromazine at nine different 5-HT receptors, six α-adrenergic receptors, five muscarinic receptors, five dopamine receptors, three histamine receptors, and the imidazoline I1 receptor (for a total of 29 different receptors). In addition, the PubChem database lists more than 20 different additional screening assays for which chlorpromazine was considered to be an "active" compound. Although D_2 antagonism is thought to reside at the core of chlorpromazine's effectiveness as an antipsychotic medication, the extent to which these other activities contribute to its effectiveness, and to its side-effect profile, remains only partially understood. The first of the atypical antipsychotics to be developed, clozapine, has a similarly promiscuous profile, with submicromolar affinities for ten 5-HT receptors, five α-adrenergic receptors, five muscarinic receptors, five dopamine receptors, and three histamine receptors (for a total of 28 different receptors) in the PDSP database. Clozapine is listed as "active" at more than a dozen additional targets in the PubChem database.[272] This theme continues today. Gray and Roth reported in 2007 that of 82 compounds with known targets that are in use or in development for the treatment of schizophrenia, 34 (41%) were designed to have multiple therapeutic targets and are likely or known to have many additional targets as well.[273]

Drug abuse, as with other psychiatric disorders, is a multimodal disorder. In other words, a complex interaction among different neurocircuits is most likely at the root of drug abuse. As such, one can question if a one-compound-one-target approach would ever be useful. It may indeed be more fruitful to return to the development of a drug that interacts in a multimodal fashion. The design of such a drug that can be directed to bind to a specific set of targets with needed efficacy while excluding binding to other targets is of course quite difficult.

What is a multi-target drug? Different expressions have been proposed: multi-target agent, dirty, promiscuous, multifunctional, multipotential, pluripotential, multiple ligand drug, multi-target-directed ligands, or polypharmacy. The 2013 IUPAC Glossary of terms used in Medicinal Chemistry now list definitions for some of these terms. For example:

- *Co-drug or mutual prodrug*

Two chemically linked synergistic drugs designed to improve the drug delivery properties of one or both drugs. The constituent drugs are indicated for the same disease, but may exert different therapeutic effects via disparate mechanisms of action.

- *Designed multiple ligands*

Compounds conceived and synthesized to act on two or more molecular targets.

- *Drug cocktail*

(1) (In drug therapy) Administration of two or more distinct pharmacological agents to achieve a combination of their individual effects.

   Note 1: The combined effect may be additive, synergistic, or designed to reduce side effects.

   Note 2: This term is often used synonymously with that of "drug combination" but is preferred to avoid confusion with medications in which different drugs are included in a single formulation.

(2) (In drug testing) Administration of two or more distinct compounds to test simultaneously their individual behaviors (e.g., pharmacological effects in high-throughput screens or drug metabolism.

- *Multi-target-directed ligand or multi-target drug*

A ligand that acts on more than one distinct molecular target. Targets may be of the same or different mechanistic classes.

Although perhaps linguistly mundane, for our purposes, the term "multi-target drug" seems appropriate.

Bivalent ligands can be taken as a type of multi-target drugs. The ideal multi-target drug, however, would resemble a typical small molecule drug that is orally bioavailable. Therefore, simply connecting drugs together by a linker, regardless of linker size, will probably not produce a viable drug. What has been described as a morphing process must occur where the important pharmacophoric regions of different drugs are identified and then reassembled into a small molecule. This process of doing so would indeed be a fascinating intellectual exercise. The concept of multi-target drugs, though not new, has been of great interest lately and several reviews describing the concept and design strategies are available.[274–278]

4.7 PHYSICOCHEMICAL PROPERTIES OF CNS DRUGS AND BLOOD-BRAIN BARRIER

Three barriers that protect the brain are considered to be present. These barriers help modulate ions, neurotransmitters, macromolecules, neurotoxins, and metabolic nutrients entering and leaving the brain. Common solutes that are present in the brain and plasma whose levels are controlled by the BBB are shown in Table 4.1.

The first barrier is the BBB. All organisms with a well-developed CNS have a BBB. In the brain and spinal cord of mammals including humans, endothelial cells

TABLE 4.1   Typical Plasma and Cerebrospinal Fluid Concentrations for Some Selected Solutes

Solute	Units	Plasma	CSF	Ratio
Na^+	mM	140	141	~1
K^+	mM	4.6	2.9	0.63
Ca^{2+}	mM	5	2.5	0.5
M^{2+}	mM	1.7	2.4	1.4
Cl^-	mM	101	124	1.23
HCO_3^-	mM	23	21	0.91
Osmolarity	mOsmol	305.2	298.5	~1
pH		7.4	7.3	
Glucose	mM	5	3	0.6
Total amino acid		2890	890	0.31
Leucine	μM	109	10.1–14.9	0.10–0.14
Arginine	μM	80	14.2–21.6	0.18–0.27
Glycine	μM	249	4.7–8.5	0.012–0.034
Alanine	μM	330	23.2–32.7	0.07–0.1
Serine	μM	149	23.5–37.8	0.16–0.25
Glutamic acid	μM	83	1.79–14.7	0.02–0.18
Taurine	μM	78	5.3–6.8	0.07–0.09
Total protein	mg/mL	70	0.433	0.006
Albumin	mg/mL	42	0.192	0.005
Immunoglobulin G (IgG)	mg/mL	9.87	0.012	0.001
Transferrin	mg/mL	2.6	0.014	0.005
Plasminogen	mg/mL	0.7	0.000025	0.00004
Fibrinogen	mg/mL	325	0.00275	0.000008
α2-macroglobulin	mg/mL	3	0.0046	0.0015
Cystatin-C	mg/mL	0.001	0.004	4

Source: Reprinted with permission from Abbott et al.[279] Copyright © 2010 Elsevier Inc.

that form the walls of the capillaries in the brain vascular system create the BBB. The combined surface area of these microvessels constitutes by far the largest interface for blood–brain exchange. This surface area, depending on the anatomical region, is between 150 and 200 cm^2/g tissue giving a total area for exchange in the brain of between 12 and 18 m^2 for the average human adult. The endothelial cells also form a tight junction with astrocytes (glia cells). The astrocytes and endothelial cells contain enzymes such as peptidases and esterases, which can metabolize drugs trying to enter the brain. This can be an advantage or a disadvantage depending on one's viewpoint. It can of course degrade drugs into inactive metabolites but it could also be used for the conversion of prodrugs into the active drug. A well-known example is the conversion of the dopamine prodrug, L-DOPA (dihydroxyphenylalanine), into dopamine for use in the treatment of Parkinson's disease.

A second interface is the blood–cerebrospinal fluid barrier that is formed by the epithelial cells of the choroid plexus facing the cerebrospinal fluid (CSF). The CSF is secreted across the choroid plexus epithelial cells into the brain ventricular system while the remainder of the brain extracellular fluid, the interstitial fluid, is derived at least in part by secretion across the capillary endothelium of the BBB.

The third interface is provided by the avascular arachnoid epithelium that underlies the dura and completely encloses the CNS, this completes the seal between the extracellular fluids of the central nervous system and that of the rest of the body. At all three interfaces, the barrier function results from a combination of physical barrier (tight junctions between cells reducing flux via the intercellular cleft or paracellular pathway), transport barrier (specific transport mechanisms mediating solute flux), and metabolic barrier (enzymes metabolizing molecules in transit). The barrier function is not fixed, but can be modulated and regulated, both in physiology and in pathology.[279]

Any drug used to treat substance abuse must be able to enter the brain from the systemic circulation through the BBB to interact with their neuronal targets. There are two main pathways for entry, passive diffusion or active transport. Most drugs are thought to enter the brain by passive diffusion. Some polar compounds such as vital amino acids pass through the BBB via transporters. A schematic of the major transport mechanisms is shown in Figure 4.11.[280]

Mechanism (a) is passive diffusion that is the most common role by which most small lipophilic compounds penetrate through the BBB. Carrier-mediated transport (b) occurs with small hydrophilic compounds and involves common transporters such as LAT1 (large neutral amino acids transporter 1) and GLUT1 (glucose transporter 1). The GLUT1 transporter is responsible for the uptake of glucose into the brain. As depicted in mechanism (c), large hydrophilic peptides and protein require receptor-mediated transport systems (e.g., TfR, transferrin receptor; LPR, LDL (low-density lipoprotein)-related protein receptor) that also play a considerable role in the transport of biopharmaceuticals. Mechanism (d) is absorptive-mediated transcytosis and is a relatively nonspecific mechanism for positively charged peptides. An important method of removing molecules from the brain is the active efflux transport system (e), which is an antiport system that attenuates the CNS concentration of many drugs by active efflux of them against concentration gradient from the brain into the

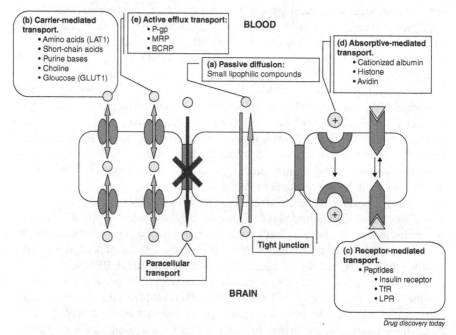

FIGURE 4.11   Diagram of transport mechanism through BBB (Reprinted with permission from Mehdipour and Hamidi.[280] Copyright © 2009 Elsevier Ltd)

blood using P-glycoprotein, multidrug resistance protein, and breast cancer resistance protein (BCRP).[280]

Being such an important component of CNS drug discovery, the development of models to help predict and determine whether a drug is bioavailable to the CNS has received an extensive amount of attention. Blood-brain barrier models ranging from purely computational to *in vivo* are available. In their simplest form, the *in vitro* models invariable consist of a layer of tightly packed endothelial cells separating two chambers. In one chamber, the drug and a buffer are added and the amount of drug that gets transferred into the other chamber is measured. There can be many variations on this theme; incorporating other cells types, transporters, enzymes, etc. Naik and Cucullo have proposed that the ideal *in vitro* BBB model should include enabling (1) the expression of interendothelial tight junctions between adjacent endothelial cells, which facilitate the formation of a very stringent and selective barrier (limited paracellular diffusion); (2) *in vivo*-like asymmetric distribution (basal vs. luminal) of relevant transporters (e.g., P-glycoprotein, glucose transporter 1), which confers polarization of the endothelial cells; (3) the expression of functional efflux mechanisms (e.g., P-glycoprotein and multidrug resistance proteins) and drug metabolism; and (4) the ability to discriminate the permeability of substances (in the absence of specific transport systems) according to their specific oil/water partition coefficient and molecular weight.[281]

Exciting *in vivo* pharmacology tools are being developed to explore the CNS bioavailability of drugs. Various *in vivo* techniques include pharmacokinetic analysis such as the *in situ* brain perfusion method, the intravenous injection/multiple time point approach, and the brain efflux index, intracerebral microdialysis, positron emission tomography (PET), magnetic resonance imaging (MRI), and histochemistry.[282] Knockout rats where the transporter genes (Mdrla, Mrpl, Mrp2, Bcrp, or Oct1) have been removed are available (e.g., NoAb Biodiscoveries). It is possible then to study how different transporters located in the BBB affect the active transport of drugs into and out of the brain. Banks have reviewed recent progress in the development of mouse models for neurological disorders of the BBB.[283]

A summary *in vitro* and *in vivo* tools to measure BBB function by Cecchelli and his research team is shown in Table 4.2.[284]

4.7.1   Receptor Occupancy

The direct measurement of receptor occupancy in the brain is one of the better solutions to the problem of measuring brain penetration. Receptor occupancy provides the pharmacologically relevant parameters and integrates all aspects that regulate brain penetration. There exist two types of methods for determining receptor occupancy: *ex vivo* and *in vivo*. Both can be done with rodents while the *ex vivo* is not suitable for humans except postmortem.

The *ex vivo* method was discussed in Chapter 2. It entails giving the rodent a drug, sacrificing the animal, and quickly removing the brain. The brain is frozen and appropriate tissue slices are removed. These slices are incubated with a radiolabeled ligand, just as in a binding assay. The thought is upon sacrifice of the animal, if the brain is processed correctly, the exact "living" state of the brain at the moment of sacrifice is preserved. The level of receptor occupancy and drug selectivity under homoeostatic conditions will be accurately represented in the tissue slices. Autoradiography is used to detect the radiolabel and the amount of receptor occupancy is calculated. The *in vivo* procedure uses PET and single photon emission tomography (SPECT) imaging techniques and will be discussed next.

Direct measurement of receptor occupancy is associated with three difficulties. Radiolabeled compounds with high affinity and selectivity for the receptor site are required and must be able to enter the brain. In rodents, the latter might be circumvented by intracerebral drug administration. This would of course not aid in determining whether the drug can enter the brain, but would allow one to measure receptor occupancy. Second, receptor occupancy measurements assess the overall brain penetration, integrated over the many mechanisms by which a drug can enter the brain. Receptor occupancy measurements alone, therefore, might not assist a medicinal chemist during the lead optimization stage of determining what is more critical, optimization of drugs properties for passive diffusion or for active transport. Third, many of the efflux transporters show species-specific selectivity and this must be taken into account.[285]

TABLE 4.2 In Vivo and In Vitro Technologies to Assess BBB Function

Techniques	Studies	Species	Pros	Cons
Intravenous injection	Rate of brain penetration. Determination of PS value from K_{in}.	Rat, mouse	Accurate measurement of PS. Applicable also for slowly penetrating compounds.	Labor intensive and low throughput—may need CD to confirm molecule reaches parenchyma. Metabolism may Confound interpretation of PS if extended experimental time periods are used
Intravenous injection	Extent of brain/CSF penetration. Determination of total brain or CSF concentration/plasma concentration ratio.	Rat, mouse	Reflects how molecules distribute between blood and brain or blood and CSF. The CSF/plasma ratio may act as a surrogate measure of brain ISF Concentrations at steady state.	Labor intensive and low throughput. Nonspecific binding in the brain and systemic PK properties of the compound complicate interpretation of the results for optimization of the compounds permeability properties.
Carotid artery single injection technique	BUI. Determination of ratio of a ^{14}C-test compound/^{3}H reference in the brain divided by the ^{14}C/^{3}H ratio	Rat, mouse	Fast, flexible, and simple technique that can be used in conscious or anaesthetized animals.	Limited sensitivity as brain extraction is measured during a single capillary transit. Needs radiolabeled compound
Microdialysis	Free drug concentration in brain determined by means of an inserted probe containing a semipermeable membrane.	Rat, mouse	In general, free drug concentrations reflect pharmacological activity better than total concentrations.	Labor intensive, low throughput, and technically demanding. Recovery problems (binding to plastics and membranes).
Autoradiography	Regional measurement of brain tracer content.	Rat, mouse	Visual demonstration of tracer distribution and regional brain tracer content in small brain regions. Good resolution.	Labor intensive and low throughput. Not suited for optimizing permeability characteristics of investigational molecules.

Method	Application	Species	Advantages	Limitations
PET	Regional measurement of brain radioactivity. Quantitative determination of cerebrovascular permeability (PS).	Large animals, humans	Rapid sequential measurements of regional brain tracer content. Noninvasive.	High cost. Labor intensive and low throughput. Each compound needs to be radiolabeled with an isotope of suitable half-life.
Intravital microscopy in combination with various staining techniques	Visualization of transported compounds across the BBB. Visualization of events in live animals such as interaction of leukocytes with the BBB.	Rat, mouse	Visualization of transport and events at the BBB in normal and pathological conditions.	Labor intensive and low throughput.
Knockout animals	Determination of involvement of transporters or receptors.	Mouse	Quite clear-cut distinction between the transporters or receptors involved in transport.	A "classification" system rather than a ranking system.
In vitro BBB models	Determination of permeability (Pe). Response of the BBB in normal and pathological conditions.	Human, cow, mouse, rat, pig	Possibility to assess permeability and involvement of transporters and receptor-mediated/adsorptive transcytosis. Can be used to estimate luminal to abluminal or abluminal to luminal transport. Cells from KO animals can be used to establish BBB models. Relatively high throughput. Suitable for optimizing BBB permeability. Low noise level and easier to elucidate molecular mechanisms at the BBB compared with *in vivo* models.	Not an intact *in vivo* system.

Source: Reprinted with permission from Cecchelli et al.[284] Copyright © 2007 Nature Publishing Group.

BBB, blood–brain barrier; BUI, brain uptake index; CD, capillary depletion; technique to differentiate endothelial cell binding and endocytosis from actual transcytosis; CSF, cerebrospinal fluid contained within the ventricles of the brain, the subarachnoid spaces and the central canal of the spinal cord; ISF, brain interstitial fluid, the extracellular fluid of the brain parenchyma; K_{in}, unidirectional transport coefficient; KO, knockout; Pe, permeability coefficient; generated by dividing the PS value by surface area of cell monolayer; PET, positron emission tomography; PK, pharmacokinetics, the study of mechanisms involved in drug absorption, distribution and elimination, and the kinetics of these processes; PS, permeability-surface area, *in vivo* permeability results are usually expressed, as the PS product as the surface area is difficult to measure *in vivo*. The PS can be derived from the transfer constant K_{in} by using the Renkin-Crone model of capillary transfer.

4.7.2   Physical Properties of CNS Drugs

Researchers at Pfizer in 2001 published a famous report that summarized physical properties of relevance to the bioavailability of drugs.[286] Now called "Lipinski's Rule of Five" or "Rule of Five," it helped define certain physical properties that the majority of bioavailable drugs possessed. Note that the observations are in relation to pharmacokinetics, not pharmacodynamics, that is, how the drug interacts with a target and the efficacy of the drug. This concept though is quite applicable to address the problem of passage of a drug through the BBB. Akin to Lipinski rules for general drug bioavailability, common physical properties of drugs for bioavailability to the CNS have been extensively studied. CNS-active drugs tend to be smaller, less polar, and more rigid than non-CNS drugs. Size and lipophilicity appear to be key parameters for passive diffusion. Numerous studies suggest that drug physical properties for good CNS penetration are:

- Molecular weight ≤ 400
- ClogP of 2-4
- Number of hydrogen bond donor groups ≤ 3
- Number of hydrogen bond acceptor groups ≤ 7
- Polar surface area ≤ 90 Å^2

A couple of recent examples of research in the area will suffice. Scientists from Cephalon have examined about 110 descriptors of CNS-active drugs versus non-CNS-active drugs.[287] They concluded that good guidelines for lead optimization of CNS bioavailability are:

- Total polar surface area < 76 Å^2
- At least one N (including one aliphatic amine)
- Fewer than seven (two to four) linear chains outside of rings
- Fewer than three (zero to one) polar hydrogen atoms
- Volume of 740–970 Å^3
- Solvent accessible surface area of 460–580 Å^2
- Positive QikProp parameter CNS

QikProp (Schrödinger) is a software that calculates the physical properties of over 20 descriptors related to ADME (absorption, distribution, metabolism, excretion) and toxicity properties based on 3D models. It predicts central nervous system activity on a -2 (inactive) to $+2$ (active) scale.

As a further example, researchers at AstraZeneca measured the solubility of 98 CNS drugs in a buffer at pH 7.4. Of these drugs, more than 90% had a solubility of $> 10\,\mu M$ while only seven had a solubility of $< 10\,\mu M$. They suggest lead development cut offs of 1 μM and 10 μM. Compounds with a solubility of < 1 μM will probably not make a good CNS drug while those of < 10 μM will be problematic. These values could serve as decision points when selecting among potential drug candidates to optimize.[288]

Further comment on physical properties of drugs will be reserved for specific examples as we discuss each therapeutic area. More detailed information can be found in the many reviews and books that are available. Some recent reviews that may be of interest to the chemist are referenced.[289–293]

4.8   BRAIN IMAGING AGENTS

A major difficulty in all psychiatric disorders is the inability to objectively diagnose and monitor the progress of the disorder via the detection and quantification of physiological biomarkers. The different techniques of imaging: MRI, PET, and SPECT, therefore, hold exciting promise to address this issue. These techniques and the ligands that have been developed allow a window into a real time observation of the workings of the brain. This has the potential of allowing a greater *in vivo* molecular understanding of drug addiction, especially in human patients.[294] The National Institutes of Mental Health maintain a CNS Radiotracer Table of all compounds containing [^{11}C], [^{18}F], or [^{123}I] that have been developed for use in humans.[295] The National Center for Biotechnology Information maintains a Molecular Imaging and Contrast Agent Database of all imaging compounds that have been reported. As of July 2013, 1444 agents were listed. The home page is http://www.ncbi.nlm.nih.gov/books/NBK5330.

The development of ligands that can be used faces the same hurdles that all CNS-active drugs face. As such many of the radioisotopic ligands used are derived from drugs that are known to be CNS active. The radioisotopes commonly used are ^{18}F and ^{11}C for PET and ^{123}I for SPECT. The greater energy of ^{124}I decay of 2.14 MeV than ^{18}F (0.64 MeV) and ^{11}C (0.97 MeV) means the positron emitted by the iodine decay will travel a greater distance in the tissue before the positron hits an electron and annihilated into γ-rays. The resolution that can be obtained using ^{124}I will then be less than the other two radioisotopes.

A positron is a particle with the same mass of an electron but with a positive charge. The concept of a positron arose from Paul Dirac's famous work on developing a relativistic model of quantum mechanism and to address the question of electron spin. Carl Anderson first discovered, and named, the particle in 1932 in his investigation of the effects of cosmic rays in a cloud chamber. The positron-electron reaction can be written as $e^+e^- \rightarrow 2\gamma$ and expressed in a Feynman diagram as shown in Figure 4.12. Here, we have an electron and a positron (both are fermions) in the initial states, separated in space, interacting with the generation of two γ-rays (photons) at a later time that represent the final state.

Many methods exist for the selective introduction of the radioisotopes into a molecule.[296] Both ^{18}F and ^{123}I are normally introduced as nucleophiles via an S_N2 reaction while ^{11}C can be introduced as an electrophile using ^{11}CH$_3$I or functional analogs of ^{11}CO$_2$.

Regardless of the method, both ^{18}F and ^{11}C must be generated on-site in a cyclotron and quickly introduced into a molecule due to their short half-lives, 110 minutes for

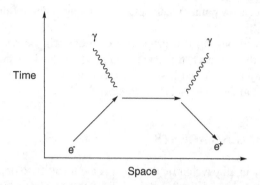

FIGURE 4.12 Feynman diagram for the electron-positron annihilation reaction

^{18}F and 20.3 minutes for ^{11}C. We can look at the synthesis of ^{11}C and its compounds as an example. ^{11}C is prepared by bombarding nitrogen gas with protons to generate ^{11}C and an α particle by the following nuclear reaction.

$$^{14}_{7}N + ^{1}_{1}H \rightarrow ^{11}_{6}C + ^{4}_{2}He$$

If the bombardment is done in the presence of 20–20,000 ppm of oxygen then $^{11}CO_2$ is produced. If 5–10% hydrogen gas is present $^{11}CH_4$ is made. With these two precursors in hand, a wide variety of reagents are available such as $^{11}CH_3I$. The total time for generation of the radioisotope, reaction with a substrate, and purification of the radiolabeled compound must not exceed 2-3 times the physical half-life. If the radioactive atom replaces the same atom in the original "cold" compound, then introduction of the radiolabel will have no effect on binding and functional activity.

A variety of imaging drugs have been developed for the dopamine, and to lesser extent, serotonin receptors due to their importance in other mental illnesses such as schizophrenia, depression, anxiety, Parkinson's disease, and Alzheimer's disease.

We can use the synthesis of $[^{11}C]$-raclopride as an example. Raclopride is a D_2/D_3 antagonist with $K_i = 1.8$ nM for D_2 and $K_i = 3.5$ nM for D_3 that is commonly used in human PET studies. Radiolabeled $[^{11}C]$-raclopride ($K_d = 1.1$ nM) was first synthesized and examined in humans in 1985.[297] It was readily synthesized by monoalkylation of the phenol with ^{11}C methyl iodide as shown in Figure 4.13.

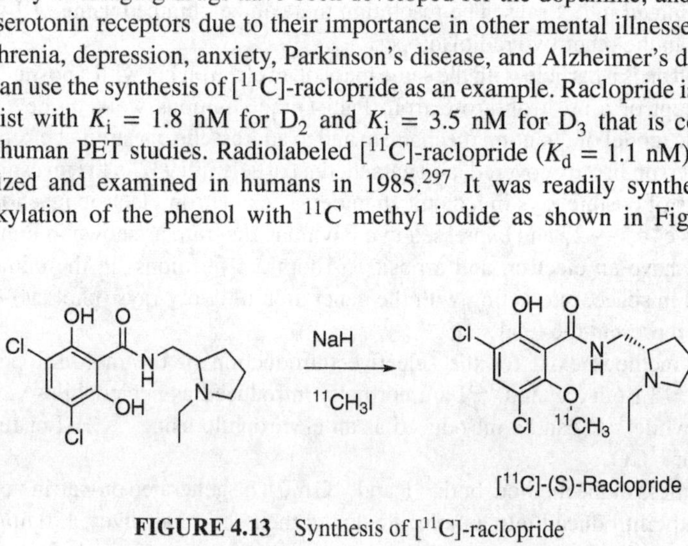

$[^{11}C]$-(S)-Raclopride

FIGURE 4.13 Synthesis of $[^{11}C]$-raclopride

The time of preparation was 50 minutes from the end of bombardment, with a radiochemical yield of 40-50%. The specific activity obtained varied from 30 to 130 Ci/mmol (Curie, Ci $= 2.77 \times 10^{12}$ disintegrations per minute). In the human study, the amount of injected radioactivity varied from 3 to 4 mCi and the specific activity was between 40 and 110 Ci/mmol. The activity was centered in the caudate/putamen region of the brain and maximal activity was observed after 20-30 minutes.

A further more quantitative study in humans showed that [^{11}C]-raclopride bound to D_2/D_3 receptors in the putaman with a $B_{max} = 14$ pmol/cm^3 and a $K_d = 3.8$ nM.[298] This is a nice example of the use of PET to investigate receptor occupancy.

Due to the short half-life of ^{11}C, a substantial amount of the radioactivity is lost during the synthesis. We can see for ^{11}C with a half-life of about 20 minutes that, in a synthesis requiring 40 minutes, half the specific activity will be lost. The syntheses therefore have to be done using larger amounts of radioactive atoms than required to compensate for this loss. Microfluidic synthesis techniques have been developed to help correct this problem.[299]

A study just reported can be used to illustrate the microfludic approach. The synthesis of [^{11}C]-raclopride using a continuous-flow microchip reactor has been reported.[300] The best method of synthesizing radiolabeled raclopride was first investigated on the cold material using three types of microchip reactors, no loop, abacus, and a full loop. The full loop design was found to give the best yields of [^{11}C]-raclopride. The microfluidic radioactive syntheses started with 0.8 mg (2.4 μmol) of desmethyl raclopride and about 50-80 mCi (5-8 nmol) of [^{11}C]-MeI using sodium hydroxide as the base in dimethyl sulfoxide as the solvent. The full loop method gave a [^{11}C]-raclopride conversion rate of about 25% at a flow rate of 2 μL/min and a residence time of 53 seconds. Material of 441 μCi can be produced which can be sufficient for some PET studies. The reported radiochemical yield was only 9%, relative to 17-45% using conventional methods. They propose that the microchip reaction chamber of polydimethylsiloxane consumes a lot of the [^{11}C]-MeI. They believe that a glass reaction loop would help prevent this.

Both human and animal PET imaging of the opioid system has received a lot of attention.[301] Several opioid ligands labeled with either [^{11}C] or [^{18}F] that are available for clinical use in humans are shown in Figure 4.14. Four of the compounds are agonists while one, ^{11}C-naltrindole, is an δ-opioid selective antagonist.

Using the μ-agonist [^{11}C]-carfentanil, it was found that up to 90% receptor occupancy by buprenorphine was achieved using clinical doses of buprenorphine (16 mg/day). In addition, to suppress withdrawal symptoms at least 50-60% μ-opioid receptor occupancy by buprenorphine is required.[302]

Missing from the list are κ-opioid selective ligands. To address this, several new κ-opioid selective ligands are being investigated. As an example, the full antagonist LY2795050 from Lilly has been labeled with ^{11}C.[303] The compound has good selectivity for the κ-opioid receptor with a $K_i = 0.72$ nM versus 26 nM and 153 nM for μ- and δ-opioid receptors, respectively. The radiolabeled compound [^{11}C]-LY2795050 was prepared in an average yield of 12% and greater than 99% radiochemical purity as shown in Figure 4.15. The key radiochemical step is a chemoselective palladium-catalyzed cyanation of the aryl iodide to produce the aryl nitrile that is then rapidly

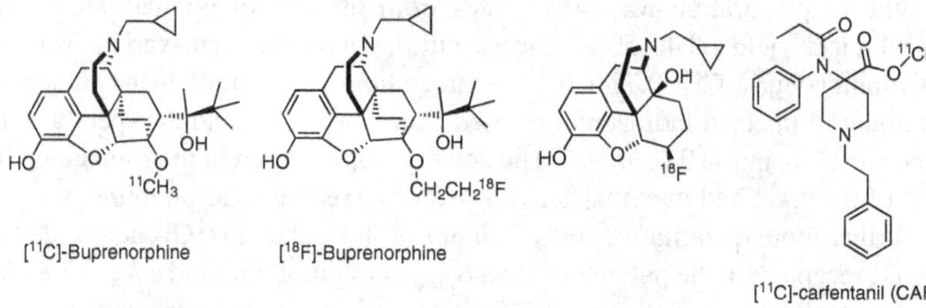

[^{11}C]-Buprenorphine

[^{18}F]-Buprenorphine

[^{11}C]-carfentanil (CAR)

K_i (nM)				
μ-opioid	0.2	0.2	2.6	0.024
δ-opioid	0.2	0.2	89	43
κ-opioid	0.2	0.2	9.3	3

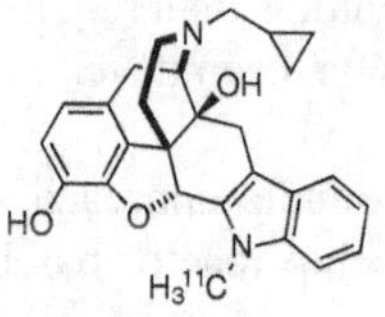

[^{11}C]-naltrindole

δ-opioid K_i = 0.02 nM

FIGURE 4.14   Opioid PET ligands

hydrolyzed to the amide. The final compound had an average specific activity of 24 GBq/μmol and the synthesis time was about 45 minutes.

The *in vivo* studies in rhesus monkeys required about 133 MBq per experiment, about 4 μg of radiolabeled LY2795050. In the rhesus monkey brain, highest uptake was observed within 20 minutes and the binding could be reduced by the addition of a large amount of another antagonist, naloxone. The highest levels of binding were

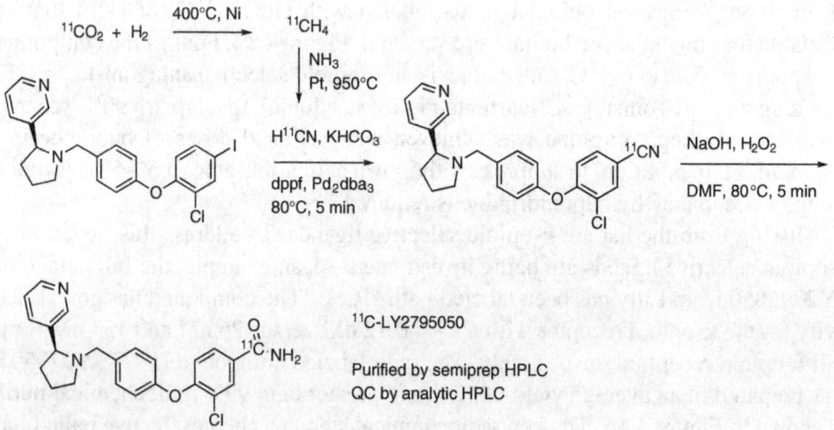

FIGURE 4.15   Synthesis of radiolabeled LY2795050

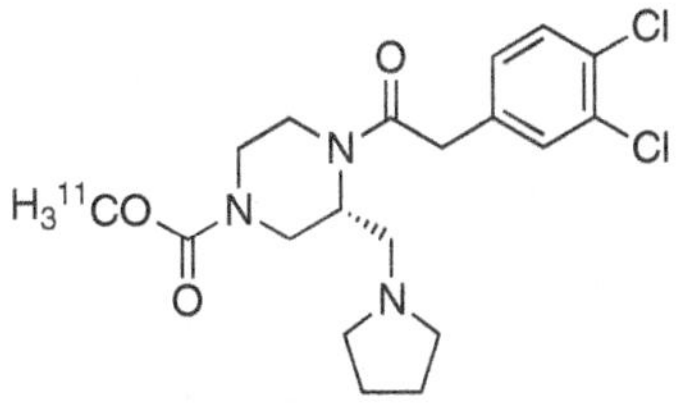

FIGURE 4.16   Structure of ^{11}C-GR103545

observed in the globus pallidus (a component of the basal ganglia) and cingulate cortex. The authors state that human trials with the compound are underway.

The same group also examined a κ-opioid selective agonist, [^{11}C]-GR103545.[304] The cold compound was first reported by Glaxo in 1993 as part of a program to develop new compounds for the treatment of pain.[305] The ^{11}C radiotracer was first synthesized in 2005 and was investigated in baboons.[306] The K_i for the κ-opioid receptor is 0.02 nM and for μ- and δ-opioid receptor is 12 and 400 nM, respectively. The compound is a potent agonist with an IC_{50} = 0.018 nM for the enantiopure compound (Figure 4.16).

From experiments in rhesus monkeys, they were able to calculate an *in vivo* global K_d = 0.48 nM (after correction for nonspecific binding) and could calculate the maximum *in vivo* binding in different regions of the brain giving a B_{max} of 0.3–6.1 nM. Highest levels of binding were observed in the cingulate cortex and the putamen (a component of the basal ganglia). From their data, they were able to calculate that a dose of 0.02 μg/kg would be appropriate for human studies. This dose produced maximum receptor occupancy in the cerebellum of rhesus monkeys of 2.6%. At this dose, a 70-kg (154 pounds) person would be injected with about 11 mCi (406 MBq) of radioactive ^{11}C.

It was observed that agonist binding was more sensitive than antagonist binding, a larger amount of radioactivity was observed with the agonists. Then, it would seem that use of an agonist would be better, but one also has to take into account any physiological effects that could occur using an agonist.

4.9 QT PROLONGATION

A potential death knell for a drug in development is its ability to initiate QT prolongation. As will be discussed in Chapter 5, the use of levo-alpha-acetylmethadol (LAAM) for the treatment of opioid addiction has fallen out of favor due to its QT prolongation effects so it is worth having a brief discussion about this topic. QT prolongation is a sudden abnormal disruption in the heart rate that can lead to death. The syndrome is called long QT syndrome (LQTS) and is often inherited and induced not only by stress or exercise, but also by drugs. LQTS is believed to be due to mutations in the *KCNH2* gene that goes by the more interesting name of *ether-a-go-go-related gene* (hERG). The gene codes for a pore-forming subunit of a delayed rectifier voltage-gated K^+ channel, Kv11.1, that passes the fast component of the two currents (fast vs. slow). This potassium channel is expressed in the heart, brain, smooth muscle tissue, endocrine cells, and some tumor cell lines. The presence in the heart though, is what causes the QT prolongation. Drug binding and blockade of this ion channel can cause the sudden cardiac arrest associated with QT prolongation.

QT refers to two of the five——P, Q, R, S, and T—electrical waves generated by the beating heart that are detected by an electrocardiogram. The electrical activity occurring between the Q and T waves is called the QT interval and usually lasts about a third of a heartbeat. Blockade of the Kv11.1 ion channel lengthens this interval, disrupting the timing of the propagating waves leading to cardiac arrest, originally called "torsade de pointe." The QT prolongation is readily seen on the electrocardiogram and *in vitro* methods to analyze for it have been developed. Any drug in clinical development that can initiate QT prolongation will most likely be removed from further consideration.[307-309]

5

MEDICATION DEVELOPMENT FOR NARCOTIC ADDICTION

"I know I'm addicted to (opioids), and it's the doctors' fault because they prescribed them. But I'll sue them if they leave me in pain."[310]

It is a twisted web we weave. Opioids and pain are inevitably linked. The sole medical use of opioid drugs is for the treatment of pain. Unfortunately, as the analgesic potency of opioids increase, for example, morphine, so do their abuse liability. A partial solution to the treatment of opioid abuse is to prevent its occurrence due to addiction derived from the use of prescription opioids for pain. Though a subject of exhaustive research, it has not been possible to effectively develop an effective and a potent analgesic without addictive effects. When one speaks of addiction to opioids, the reference is to heroin (3,6-diacetylmorphine), morphine, and prescription analgesics such as oxycodone (Oxycontin®, Percocet®).

The term "opioid" is defined for compounds that produce pharmacological properties similar to the opium alkaloids such as morphine and can be antagonized by an opioid antagonist. A vast, structurally diverse number of compounds ranging from peptides to small molecules are considered as opioids, with morphine the standard for comparison of activity. The majority of the analgesic and addiction literature uses morphine as the standard. Therefore, it is worth describing in some detail the chemical and pharmacological properties of morphine.

Morphine is one of the many alkaloids produced by the poppy *Papaver somniferum*. The juice or latex of the poppy is referred to as opium. The main source of opium is currently from Afghanistan, Southeast Asia (Lao People's Democratic

Drug Discovery for the Treatment of Addiction: Medicinal Chemistry Strategies, First Edition. Brian S. Fulton.
© 2014 John Wiley & Sons, Inc. Published 2014 by John Wiley & Sons, Inc.

Republic and Myanmar) and Latin America (Mexico and Columbia). In 2009, the United Nations Office on Drugs and Crime (UNODC) estimated that 85% of the world production of opium originates in Afghanistan.[311] In November 2013, the Ministry of Counter Narcotics (Kabul Afghanistan) and UNODC reported in the 2013 Afghanistan Opium Survey that poppy cultivation in Afghanistan rose 36% in 2013, a high record. Meanwhile, opium production amounted to 5500 tons, an increase of almost 50% from 2012. For poorer Afghani farmers, there is a strong financial motivating force for the cultivation of poppies. Afghani farmers can earn up to $145 per kg of opium where per capita income is only about $1100, 36% of the population is below the poverty level, and an unemployment rate of 35% (CIA World Factbook). Perhaps not surprisingly, almost 90% of opium poppy cultivation in 2013 remained confined to nine provinces in the southern and western regions, which include the most insurgency-ridden provinces in the country. Helmand, Afghanistan's principal poppy-producer since 2004 and responsible for nearly half of all cultivation, expanded the area under cultivation by 34%, followed by Kandahar, which saw a 16% rise.

In terms of quantities, 7 kg of Afghan opium can produce 1 kg of Afghan heroin. Afghan opium generally has higher morphine content than the opium produced in Myanmar, which requires approximately 10 kg of opium for each kilogram of heroin processed. Karch has presented in some detail the methodology for the clandestine production of morphine from opium.[312] Briefly, the opium is dissolved in a basic solution and then the free base of morphine is precipitated by the addition of ammonium chloride. The free base is filtered and purified by the formation of the hydrochloride or hydrosulfate salt and then dissolved in water. Solids are filtered off and the free base is obtained by the addition of ammonium hydroxide. The resultant precipitate of morphine is isolated by filtration. Heroin is produced by acetylation of morphine with acetic anhydride or acetyl chloride. As one might imagine, the morphine produced is not pure and contains other basic alkaloids from opium, especially codeine and thebaine. Street heroin is sold as "bags" of 30–50 mg of which only 1–10% may actually be heroin.[313]

5.1 PHARMACOLOGY OF NARCOTIC ADDICTION AND PAIN

Opioids cause addiction due to their binding to the μ-opioid receptor. When opioids bind to the μ-opioid receptor located on GABA neurons in the ventral tegmental area (VTA), an inhibitory action occurs. GABA is not released into the synapse where it can bind to $GABA_A$ receptors on dopamine neurons originating in the VTA. GABA is an inhibitory neurotransmitter, the brake is removed and dopamine is released by the dopaminergic neuron from the axonal end located in the nucleus accumbens. It is noteworthy that pure δ-opioid receptor and κ-opioid receptor agonists do not show an abuse liability but do, however, have analgesic properties. However, their side effects do not allow them to be useful analgesics.

It is worth taking time to briefly explore the complex mechanism of pain and analgesia.[314] We all know what pain is. If one sticks your hand into a campfire then a rather distasteful feeling of pain is sensed. The sensation of pain results from peripheral activation of receptors on the surface of organs such as the skin or central sensitization of organs in the body. A signal, nerve impulse, is transmitted to the spinal cord. More specifically, pain neurons (nociceptors) in the peripheral central nervous system leading to the dorsal horn of the spinal cord transmit the impulse. Upon excitation, the neuron releases the excitatory amino acids, glutamic acid and aspartic acid (along with neuropeptides), that interact with presynaptic and postsynaptic NMDA and AMPA receptor-coupled ion channels. Activation of the ion channels causes an influx of Na^+ and Ca^{2+} ions, depolarizing the neuron and sending a signal via the ascending spinothalamic tract neuronal system to the limbic system and parts of the cortex where the sensation of pain is developed. A descending neuronal pathway from the brain to the site of stimulus, again via the spinal cord and dorsal horn, transmits the response. Get your hand out of the fire!

Opioids inhibit the nociceptive signal leading to the brain. A key point for the development of analgesic development is that this signal disruption can occur outside the CNS in peripheral tissue. Thus, the development of analgesic drugs that cannot cross the blood-brain barrier and act only at peripheral μ-, δ-, or κ-opioid receptors have the potential of blocking pain without the centrally induced side effect of addiction.[315]

5.2 PRESCRIPTION DRUG ADDICTION

Addiction and abuse of prescription narcotics such as OxyContin® has become a major health crisis and has been widely documented. Commonly abused prescription analgesics are hydrocodone (Vicodin®), oxycodone (OxyContin®), oxymorphone (Opana®), hydromorphone (Dilaudid®), propoxyphene (Darvon®), meperidine (Demerol®), fentanyl (Duragesic®), and diphenoxylate (Lomotil®). The first four drugs are based on the morphine structure while the latter four are synthetic compounds. These drugs come in tablet formulations but are often abused by crushing the tablet and snorting or injecting the powder. It has been recognized that a factor in their abuse is their ready availability as a result of physicians' aggressive dispensing of narcotic analgesics and insufficient means of tracking the numbers of prescriptions an individual can obtain. Unfortunately, for legitimate treatment of severe chronic pain, there is often little choice than the use of narcotics.

Common commercial analgesics besides OxyContin® are Percocet® and Vicoden®. OxyContin® can contain from 10 to 80 mg of the μ-opioid agonist oxycodone. Percocet® is marketed by Endo Pharmaceuticals and is formulated as a tablet containing 2.5–10 mg of oxycodone and 325 mg of the non-opioid analgesic acetaminophen. Endo also markets Percodan® that contains 4.8 mg of oxycodone and 325 mg of aspirin (acetyl salicylic acid). Vicoden® is marketed by Abbvie, which was formed in 2013 when Abbott Laboratories separated into two independent companies. Abbvie retained the proprietary pharmaceutical business of Abbott Laboratories. Vicoden® contains 5–10 mg of oxycodone and 300 mg acetaminophen. Another product from Abbvie is Vicoprofen® that contains 7.5 mg of hydrocodone and 200 mg of ibuprofen.

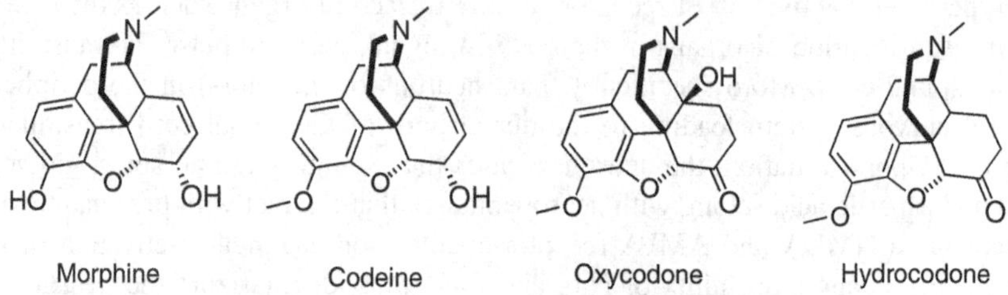

FIGURE 5.1 Structures of common opioids

It is clear that there is a large level of concern developing within the general public on the abuse of prescription drugs. A new extended-release formulation of oxycodone called Zohydro from Zogenix is due to be available in 2014. The gelatin capsules will contain from 10 to 50 mg of oxycodone and are indicated for severe pain. It is being reported though that the capsules can be easily crushed for nasal or intravenous use. The amounts of oxycodone for this type of formulation are quite large leading to concern about nonprescription abuse or perhaps even death from overdose. In fact on February 26, 2014, a letter with over 40 public health group signatories was sent to the FDA urging the FDA to withdraw support for the drug. On December 10, 2013, State Attorney Generals from 28 states sent a letter to the FDA also asking them to reconsider the approval of the drug.

The structures of the common analgesic opioids derived from morphine are shown in Figure 5.1. All are controlled substances, morphine and oxycodone are Schedule II controlled substances, hydrocodone is a Schedule III drug, and codeine is a Schedule V drug.

To make it more difficult for oxycodone and other opioid analgesics to be abused, manufactures are developing tamper-resistant formulations.[316] As an example, Pain Therapeutics has introduced a tamper-resistant formulation of oxycodone called Remoxy. The formulation is based on their ORADUR® extended-release technology. The formulation is a viscous material that in the stomach becomes a matrix in which the drug is trapped. In this case, oxycodone is at the same amount (40 mg) as in an oxycontin pill. As a comparison, the formulation is much more viscous than honey; 60,000 versus 1760 cp, respectively.[317] The development of tamper-resistant formulations has, however, not completely alleviated the problem of prescription drug abuse. A July 2012 broadcast on National Public Radio reported that as it becomes more difficult to obtain prescription oxycodone, other analgesics such as oxymorphone (Opana®) are becoming the drug of choice in parts of the United States.

5.3 APPROVED MEDICATIONS

Approved medications for opioid abuse are methadone, levo-alpha-acetylmethadol (LAAM), buprenorphine, naltrexone, and lofexidine (UK). The first three act as μ-opioid partial agonists and have proven effective in long-term maintenance

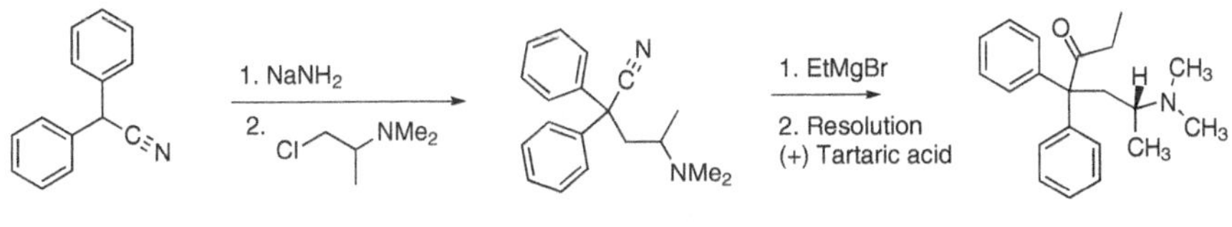

FIGURE 5.2 Synthesis of methadone

treatment for preventing relapse. Naltrexone is an opioid antagonist useful for opioid withdrawal and perhaps long-term maintenance treatment for preventing relapse. Lofexidine is quite different acting as a α_{2A} adrenergic agonist and is used primarily for opioid withdrawal. Several recent reviews have looked at the treatment of opioid addiction.[318,319]

5.3.1 Methadone-Opioid Partial Agonist

Methadone was first reported in the patent literature in 1944 by Von Max Bockmühl of Farbwerke Hoechst. He later published a full report on the synthesis and pharmacological properties of methadone with coauthor Gustav Ehrhart in 1948. The production process is shown in Figure 5.2.

Methadone is prescribed for maintenance therapy to help prevent relapse. The properties of the drug have been extensively studied. The pharmacokinetics and metabolism of methadone have been reviewed.[321] It has variable bioavailability (41–95%) with an onset of action time (T_{max}) of 1–6 hours and a long half-life of up to 50 hours. Methadone is less polar than morphine with a log $D_{7.4} = 1.65$ versus a log $D_{7.4} = -5.0$ for morphine and can be given orally. At a pH 7.4 both compounds will exist in water as the ammonium salt. The solubility of the hydrochloride salt of methadone in water is 12 g/100 mL while for morphine hydrochloride the solubility is 5.7 g/100 mL. Typical doses of methadone are 25–60 mg per day.[322] The maintenance program typically lasts for 1 year but can extend up to 10 years or more. The drug has abuse liability at oral doses greater than 100 mg.[323] Methadone is administrated as the racemic mixture but the R–(–) enantiomer is 18 times more potent that the S–(+) enantiomer.

Morphine and R–(–)-methadone both bind to the μ-opioid receptor with similar affinity ($IC_{50} = 2.5$ nM and 3.0 nM, respectively) with weaker binding to the δ- and κ-opioid receptors (IC_{50}'s of R-methadone > 370 nM).[324] Methadone tends to function as a full agonist, depending on the functional assay used. Major differences can exist though in other functional activities such as affecting the activity of intracellular kinases, recruitment of β-arrestin2, and receptor desensitization and internalization.[325] Differential modulation of these functional activities is thought to be important in the abuse liability of an analgesic.

The structure–activity relationship (SAR) of methadone is fully developed and the results for efficacy are summarized in Figure 5.3.[326]

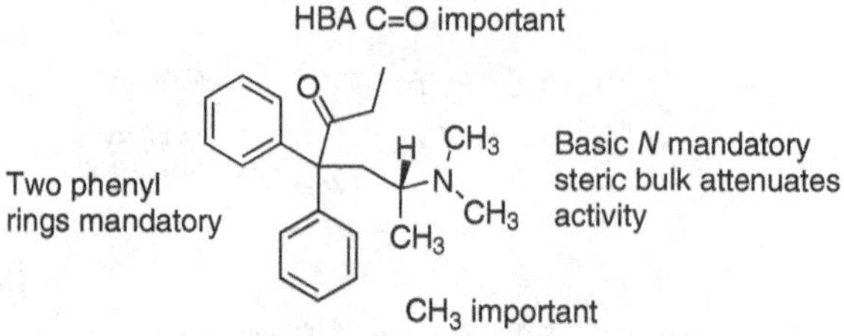

HBA C=O important

Two phenyl
rings mandatory

H CH$_3$
N
CH$_3$
CH$_3$

Basic *N* mandatory
steric bulk attenuates
activity

CH$_3$ important

FIGURE 5.3   SAR of methadone

Some of the effects of methadone may be due to its antagonist activity at the NMDA receptor channel. Methadone has weak binding affinity at the NMDA receptor with the R−(−) enantiomer slightly more potent than the S−(+) enantiomer (K_i = 3.4 vs. 7.4 μM, respectively, in rat forebrain synaptic membranes) while morphine has no detectable binding (K_i > 100 μM).[327] Binding appears to be at a noncompetitive site on the receptor as methadone was able to displace the noncompetitive antagonist [^{3}H]-MK-801 but not the competitive NMDA receptor antagonist [^{3}H]-CGS-19755. The S enantiomer of methadone has been shown to antagonize the effects of NMDA in rats at doses that also block the induction of morphine tolerance. Administration of NMDA to rats induces a hyperalgesic effect where the rat is more sensitive to pain. Pretreatment of the rat with S-methadone (in this study), before injection of NMDA, completely blocks the hyperalgesic effects of NMDA.[328] It is not clear if the effect on morphine tolerance was due to binding to the μ-opioid and/or the NMDA receptors. Most likely the interaction of methadone with μ-opioid receptor *in vivo* is more important as much higher concentrations of methadone are required to block the NMDA receptor channel than required to activate the μ-opioid receptor, at least in whole-cell recordings of changes in membrane potentials.[329]

Other pharmacological information is being obtained suggesting that the mechanism of action may be more complex. Methadone has pharmacological activity as an $\alpha_3\beta_4$ nAChR antagonist.[330] Methadone inhibited nicotine-stimulated ^{86}Rb$^+$ efflux from cells expressing the $\alpha_3\beta_4$ nAChR with an IC$_{50}$ = 1.9 μM. Very little difference in antagonist activity was measured between the R and S enantiomers. Studies in both transfected cells and *in vivo* in *Abcb1a* gene knockout mice have shown that methadone is a substrate for the P-glycoprotein transporter.[331,332]

5.3.2 LAAM-Opioid Agonist

As part of the methadone SAR studies, Bockmühl and Ehrhart also reported an analog where the ketone was reduced to an alcohol but the analgesic activity was not reported. In the following year, a report from Bristol Laboratories presented the synthesis and analgesic activity of the acetyl ester of hydroxyl methadone, α-acetyl methadol (i.e., LAAM).[333] The ester, LAAM (Figure 5.4), showed analgesic activity about six times greater than that of methadone though it was also about twofold more toxic in mice; methadone had an LD$_{50}$ = 28.6 mg/kg with an LD$_{50}$ = 13.8 mg/kg for

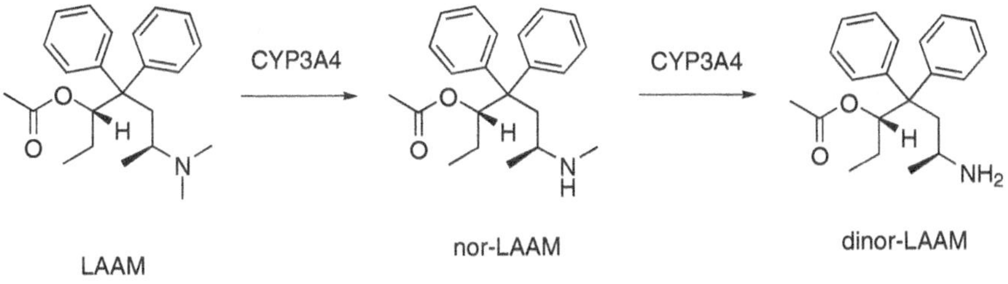

FIGURE 5.4   Active metabolites of LAAM

LAAM. Of note the alcohol was less toxic (LD_{50} = 76 mg/kg) and about 2.6 times as potent as methadone.

Clinical trials in the 1970s established a favorable efficacy of LAAM relative to methadone in the treatment of opioid addiction and the FDA approved LAAM in 1993 for maintenance therapy in opioid addiction. Roxane Laboratories marketed LAAM under the name ORLAAM™. Clinical studies have shown LAAM as having good efficacy and being well tolerated in the treatment of opioid dependence,[334] however, LAAM has also been shown to induce QTc-prolongation to a higher degree than methadone.[335] In 2003, Roxane Laboratories announced it was discontinuing the sale and distribution of ORLAAM™ in the United States. LAAM is a longer-acting drug than methadone and the metabolism of LAAM by CYP3A4 produces active metabolites that are shown in Figure 5.4.

The primary pharmacological action of LAAM is as a μ-opioid agonist (K_i = 15 nM),[336] though, as with methadone, it also acts as a noncompetitive α3β4 nAChR antagonist (IC_{50} = 2.5 μM).[330]

5.3.3   Buprenorphine-Opioid Partial Agonist

Buprenorphine (Suboxone®, Subutex®) was discovered at the English company Reckitt & Coleman in 1966 and was approved for use in 1978. J.W. Lewis has written a fascinating history of the discovery of buprenorphine during which the extremely potent μ-opioid agonist etorphine was also discovered.[337] Buprenorphine is structurally novel in that it is derived from a Diels-Alder reaction of thebaine with methyl vinyl ketone as shown in Figure 5.5.

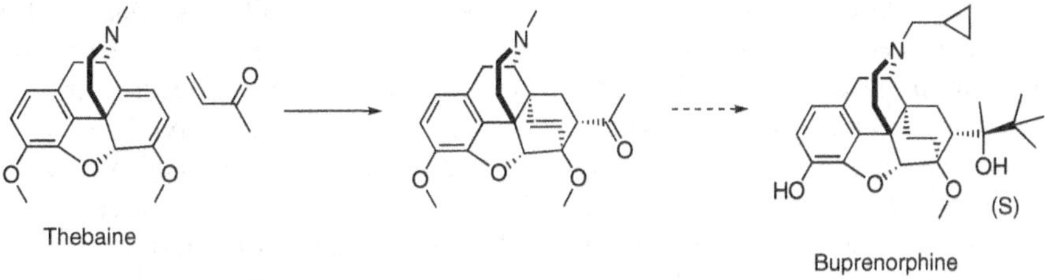

FIGURE 5.5   Synthesis of buprenorphine

The functional activity of buprenorphine is somewhat tricky, dependent in part on the type of functional assay used. For example, in a cell line using the $[^{35}S]GTP\gamma S$ assay buprenorphine behaves as a μ-opioid agonist with an $EC_{50} = 0.2$ nM and is about equal potent as morphine.[338] However, in another $[^{35}S]GTP\gamma S$ functional binding assay, this time using membranes obtained from the caudate putamen of guinea pigs, buprenorphine did not stimulate $[^{35}S]GTP\gamma S$ binding and inhibited agonist-stimulated $[^{35}S]GTP\gamma S$ binding. Antagonists K_i values of 0.088, 1.15, and 0.072 nM at μ-, δ-, and κ-opioid receptors, respectively, were measured. Based on this, buprenorphine is a more potent antagonist than the nonselective opioid antagonist naltrexone that has antagonist K_i values of 1.4, 25, and 11 nM at μ-, δ-, and κ-opioid receptors, respectively.[339]

In agreement with the cell-based assay using cloned μ-opioid receptor mentioned above, buprenorphine in HN.10 cells expressing the cloned μ-receptor was a potent ($K_d = 0.52$ nM) but low efficacy ($E_{max} = 33\%$ stimulation) agonist. However, when the potent and selective μ-opioid agonist DAMGO was added, buprenorphine potently inhibited DAMGO-stimulated $[^{35}S]$-GTPγS binding with a $K_i = 0.10$ nM. Then, buprenorphine is a more potent antagonist than agonist though it can be considered a partial agonist.

Paronis and Bergman have recently examined the pharmacology of buprenorphine in rhesus monkeys.[340] They investigated the effects of different opioid agonists and antagonists on food responding during both acute and chronic treatment with buprenorphine. The right-ward shift in the dose-response curves during chronic treatment with buprenorphine was much larger with a μ-opioid agonist (heroin) than with the κ-opioid agonist U50,488 demonstrating the effects of buprenorphine on the μ-opioid is more important than on the κ-opioid receptor. Introduction of naltrexone was found to initiate precipitated withdrawal. They concluded that the buprenorphine antagonist effect on the μ-opioid receptor is potentiated during chronic treatment, can lead to the development of tolerance to buprenorphine, and could produce dependence in previously nondependent monkeys.

Mello and Mendelson first showed that buprenorphine could control opioid addiction in clinical trials that were reported in 1980.[341] Its main use is similar to that of methadone as a maintenance medication to prevent relapse. Buprenorphine is available as an oral drug under the brand name Subutex® and Suboxone®, which is a mixture of buprenorphine and naloxone. Suboxone was approved by the FDA in 2002 for the treatment of opioid addiction and is made by the Reckitt Benckiser (RB) Group though they lost exclusive rights in 2009 and generic formulations are available. The drug is currently sold only as a sublingual film. In March 2013, RB Pharmaceuticals announced it was voluntarily withdrawing their tablet formulation from the US market. Suboxone can be considered to be a blockbuster drug. According to Drugs.com, total sales in 2012 of $1.5 billion were reported (*Source*: http://www.drugs.com/stats/suboxone). The use of Suboxone has proven effective for the treatment of prescription opioid abuse. In a 600 outpatient 12-week clinical trial, 49% of the patients had a beneficial effect (negative urine tests) from taking Suboxone. Once the medication was stopped though there was a high rate of relapse.[342]

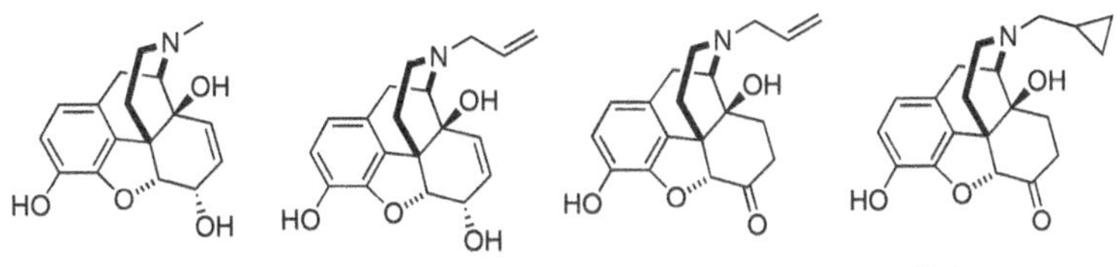

K_i (nM)	μ-opioid	2.2	11	3.7	0.74
	δ-opioid	68		32	44
	κ-opioid	26	1.6	66	0.3

FIGURE 5.6   Structures of common opioid antagonists

As with prescription analgesics, both methadone and buprenorphine are suscepti-ble to misuse. One effort to circumvent misuse is to develop drug delivery systems that make it impossible or very difficult to obtain the drug in a pure, powder form that can be injected or snorted. An extended-release form of buprenorphine called Probuphine® is being developed by Titan Pharmaceuticals. The implant is composed of an ethylene-vinyl acetate matrix that can deliver a 6-month dose of buprenorphine hydrochloride with a single injection. A New Drug Application (NDA) was submitted to the FDA in October 2012 and an Advisory Committee voted in favor of approval in March 2013. As of December 2013, Titan was in discussions with the FDA on the need to develop additional clinical data for dosing levels. The dosage levels being examined range from 8 to 16 mg of buprenorphine per day.

5.3.4   Naltrexone-Opioid Antagonist

Opioid antagonists were first used clinically in the 1940s with the development of nalorphine. The structures and binding affinities at the opioid receptors of common opioid antagonists are shown in Figure 5.6.

Nalorphine was later determined to be a partial agonist with IC_{50}'s of 2.7 and 483 nM at μ- and κ-opioid receptors, respectively. For comparison, morphine has IC_{50}'s of 17 and 483 nM. Both nalorphine and morphine have similar agonist efficacy of about 50% inhibition of Forskolin-stimulated adenylyl cyclase activity.[343,344] Not surprisingly, nalorphine was found to have analgesic properties.

Oxidation of the 6-OH group and reduction in the C7-C8 double bond of nalorphine led to the full antagonist naloxone. Replacement of the N-allyl group of naloxone with a cyclopropyl methyl group produced naltrexone. The full synthesis of naltrexone from thebaine is shown in Figure 5.7.

Within this series is the remarkable change in functional activity upon seemingly simple changes to the N-alkyl group. For example, taking morphine, increasing the size of the N-alkyl group from methyl to butyl, increases the level of μ-opioid antagonism till no agonist activity at all is observed with N-butyl morphine. However, going further and substituting the methyl group with an n-pentyl or n-hexyl alkane fully restores agonist activity. The N-(2-phenethyl) derivative is 14 times more potent than morphine!

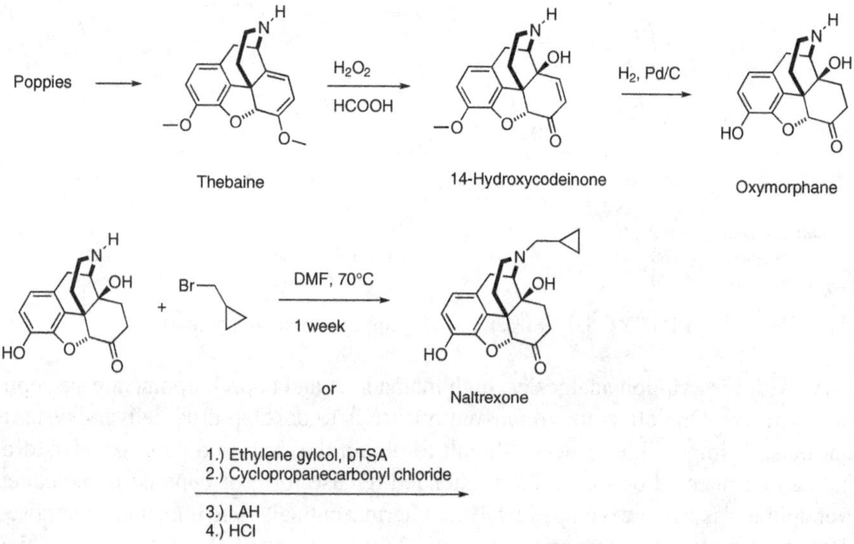

FIGURE 5.7 Synthesis of naltrexone

Meanwhile we can see where redox changes in the C ring with simultaneous modification of the *N*-alkyl group produces full antagonists.

The reasons for these dramatic changes in functional activity have not been rigorously explained. A molecular dynamics study (complete with ".avi" video simulations) suggests that the difference can be due to different modes of binding of the compounds to the receptor.[345] The molecular dynamics simulations revealed distinct binding modes of analyzed antagonists and agonists to the μ-opioid receptor. They all interacted with aspartate D3.32 on TM3 of the μ-opioid receptor but while the antagonists formed a bond with tyrosine Y3.33 on TM3, the agonists bound to histidine H6.52 on TM6. Furthermore, it was possible to observe a break of an intramolecular hydrogen bond from D3.32 to Y7.43 between TM3 and TM7 (3-7 lock) during simulations of agonist complexes but not those of antagonist. Interdependence between the 3-7 lock and the rotamer toggle switch of tryptophan W6.48 is proposed based on simulation of the naltrexone complex that was restrained to force an agonistic action.

Naltrexone is a μ-opioid receptor neutral antagonist with a binding affinity $K_i = 0.46$ nM and functional activities of $pA_2 = 9.35$ ([^{35}S]GTPγS) and $pK_B = 9.28$ (cAMP).[346] Naltrexone also binds strongly to the δ- and κ-opioid receptors with K_i's of 44 and 0.3 nM, respectively. As naltrexone can readily displace morphine from the opioid receptors (morphine K_i's μ = 2.2 nM, δ = 68 nM, κ = 26 nM), it is used in emergency room settings to treat narcotic overdose patients.

Alkermes has developed a sustained-release version of naltrexone called Vivitrol^{TM}. It is intended as a once-a-month intramuscular injection of 380 mg of naltrexone administered by a physician. The naltrexone is encased in microspheres composed of polylactide-*co*-glycolide. It is reported that following IM injection, naltrexone plasma concentration first peaks after 2 hours, followed by a second peak about 2–3 days later. Beginning approximately 14 days after dosing, concentrations of naltrexone slowly decline, with measurable amounts still existing after 1 month (prescribing insert www.vivitrol.com).

While the use of oral naltrexone is useful for opioid withdrawal, its efficacy in maintenance treatment for the prevention of relapse is said to be more questionable.[318]

The development of other sustained-release formulations is being investigated. Two studies will be examined to illustrate this point. An academic study in Australia conducted a 6-month randomized, double-blind, double-dummy trial involving 69 subjects to investigate whether a sustained-release naltrexone implant was better at reducing craving and relapse than oral naltrexone. The implant contained approximately 2.3 grams naltrexone loaded in poly(*trans*-3,6-dimethyl-1,4-dioxane-2,5-dione) lactide microspheres compressed into tablets and surrounded by a poly(*trans*-3,6-dimethyl-1,4-dioxane-2,5-dione) lactide coating. Implant treatment involved minor surgery under local anesthetic with the subcutaneous insertion of tablets into the side of the lower abdomen. The oral treatment was 50 mg naltrexone per day. They found that implanted naltrexone was much better at reducing craving and preventing relapse than the oral group. After 6 months, there was an approximately 42% relapse rate for oral naltrexone users compared with a 25% relapse rate for those receive the implant. Naltrexone plasma levels of >0.5 ng/mL were required with best results at plasma levels of 3 ng/mL.

Another double-blind placebo-controlled study involved 60 heroin-dependent subjects who were treated with either 192 mg or 384 mg of depot naltrexone microcapsules. The exact composition of the microcapsules was not mentioned. The study lasted 8 weeks and the subjects received depot injections at the beginning of week 1 and week 5.[347] They found that retention was dose-dependent with 68% of the patients, remaining in the program versus only 39% for placebo. The effect on opioid use was mixed depending on how it is calculated. Initially, it was assumed that missed clinic visits and urine samples would have resulted in a positive urine test. Using this assumption, only 25% of the placebo group gave urine samples negative for opioids while 62% of the subjects given the 384 mg dose tested negative. If the assumption was not used, then there was no significant difference between placebo and the naltrexone-treated patients. This simply reflects the common difficulty of outpatient compliance in clinical tests. As in the first study described, the peak naltrexone plasma levels after 1 week of treatment were 1.9 ng/mL for the 192 mg depot dose and 3.2 ng/mL for the 384 mg depot dose.

At present, methadone and buprenorphine, despite a potential abuse liability, are still the drugs of choice for long-term maintenance programs. They appear to offer the best compromise especially in terms of patient compliance.

5.3.5   Lofexidine-Adrenergic Agonist

Lofexidine is an adrenergic α_{2A} agonist that is approved in the United Kingdom as BritLofex® (Britannia Pharmaceuticals Limited) for opioid detoxification but not yet approved by the FDA. Both clonidine and lofexidine are prescribed for off-label

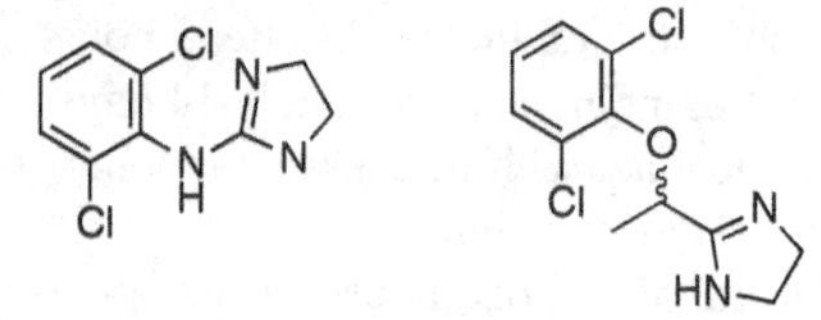

	Clonidine	Lofexidine
	K_i (nM)	K_i (nM)
α_{2A}	31	4.4
α_{2B}	40	68
α_{2C}	63	69

FIGURE 5.8   Structures and binding affinities of clonidine and lofexidine

use in the treatment of substance abuse. Their mechanism of action is considered to be the reduction in anxiety that occurs during opioid withdrawal. Anxiety is associated with high levels of norepinephrine; clonidine and lofexidine, upon binding to α_{2A} presynaptic autoreceptors located on adrenergic neurons, inhibit the firing of the neurons, thus decreasing the release of norepinephrine. A further mechanism of lofexidine is to alleviate some of the physical symptoms of opiate withdrawal. Increased levels of norepinephrine are observed during opioid withdrawal. By the action of norepinephrine on adrenergic receptors, blood pressure and heart rate will increase during withdrawal. Lofexidine, by reducing norepinephrine levels, helps to alleviate the physical symptoms of withdrawal. Lofexidine has not been shown to affect the sensation of craving.

The α_{2A} agonist's clonidine and lofexidine arose from the imidazoline and guanidine class of α_1 and α_2 agonists. Clonidine was developed in the early 1960s at Boehringer for use as a centrally active hypotensive drug, while lofexidine was developed in the late 1960s. The structures and binding affinities of clonidine and lofexidine are shown in Figure 5.8.[348,349] The binding affinity of (S)-(−) enantiomer ($IC_{50} = 3.6$ nM) of lofexidine is 10 times than that of the (R)-(+) enantiomer ($IC_{50} = 43$ nM) to the α_2 adrenergic receptor.[350]

Synthesis of Clonidine   Clonidine is readily synthesized as shown in Figure 5.9 by formation of the thiourea of 2,6-dichloroaniline with ammonium thiocyanate. Methylation of the sulfur atom with methyl iodide generates the imino-methyl-thioether that is converted into the imidazoline with ethylene diamine.

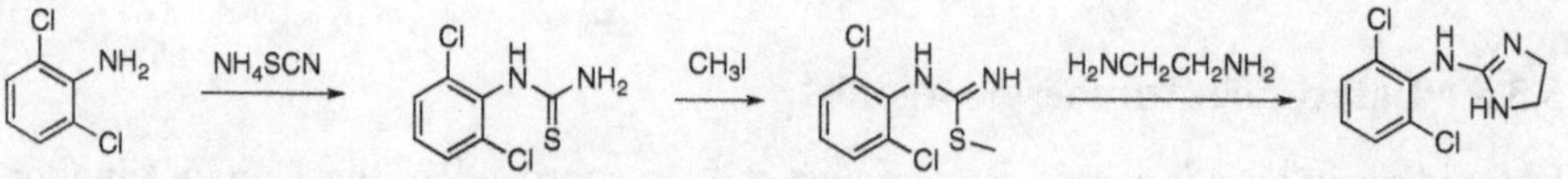

FIGURE 5.9   Synthesis of clonidine

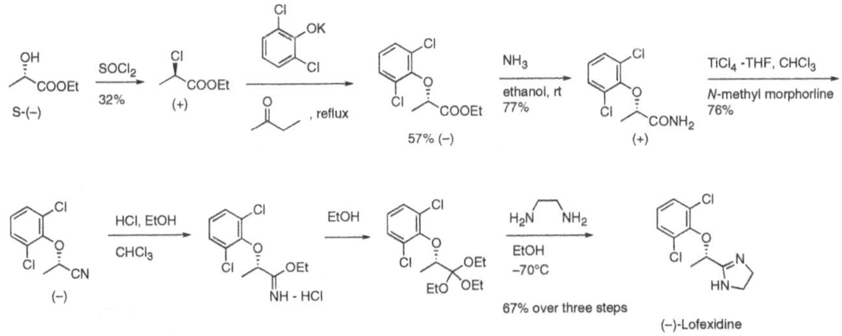

FIGURE 5.10   Synthesis of (S)-(−)-lofexidine

Synthesis of Lofexidine   Lofexidine has one chiral center and the different enantiomers of Lofexidine were first synthesized from ethyl lactate as shown in Figure 5.10 for the more active S-(−) enantiomer.[350] Conversion of the nitrile to the imidazoline was performed with isolation of the intermediates by simple filtration but without separate purification steps.

A modified procedure has been developed leading to R-(+)-Lofexidine from ethyl L-(−)-lactate (Figure 5.11) in an overall yield of 75–80% without chromatographic or distillation purification steps.[351]

It is recommended that the dosage of lofexidine should be titrated according to the patient's response. Initial dosage should be 0.8 mg per day in divided doses. The dosage may be increased by increments of 0.4–0.8 mg per day up to a maximum of 2.4 mg daily. Maximum single dose should not exceed 4 × 0.2 mg tablets (0.8 mg). Each patient should be assessed on an individual basis; those undergoing acute detoxification will usually require the highest recommended dose and dosage increments to provide optimum relief at the time of expected peak withdrawal symptoms.

The pharmacokinetics of lofexidine in humans has been reported. In a single dose study in healthy individuals using 1.2 and 2.0 mg, the average time to maximum concentration (C_{max}) was observed at approximately 3 hours with a C_{max} = 1755 ng/L for 1.2 mg and C_{max} = 2795 ng/L for 2.0 mg. The rates of elimination (K_e) at each dose were similar, 0.063 h^{-1} at 1.2 mg and 0.065 h^{-1} at 2.0 mg and the half-live of the drug was 11.16–11.44 hours with 2,6-dichlorophenol as the main metabolite.[352]

The above single dose study was modified to mimic lofexidine clinical trials in addicts that usually composed of three phases. Phase I (opioid agonist stabilization phase), Phase II (treatment phase), and Phase III (post-detoxification phase). The

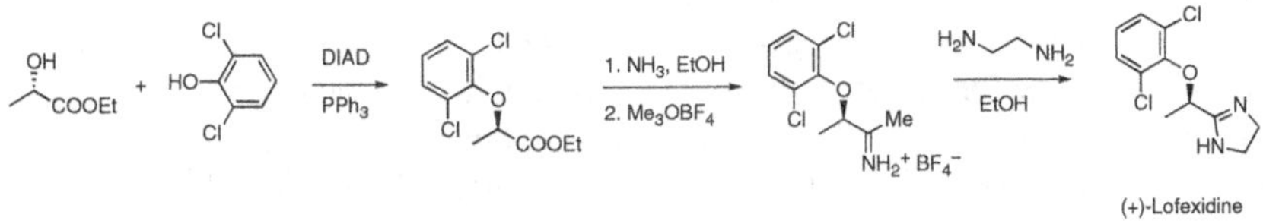

FIGURE 5.11   Synthesis of (R)-(+)-lofexidine

study used an open-label, randomized design. Three healthy male volunteers received multiple doses of lofexidine between study days 9 and 16. On study days 1 through 8, the subjects did not receive any medications and this period was used to mimic Phase I in lofexidine clinical trials in addicts. On the first day of dosing (study day 9), the subjects received 0.4 mg lofexidine twice daily (BID). From days 10 to 15 of the study, they received 0.8 mg BID. The dosing schedule ended on day 16 of the study, when the subjects received a 0.6-mg dose BID. On day 9 after the 0.4 mg dose, the C_{max} detected at 3.3 hours was 433 ng/L. This maximum concentration compares relatively well with the 2.0 mg dose from the single dose study when it is normalized to a C_{max} of 560 ng/L for a 0.4 mg dose. On day 10, the C_{max} had increased to 1450 ng/L.

We can compare these results with those involving opioid-dependent patients undergoing detoxification. During an inpatient randomized double-blind study involving 35 patients taking lofexidine and 33 patients on placebo the amount of lofexidine in the blood was determined at day 7. From days 4 to 7, the patients were given a dose of 3.2 mg lofexidine. A C_{max} of 3442 ng/L was determined and the authors concluded that steady-state concentrations were reached.[353] It is difficult to compare this with the study on non-opioid taking subjects due to the different dosage schedules, but the authors conclude that they are in agreement. This suggests that opioid-dependent patients will not metabolize lofexidine differently than non-opioid-dependent patients.

A Phase III trial for opiate detoxification (Study NCT00235729) sponsored by the Department of Veterans Affairs has been completed but results are not yet available. Lofexidine has no abuse liability and a favorable safety profile so if it was approved it would become a valuable drug for opioid detoxification and could be used to supplement the use of methadone and buprenorphine.

A second Phase II clinical trial (NCT00142909) to study "Effectiveness of Lofexidine to Prevent Stress-Related Opiate Relapse During Naltrexone Treatment" was sponsored by Yale University in collaboration with NIDA. As reported in *NIDA Notes 12/2008* "the researchers stabilized 18 opioid-detoxified men and women on naltrexone (50 mg) and lofexidine (2.4 mg) daily for 1 month. They then retained all the patients on naltrexone for 4 more weeks, but kept 8 on lofexidine and gave 10 others identical-looking pills containing lofexidine doses that—unbeknownst to the patients—tapered to zero over several days. Of the 13 patients who completed the study, 80% of those who continued to receive combination therapy submitted opiate-free urine samples throughout the 4-week period, compared with 25% of those tapered to placebo. A follow-up laboratory session that exposed 10 of the patients to stressful and opiate-related stimuli showed that lofexidine—but not placebo—reduced the patients' reaction to stress, stress-induced opiate craving, and negative emotions (such as anger), all of which can trigger relapse."

A Cochrane review investigated the clinical uses of clonidine and lofexidine versus methadone for opioid withdrawal. Although limited data prevented a rigorous statistical analysis, it was concluded that little significant differences exist with respect to reducing the effects of withdrawal.[354] Perhaps due to the known adverse effects of clonidine and to a lesser extent, lofexidine, patients on methadone treatment stayed in treatment longer.

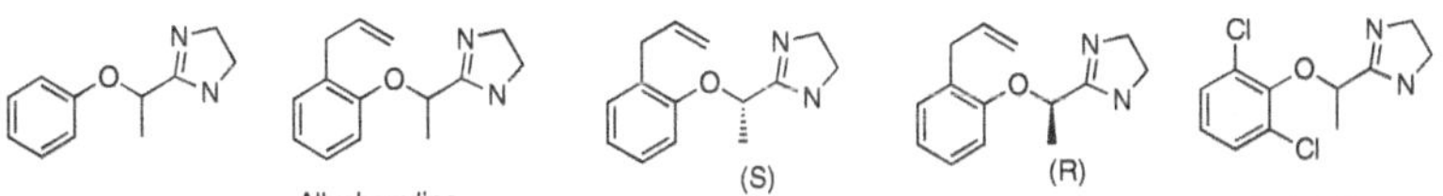

FIGURE 5.12   SAR of α_{2C}-agonists/α_{2A}-antagonists

K_i (nM)		Allyphenyline	(S)	(R)	Lofexidine
	α_{2A}	27	57	52	100	4
	α_{2B}	166	388	398	562	67
	α_{2C}	263	85	71	177	69
EC_{50} (nM)	α_{2A}	Antagonist	Antagonist	Antagonist	Antagonist	6
	α_{2B}	Antagonist	No activity	1000	No activity	126
	α_{2C}	Antagonist	50	25	162	1

While on the subject of adrenergic receptors, the α_{2C} subtype is known to be important in CNS processes. An Italian group has developed a series of α_{2C}-agonists/α_{2A}-antagonists based on the aryl imidazoline structure of clonidine and lofexidine that enhance morphine efficacy and reduce the acquisition of tolerance dependence to morphine in mice.[355]

Structures of key compounds and their binding affinities and functional activity are shown in Figure 5.12. While allylphenyline has no inherent antinociceptive effects, when 0.05 mg/kg was co-administered with morphine (5 mg/kg subcutaneous) the analgesic effect of morphine was enhanced in both magnitude and especially, time. For example, at 60 minutes, the maximum possible effect for morphine was about 60% while for the combination dosage of 0.05 mg/kg allylphenyline and 5.0 mg/kg morphine the effect was 100%. At the same dosage, naloxone-induced withdrawal symptoms were reduced and the ability of the mice to develop tolerance to the morphine was also reduced by about 50%.

The antagonist to agonist switch at the α_{2C} receptor (e.g., allylphenyline vs. lofexidine) was ascribed to an a hydrophobic interaction of the allyl group with a tryptophan and two phenylalanine amino acids in the binding pocket of the receptor. Interestingly, the presence of two ortho chlorines on the phenyl ring (lofexidine) results in agonism at all three receptors but the removal of a chlorine atom results in antagonism at the α_{2A} receptor. Although the stereoelectronic effects are complex, an SAR and a molecular study of 26 compounds suggest that lofexidine is an agonist at all three adrenergic receptors due to having two ortho lipophilic ($+\pi$) substituents of moderate size (molar refractivity (MR) < 16 cm^3/mol).[348] The α_{2A} receptor appears to be especially sensitive to the structural changes in the ligands studied. In all it appears that this unique set of α_{2C}-agonism/α_{2A}-antagonism may be of interest.

5.4   MEDICATION DEVELOPMENT

5.4.1   μ-Opioid Agonist and NET/SERT Reuptake Inhibiter: Tramadol

Tramadol (Rybiz®, Ryzolt®, Ultram®) is used to relieve moderate to moderately severe pain and was developed by Janssen Pharmaceuticals. The drug is now available

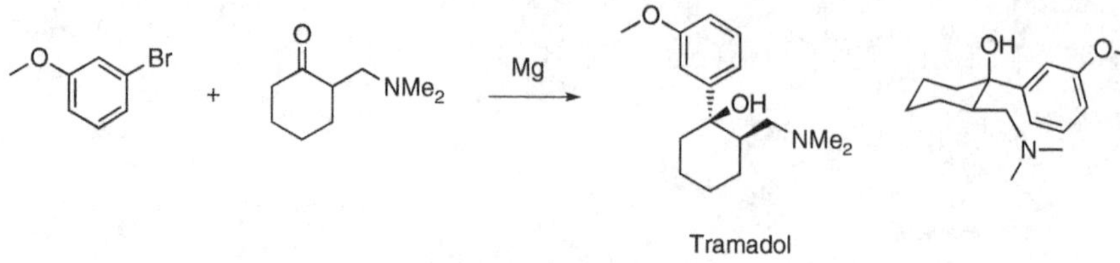

Tramadol

FIGURE 5.13   Synthesis of tramadol

as a generic drug product. An extended-release form of tramadol is available for people who are expected to need medication to relieve chronic pain. The clinical pharmacology of tramadol with regard to pain has been reviewed.[356] Tramadol is a weak μ-opioid receptor agonist (K_i = 2.1 μM), a weak NET reuptake inhibiter (IC_{50} = 1.4 μM), and a weak SERT reuptake inhibitor (IC_{50} = 2.2 μM).[357] In addition to tramadol, decreasing pain through its action on the μ-opioid receptor, additional analgesic effects are obtained by increasing the levels of serotonin and norepinephrine. Both neurotransmitters are known to be important in the descending neuronal pathway from the brain to the dorsal horn in the spinal cord. In human clinical trials at doses of 200 and 400 mg (qd, po), it showed evidence of attenuating opioid withdrawal symptoms without any significant agonist effects.[358]

Tramadol contains two chiral centers and is sold as the cis-(±) form for the treatment of pain. The compound is generally prepared as shown in Figure 5.13 following the procedure of Flick in US Pat. No. 3652, 589.

The α-methylenedimethylamine cyclohexanone base is prepared by a Mannich reaction using cyclohexanone, formaldehyde, and dimethylamine. The Mannich reaction generates a chiral center and two enantiomers are produced as a racemic mixture. Addition of the Grignard reagent from 3-bromoanisole to the carbonyl carbon generates a second chiral center leading to a diastereomeric mixture of four compounds in a cis-to-trans ratio of about 80:20. All four potential compounds are shown in Figure 5.14

The cis-(−)-1S,2S enantiomer of the cis pair has greater norepinephrine reuptake inhibition potency than the cis-(+)-1R,2R enantiomer (K_i = 0.43 μM vs. 2.51 μM), while the opposite is true for the inhibition of serotonin reuptake; K_i cis-(+)-1R,2R enantiomer = 0.53 μM, K_i cis-(−)-1S,2S enantiomer = 2.35 μM. The cis-(+)-1R,2R enantiomer binds more strongly to the opioid receptors with K_i's of 1.33, 62, 54 μM at μ-, δ-, and κ-opioid receptor, respectively. The cis-(−)-1S,2S enantiomer has lower

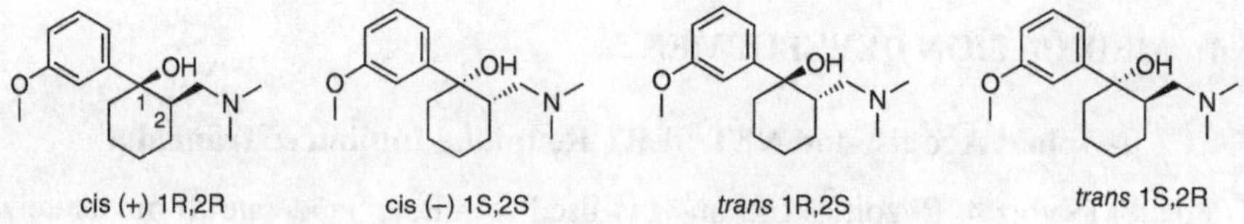

cis (+) 1R,2R   cis (−) 1S,2S   trans 1R,2S   trans 1S,2R

FIGURE 5.14   Structures of tramadol stereoisomers

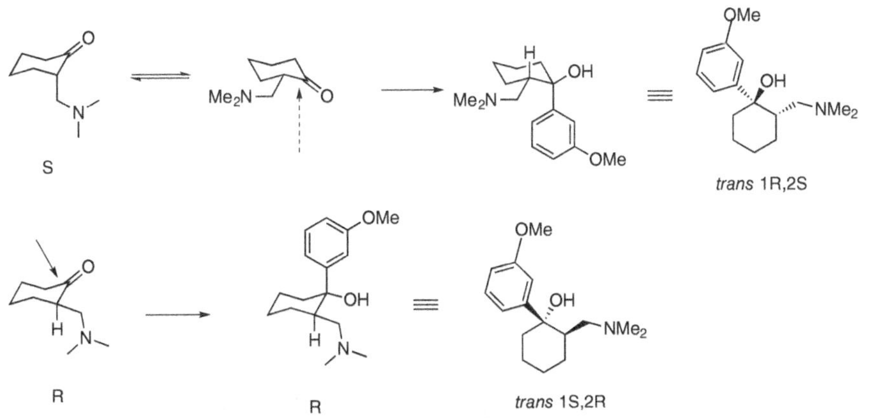

FIGURE 5.15 Diastereoselective synthesis of *trans*-tramadol

affinity at the μ-, δ-opioid receptors with K_i = 25 and 213 μM, respectively, and is equipotent at the κ-opioid receptor (K_i = 53 μM). In rat antinociception studies *cis*-(+)-1R,2R tramadol was more potent than *cis*-(−)-1S,2S tramadol.[357]

SAR studies that focused on optimization of the analgesic properties of tramadol have been reported but not with respect to the treatment of opioid withdrawal.[359] It is not clear what the relative importance of potency at the opioid receptors and the two monoamine transporters are with regard to use as an opioid withdrawal drug.

Diastereoselective syntheses of both the *cis* and *trans* racemic mixtures have been reported in the patent literature. During the Grignard step, addition of 0.75 equivalents of lithium chloride, and using the bidentate solvent 1,2-dimethoxyethane, gives a 92:8 ratio of the *trans/cis* isomers in 68% yield. Using the monodentate solvent tetrahydrofuran reduces the ratio to *trans/cis* of 83:17 (US2005215821). A possible mechanism for the diastereoselectivity is shown in Figure 5.15. The nucleophilic anisole magnesium bromide approaching the carbonyl carbon from the axial position to give an equatorial hydroxy group can explain the predominance of the *trans* isomer. The use of both a coordinating bidentate solvent and lithium chloride would decrease the aggregation state of the anisole magnesium bromide increasing its reactivity. Why this would favor axial attack is not clear unless the presence of lithium affects the chelation of magnesium to the carbonyl carbon.

The *cis* racemic mixture can be prepared stereoselectively by first forming a complex of the (±)-2((dimethylamino)methyl)cyclohexanone with catalytic amounts of iron (III) chloride of about 1 mol%. The resultant violet solution is slowly added at −50°C to a suspension of the lithium salt of 3-bromoanisole in THF. A 94:6 ratio of *cis/trans* tramadol is produced in yields of 89% (US2003065221). A logical explanation as depicted in Figure 5.16 is the ferrous ion acts as a Lewis acid and coordinates with the carbonyl oxygen atom and the amino group. This apparently forces the anisole lithium bromide to approach from the equatorial position producing an axial hydroxyl group and giving the *cis* isomer.

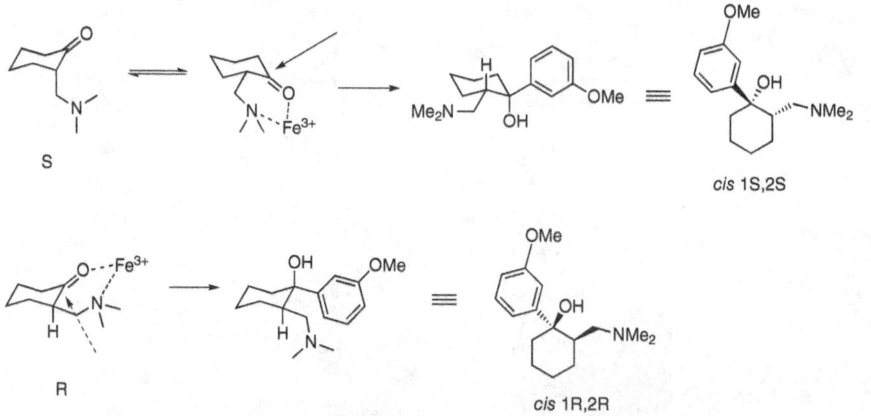

FIGURE 5.16   Diastereoselective synthesis of *cis*-tramadol

A continuous flow synthesis of the Grignard step produced tramadol in 96% yield with a *cis:trans* diastereomeric ratio of 80:20, as originally reported. The continuous flow Grignard reaction was accomplished by injecting a 1.55 M solution of the anisole magnesium bromide and a 0.25 M solution of the ketone into a 10-mL reactor at 0.30 mL/min at room temperature.[360]

Enantiopure drug may be obtained by resolution of the *cis* or *trans* compounds. Two detailed methods for the resolution of (±)-*cis*-tramadol using mandelic acid are available.[361,362] Interestingly, an enantioselective synthesis of tramadol has not been reported.

5.4.2   NMDA Antagonist: Memantine

Memantine (Namenda®) was approved in 2003 for the treatment of moderate to severe Alzheimer's disease and is marketed by Forest Laboratories. One of its main benefits is the lack of significant side effects. Memantine is an uncompetitive NMDA antagonist ($IC_{50} = 0.54\ \mu M$) that is thought to function by preferentially blocking open NMDA channels located on postsynaptic neurons. This will inhibit the migration of Ca^{2+} ions into the neuron and attenuate depolarization of the neuron, reducing the development of an action potential. The exact binding mechanism of memantine in the NMDA ion channel is not known, but the group of Dougherty has recently produced a homology model that is thought to mimic the GluN1 and GluN2B transmembrane regions of NMDA where memantine binds.[363]

The primary interaction of memantine with NMDA ion channel is by the ammonium salt interacting with an asparagine side chain carboxamide group on the GluN1 subunit via hydrogen bonding.[364] By changing the amino acids in the binding sites by point mutations and then measuring binding affinity, they identified a hydrophobic pocket consisting of two alanine amino acids that interact with the methyl groups of memantine as shown in Figure 5.17. Based on this model, memantine binds to

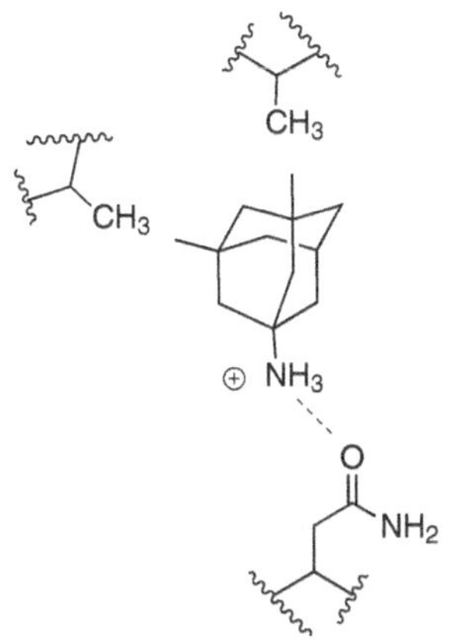

FIGURE 5.17   Binding of memantine to NMDA

the NMDA ion channel with the amino group oriented toward the cytoplasm of the neuron.

Amantadine, a structural similar antiviral drug was also examined. Amantadine is identical to memantine except that it lacks the two-methyl groups. The lower binding affinity of amantadine ($IC_{50} = 41$ μM) can be explained then to the loss of the two hydrophobic, van der Waals interactions.

Lilly Research Laboratories first reported memantine in 1965 as part of a program developing hypoglycemic drugs. It is a structurally unusually CNS drug as it contains an adamantane ring and a single functional group, an amino group. Figure 5.18 outlines the synthesis of memantine. Following the procedures developed by Stetter, the compound is prepared from 1,3-dimethyladamantane by bromination to give a 3° bromide. The bromide then undergoes an unusual reaction with acetonitrile and aqueous acid to produce the amide in high yield. Hydrolysis of the amide produces memantine in an overall yield of 63%.[365,366]

As we have seen, the glutaminergic neurocircuit is important in drug addiction so modulation of it would be expected to have some effect. Memantine has been investigated for its effects on alcohol, cocaine, and opioid dependence.[367] It produced modest reductions in subjective aspects (craving) in heroin-dependent patients. Of interest is that one study reported that memantine at doses of 30 mg per day could reduce

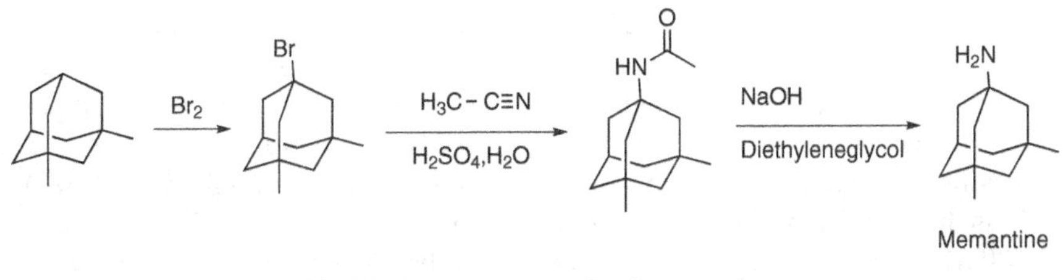

FIGURE 5.18   Synthesis of memantine

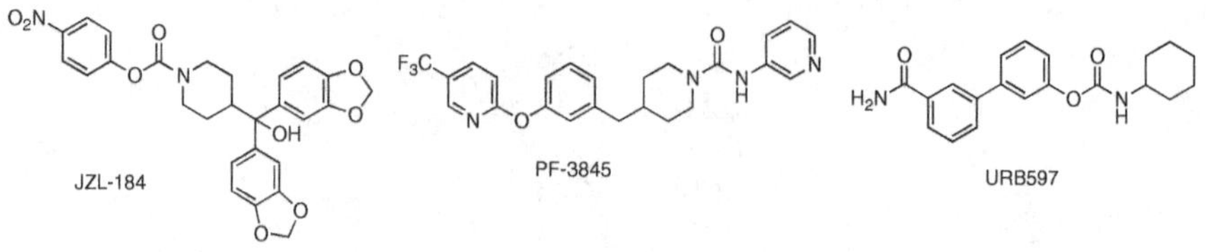

FIGURE 5.19 FAAH and MAGL inhibitors

alcohol craving but it also enhanced the dissociative effects of alcohol. Ketamine, a widely abused club drug taken for its dissociative effects, is also an NMDA antagonist. One of the reported side effects on memantine during the treatment of Alzheimer's patients is hallucinations.

5.4.3 Fatty Acid Amide Hydrolase Inhibitors

The enzymes, fatty acid amide hydrolase (FAAH) and monoacylglycerol lipase (MAGL), metabolize several endogenous fatty acid amides such as anandamide, which activate CB_1 receptors. Inhibition of these enzymes would increase endogenous anandamide and 2-archidonylglycerol levels mimicking the effect of exogenous tetrahydrocannabinol (THC). Rodent studies have found that inhibition of FAAH with the irreversible inhibitor PF-3845 (K_i = 230 nM) or inhibition of MAGL with the irreversible inhibitor JZL184 (IC_{50} = 8 nM) reduced naloxone-precipitated morphine withdrawal symptoms.[368] How the increase in the concentration of the endogenous endocannabinoids anandamide and 2-archidonylglycerol by inhibition of their metabolism modulates opioid withdrawal is unknown. It was proposed that a possible mechanism might be that stimulation of CB_1 receptors colocalized with the μ-opioid receptor (both $G_{i/o}$ coupled) would attenuate the increase in adenylyl cyclase and cAMP production that occurs during opioid withdrawal. The different inhibitors are structurally diverse though the active moieties, carbamate or urea, are the same as shown in Figure 5.19.

Further evidence of an FAAH/opioid connection is evidenced that the antinociceptive effects produced by an FAAH irreversible inhibitor URB597 (IC_{50} = 5 nM)[369] and anandamide combination could be blocked by the administration of an μ-opioid receptor antagonist (naloxone) or an κ-opioid receptor antagonist (nor-binaltorphimine).[370] This suggests that activation of the CB_1 receptor via anandamide causes the release of endogenous opioid peptides.

The covalent MAGL inhibitor JZL184 was discovered by Cravatt and a detailed account of the synthesis, SAR, and enzymatic inhibition activity is available.[371] Pfizer developed the FAAH inhibitor PF-3845 (Figure 5.20) and they have also given a detailed account of the discovery and properties of the inhibitor.[372] Further, SAR studies by Pfizer have led to a potent and selective irreversible inhibitor of FAAH with an IC_{50} = 7.2 nm and a second-order rate constant of inhibition k_{inact}/K_i = 40,300 $M^{-1} s^{-1}$. The compound was reported to be completely selective for FAAH in a serine hydrolase functional proteomic-screening assay at 100 μM and in a broad

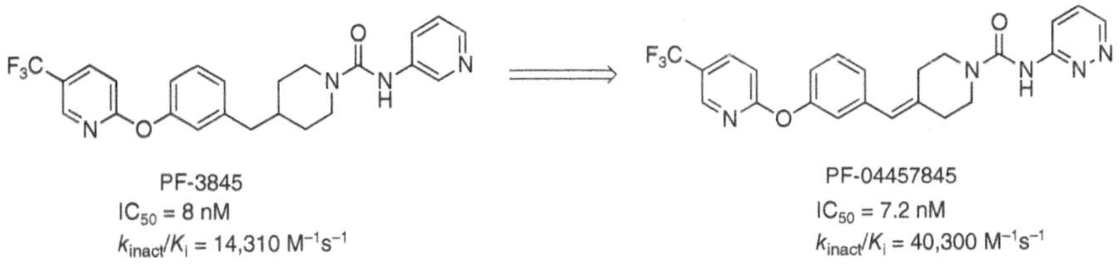

FIGURE 5.20   Optimization of FAAH inhibitor PF-3845

68 target-screening panel.[373] Pfizer has completed a clinical trial (NCT00981357) with PF-04457845 for the treatment of pain in patients with osteoarthritis of the knee.

It remains to be seen if this approach will be useful in the treatment of drug addiction. Yale University and NIDA are currently recruiting participants for a Phase II study (NCT01618656) to investigate the safety and efficacy of PF-04457845 to treat cannabis withdrawal.

5.4.4 5-HT$_{3C}$ Antagonist: Ondansetron

Ondansetron (Zofran®, Zudan®) is an approved drug that is used in cancer chemotherapy for the prevention of nausea and vomiting. GlaxoSmithKline first reported the compound as GR38032F in the early 1980s. The drug contains one chiral center and is sold as the racemate.

A genetic mapping experiment in mice revealed that the *Htr3a* gene was important in opioid withdrawal. This gene produces the 5-HT$_3$ receptor and chronic morphine use in mice was found to lead to a fivefold reduction in the expression of the gene in the amygdala and dorsal raphe. Behavioral studies using the 5-HT$_3$ receptor antagonist ondansetron showed that the antagonist was effective in attenuating the physical dependence and the reinforcing effects of morphine. Based on these animal model studies, 8 mg of ondansetron was preadministered to eight nonuser human patients before injection of 10 mg/70 kg of morphine over a 10-minute period. After 2 hours, precipitated withdrawal was initiated with 10 mg/70 kg naloxone and withdrawal symptoms were recorded. Withdrawal symptoms were measured using the Subjective Opioid Withdrawal Scale and Objective Opioid Withdrawal Scale. Withdrawal symptoms were reduced by 76% in all patients.[374]

A clinical trial (NCT01549652) using the 5-HT$_{3C}$ antagonist ondansetron is scheduled to occur to test the ability of the drug to reduce symptoms associated with opioid withdrawal and to prevent the progression of opioid physical dependence. The trail is being conducted by Stanford University in collaboration with NIDA and will involve 133 patients.

As we discussed in Chapter 2, the 5-HT$_3$ receptor is a ligand-gated ion channel and antagonism will reduce the concentrations of extracellular dopamine. The functional interactions between ondansetron antagonism of 5-HT$_{3A}$R, opioid use, and dopamine release are not clear. Microdialysis experiments have shown that ondansetron decreases dopamine levels in the striatum and that 5-HT$_3$ receptor antagonists reduce dopamine levels in the nucleus accumbens by inhibiting the increase

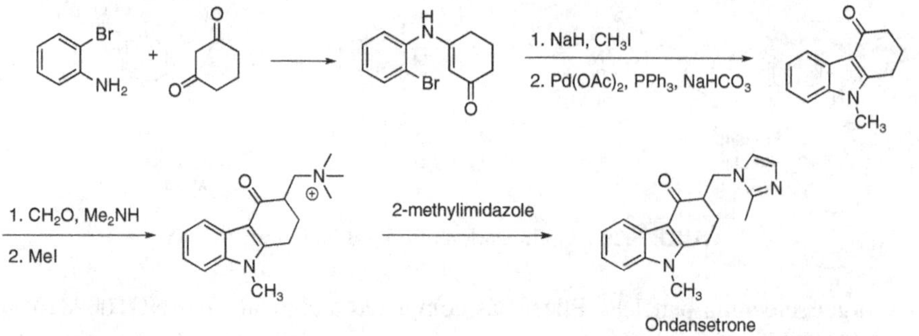

FIGURE 5.21   Synthesis of ondansetron

in firing of dopamine cell bodies in the VTA induced by peripheral administration of morphine.[375,376]

Ondansetron is highly selective for the 5-HT$_{3A}$ receptor with a $K_i = 1.6$ nM. In a binding profile assay, it showed little binding to 38 other receptors with the exception of the σ-receptor with a $K_i = 680$ nM.[377] The drug is very potent in tissue assays giving K_B's of 0.4–2.5 nM.[378] The potency of each enantiomer appears to be about the same except in a guinea-pig ileum longitudinal assay the (R) enantiomer was about 10-fold more potent than the (S) enantiomer. A pharmacophore model has been proposed consisting of a heteroaromatic ring system, a coplanar carbonyl group, and a nitrogen center at well-defined distances. Introduction of a six-membered hydrophobic ring connecting the indole N with the C7 position of the indole ring enhances binding affinity.[377]

The compound consists of a carbazole core linked to 2-methylimidazole. The general synthesis as shown in Figure 5.21 starts with the condensation of 2-bromoaniline and 1,3-cyclohexanedione to give the β-aminoaryl cyclohexanone. The indole ring was constructed by N-methylation of the aniline followed by a Heck coupling. A Mannich reaction serves a dual role by introducing the keto α-methylene group and subsequent quaternization of the amine produces a good leaving group. A substitution reaction by 2-methylimidazole completes the synthesis.

5.4.5   Ibogaine Alkaloids

A very interesting compound that has been investigated is the indole alkaloid ibogaine. Ibogaine is a structurally interesting indole alkaloid extracted from the root bark of the African shrub *Tabernanthe iboga*. It has a complex pharmacological profile, and interacts with multiple systems of neurotransmission.[379] Due to its respiratory and cardiac problems (including lengthening of the QT interval), the drug does not show promise in the treatment of drug addiction yet the pharmacological properties are of interest. Ibogaine has psychoactive properties and appears to modulate tolerance to opiates and other drugs of abuse.

Ibogaine has gained popularity as a medication at some substance abuse treatment centers (e.g., www.ibogainesanctuary.com). Beyond individual testimonials, there

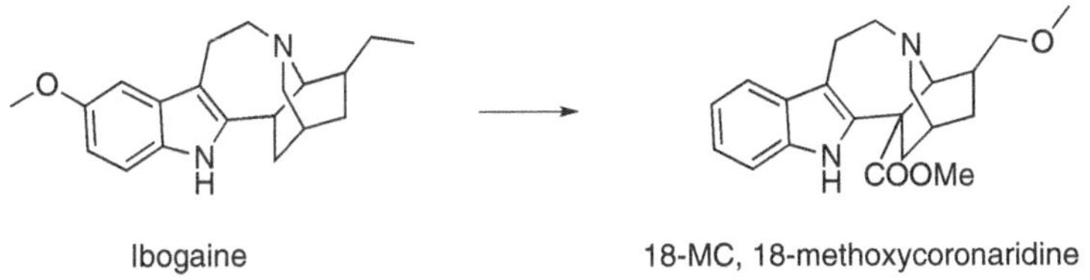

Ibogaine   18-MC, 18-methoxycoronaridine

FIGURE 5.22 Structures of ibogaine and 18-MC

are no validated clinical trials though that has proven its usefulness. In fact, the highly touted hallucinogenic effects would seem to make it improbable for use as a medication as while as the fact it has DEA Schedule I classification.

The Glick group has studied an analog of ibogaine, 18-methoxycoronaridine (18-MC) shown in Figure 5.22.

At doses of 10–40 mg/kg, 18-MC can attenuate the self-administration of heroin and cocaine in a dose-dependent fashion and at 40 mg/kg practically completely block the self-administration of both heroin and cocaine.[380] The drug has some level of bioavailability into the brain as it was delivered as the salt intraperitoneally. The free base has a calculated log P of 1.85 and a polar surface area of 50.8 Å^2. *In vivo* microdialysis measurements of dopamine and the dopamine metabolites DOPAC and HVA showed that extracellular dopamine levels in the nucleus accumbens was reduced 20–30% after 120 minutes.

The attenuation of heroin and cocaine self-administration may be due to the antagonist properties of 18-MC at the $\alpha_3\beta_4$ nAChR.[381] As mentioned, ibogaine and 18-MC bind to multiple targets. In brief, 18-MC binds to the μ, δ, and κ-opioid receptors with $K_i = 1$–5 μM and the nACh $\alpha_3\beta_4$ receptor with $IC_{50} = 0.75$ μM.[381,382] Using rat, thalamic membranes, ibogaine and 18-MC were μ-opioid antagonists with functional Ke values of 3 μM for ibogaine and 13 μM for 18-MC. Furthermore, 18-MC did not stimulate [^{35}S]GTPγS binding in Chinese hamster ovary cells expressing human or rat μ-opioid receptors and had only limited partial agonist effects in human embryonic kidney cells expressing mouse μ-opioid receptors.[383] Ibogaine's binding profile showed μM binding to 16 receptors, transporters, and ion channels.[384]

18-MC does not appear to have been tested in humans; while the respiratory and cardiac problems of ibogaine have been documented in humans[385]; studies of 18-MC in rodents suggest adverse side effects might be less than with ibogaine. Regardless, antagonism of $\alpha_3\beta_4$ nAChR has not received a lot of attention; 18-MC suggests that this might be a valid target to pursue. Brown has recently reviewed the properties, preclinical and clinical drug investigations, and the possible role of ibogaine in the treatment of drug addiction.[386]

6

MEDICATION DEVELOPMENT FOR STIMULANT ADDICTION

If chemists were charged with developing an addictive drug that was readily available with incredible profit margins, they would hardly do better than that with cocaine. Unfortunately, clandestine chemists may have succeeded with methamphetamine.

6.1 PHARMACOLOGY OF COCAINE ADDICTION

Cocaine is an intense, euphoria-producing stimulant drug with strong addictive potential. Cocaine is usually distributed as a white, crystalline powder, and to increase profits, cocaine is often diluted ("cut") with a variety of substances, the most common of which are sugars and local anesthetics. In contrast, cocaine free base (crack) looks like small, irregularly shaped chunks (or "rocks") of a whitish solid. Powdered cocaine can be snorted or injected into the veins after dissolving in water. Cocaine free base (crack) is smoked, either alone or on marijuana or tobacco.[387]

Cocaine is derived from the leaves of the coca tree (*Erythroxylum coca*) that is indigenous to the Andes Mountain areas of South America, particularly Bolivia and Peru. *Erythroxylum coca* is a shrubby tree that grows to heights of up to 18 feet. The leaves are of a brownish-green color, stiff, and are bitter tasting. The leaves have been used for thousands of years by the indigenous population as a tea or by chewing the leaves. Beyond their stimulatory effect, the leaves contain valuable minerals and vitamins A, B_2, and E.[388]

The majority of coca grown for use in the production of cocaine is grown in Columbia. The cocaine manufacturing process takes place in remote jungle

Drug Discovery for the Treatment of Addiction: Medicinal Chemistry Strategies, First Edition. Brian S. Fulton.
© 2014 John Wiley & Sons, Inc. Published 2014 by John Wiley & Sons, Inc.

laboratories where the raw product is extracted from the coca leaves. The coca leaves are first placed in a pit lined with plastic. The leaves are then soaked in a dilute solution of water and a strong alkali such as lime. After 2–3 days, a volatile organic solvent is added such as gasoline, methyl isobutyl ketone, ethyl acetate. The leaves are removed, and sulfuric acid or hydrochloric acid is added to the mix. The acid protonates any basic alkaloids causing them to be soluble in the acidic water layer. The organic solvent presumably removes any nonbasic organic materials. The acidic water layer is made basic using lime or ammonia that precipitates the more basic alkaloids. The solid material, cocaine paste, is removed and allowed to dry in the sun. At this stage, the paste is a mixture of cocaine and other alkaloids along with any other materials used in the processing. The paste can be used as is though it is quite harmful. It can take 100–150 kg of dry leaf to produce 1 kg of cocaine paste.

The paste can be further processed at a base laboratory or it can go directly to a crystal laboratory. At a base laboratory, the impurities (cinnamoyl esters of cocaine) in the coca paste are removed by oxidation in an acidic solution with potassium permanganate, potassium dichromate, or sodium hypochlorite to water-soluble products. The solution is made basic and the purified cocaine precipitates. This material is taken into the crystal laboratory up to 50 kg batches, and it is slurried in ether, treated with hydrochloric acid, and filtered to give the cocaine hydrochloride salt as a white crystalline powder.[312]

Treating cocaine with sodium bicarbonate and water makes the free base of cocaine known as crack. The free base precipitates as small, irregularly shaped chunks (or "rocks") of cocaine. Cocaine hydrochloride is ingested by snorting or by injection. The free base of cocaine is smoked.

Cocaine is also abused in combination with an opiate, like heroin, a practice, known as "speedballing." Although injecting into veins or muscles, snorting and smoking are the common ways of using cocaine, all mucous membranes readily absorb cocaine. Cocaine users typically binge on the drug until they are exhausted or run out of cocaine. The intensity of cocaine's euphoric effects depends on how quickly the drug reaches the brain, which depends on the dose and method of abuse. Following smoking or intravenous injection, cocaine reaches the brain in seconds, with a rapid buildup in levels. This results in a rapid onset, intense euphoric effect known as a "rush" that lasts about 20–30 minutes. By contrast, the euphoria caused by snorting cocaine is less intense and does not happen as quickly due to the slower buildup of the drug in the brain. Other effects include increased alertness and excitation, as well as restlessness, irritability, and anxiety. Tolerance to cocaine's effects develops rapidly, causing users to take higher and higher doses. Taking high doses of cocaine or prolonged use, such as binging, usually causes paranoia. The crash that follows euphoria is characterized by mental and physical exhaustion, sleep, and depression lasting several days. Following the crash, users experience a craving to use cocaine again. Physiological effects of cocaine include increased blood pressure and heart rate, dilated pupils, insomnia, and loss of appetite. The widespread abuse of highly pure street cocaine has led to many severe adverse health consequences such as cardiac arrhythmias, ischemic heart conditions, sudden cardiac arrest, convulsions, strokes, and death. In some users, the long-term use of inhaled cocaine has led to a unique respiratory syndrome, and chronic snorting of cocaine has led to the erosion of the upper nasal cavity.[387]

Cocaine is a Schedule II drug under the Controlled Substances Act, meaning it has a high potential for abuse and limited medical usage. Cocaine hydrochloride solution (4% and 10%) was used primarily as a topical local anesthetic for the upper respiratory tract. It was also used to reduce bleeding of the mucous membranes in the mouth, throat, and nasal cavities. However, better products have been developed for these purposes, and cocaine is rarely used medically in the United States.

Although controversial, oral cocaine has been proposed as a substitution therapy for cocaine addiction. An interesting monograph by Llosa is available on the history and the use of oral cocaine with respect toward addiction.[389]

According to the 2005 Colombia Threat Assessment, 90% of the cocaine shipped to the United States originated in Columbia and entered through a Central America-Mexico corridor. The 2013 National Drug Threat Assessment Summary states that the trend of lower cocaine availability that began in 2007 continued into 2012. Seizures at the Southwest Border and price and purity data also indicate decreased availability of cocaine. Southwest Border cocaine seizures were markedly down the first quarter of 2012 as compared with 2011. This trend continued over the first half of 2012. According to National Seizure System data, approximately, 16,908 kg of cocaine was seized at the Southwest Border in 2011. During 2012, only 7143 kg of cocaine was seized, a decrease of 58%.

According to the 2013 National Drug Threat Assessment, the decline in cocaine availability occurring in various areas throughout some domestic drug markets appears to be the aggregate result of various factors.

- Counterdrug efforts may be sufficiently disrupting Colombian traffickers' ability to increase cocaine transportation. Reporting indicates that the combined effect of several large seizures and the arrests of several high-level traffickers have made Transnational Criminal Organizations (TCOs) reluctant to transport large shipments of cocaine.
- Conflict between and within TCOs in Mexico, the transit area for most of the cocaine entering the United States is also a significant factor impacting cocaine flow to the United States. Clashes for the control of lucrative smuggling routes have frequently led to increased violence between, and among, TCOs. These conflicts may also affect the amount of cocaine moved, as groups scale back their smuggling efforts until disputes abate.
- Cocaine production rates in Colombia—the source of most of the cocaine distributed in the United States—have declined in recent years. Available data on cultivation, yield, and trafficking indicate that cocaine production in Colombia declined in 2012 from the high levels seen in the period 2005–2007.

6.1.1 Mechanism of Action

The primary action of cocaine in inducing addiction is binding to the dopamine transporter (DAT) and preventing reuptake of synaptic dopamine although cocaine has about equal functional activity at all three monoamine transporters (DAT, NET, SERT IC_{50} = 310, 221, and 260 nM, respectively).[390] Cocaine binds somewhat to the NET (K_D = 1.4 μM) and binds to the SERT with a K_D = 340 nM and to the DAT with a K_D = 220 nM. The relative ratio of binding to the monoamine transporters is therefore DAT:SERT:NET = 1:1.5:6.4.[43] The pharmacokinetics of cocaine has been extensively studied.[312,391]

6.2   PHARMACOLOGY OF METHAMPHETAMINE ADDICTION

Methamphetamine is a Schedule II stimulant under the Controlled Substances Act, which means that it has a high potential for abuse and limited medical use. It is available only through a prescription that cannot be refilled. Today, there is only one legal product, Desoxyn®, containing methamphetamine as the active ingredient. Desoxyn® is manufactured by AbbVie in 5-mg tablets and marketed by Recordati Rare Diseases, Inc. for very limited use in the treatment of obesity and attention deficit hyperactivity disorder (ADHD). In use for the treatment of ADHD, the medication guide from Recordati Rare Diseases specifies the drug for the treatment of children 6 years or older with a behavioral syndrome characterized by moderate to severe distractibility, short attention span, hyperactivity, emotional lability and impulsivity. An initial dose of 5-mg Desoxyn once or twice a day is recommended, and the daily dosage may be raised in increments of 5 mg at weekly intervals until an optimum clinical response is achieved. The usual effective dose is 20–25 mg daily. The total daily dose may be given in two divided doses daily.

Methamphetamine can be swallowed, injected, inhaled, or smoked. Direct introduction into the brain by smoking or injection produces a brief, intense sensation or rush. The hydrochloride salt of methamphetamine "ice," like "crack" with cocaine, is smoked. Crystal meth resembles glass fragments or shiny blue-white "rocks" of various sizes. Less direct delivery orally or nasally produces a long-lasting high instead of the rush. Long-term, chronic methamphetamine use leads to a multitude of physical and psychological negative changes. Physical changes include neuronal cell death of dopamine- and serotonin-producing cells.[387]

The stimulatory effects of amphetamines and methamphetamine are similar to cocaine, but their onset is slower and their duration is longer. In contrast to cocaine, which is quickly removed from the brain and is almost completely metabolized, methamphetamine remains in the central nervous system longer, and a larger percentage of the drug remains unchanged in the body, producing prolonged stimulant effects. The half-life of methamphetamine elimination is about 12 hours and smoking can produce a high that lasts as long as 24 hours. Chronic abuse produces a psychosis that resembles schizophrenia and is characterized by paranoia, picking at the skin, preoccupation with one's own thoughts, and auditory and visual hallucinations. Violent and erratic behavior is frequently seen among chronic abusers of amphetamines and methamphetamine.

Methamphetamine abuse raises the risk of contracting or transmitting HIV and hepatitis B and C—not only for individuals who inject the drug, but also for noninjecting methamphetamine abusers. Among injecting drug users, HIV and other infectious diseases are spread primarily through the reuse or sharing of contaminated syringes, needles, or related paraphernalia. But regardless of how methamphetamine is taken, its intoxicating effects can alter judgment and inhibition and lead people to engage in unsafe behaviors like unprotected sex.

Methamphetamine abuse is associated with a culture of risky sexual behavior, both among men who have sex with men and in heterosexual populations, a link that may be attributed to the fact that methamphetamine and related stimulants can increase libido. (Although paradoxically, long-term methamphetamine abuse may be associated with decreased sexual functioning, at least in men.) The combination of injection practices and sexual risk taking may result in HIV becoming a greater problem among methamphetamine abusers than among other drug abusers, and some epidemiologic reports are already showing this trend. For example, while the link between HIV infection and methamphetamine abuse has not yet been established for heterosexuals, data show an association between methamphetamine abuse and the spread of HIV among men who have sex with men.

Methamphetamine abuse may also worsen the progression of HIV disease and its consequences. In animal studies, methamphetamine has been shown to increase viral replication. Clinical studies in humans suggest that current methamphetamine users taking highly active antiretroviral therapy to treat HIV may be at greater risk of developing AIDS than nonusers, possibly as a result of poor medication adherence. Methamphetamine abusers with HIV also have shown greater neuronal injury and cognitive impairment due to HIV, compared with those who do not abuse the drug.[392]

Most of the methamphetamine abused in the United States is manufactured on an industrial scale in "superlabs" here or, usually, in Mexico. Within the past several years, there have been news reports of seizures of tons of methamphetamine and hundreds of tons of precursor chemicals by law enforcement authorities in Mexico. Based on these reports, the center of manufacturing is in Mexico and Central America with precursor chemicals shipped from China. The manufacture and distribution of meth is controlled by Mexican drug cartels.

Methamphetamine is also produced in Southeast Asia and is becoming a major health issue. A United Nations Office on Drugs and Crime report: Patterns and Trends of Amphetamine-Type Stimulants and Other Drugs—Challenges for Asia and the Pacific 2013, described the growing use of methamphetamine in Asia. Methamphetamine is now the first or second most used illicit drug in 13 of the 15 Asia Pacific countries surveyed with the use of methamphetamine increasing in Cambodia, China, Japan, Lao People's Democratic Republic, Myanmar, the Republic of Korea, Thailand, and Vietnam. In 2012, 227 million methamphetamine pills were seized in East and Southeast Asia which was a 50% increase from 2011. In addition to the pills, 11.6 tons of crystal methamphetamine was also seized in 2012. Myanmar was recognized as a major site of illicit methamphetamine production. Production in Southeast Asia is facilitated as precursor chemicals used in illicit methamphetamine manufacture are often easily obtainable due, in part, to their production and use in the manufacture of pharmaceutical preparations in the region and in the neighboring regions of South Asia. According to recent news reports, North Korea is a significant producer of methamphetamine in state-run laboratories with methamphetamine addiction becoming a major health issue in some provinces.

The drug is also easily made in small clandestine laboratories, with relatively inexpensive over-the-counter ingredients such as pseudoephedrine, a common ingredient in cold medications. Several clandestine synthetic methods are available using ingredients and equipment readily available at hardware and grocery stores. To curb production of methamphetamine, the US Congress passed the Combat Methamphetamine Epidemic Act in 2005, which requires that pharmacies and other retail stores keep logs of purchases of products containing pseudoephedrine and limits the amount of those products an individual can purchase per day. A few states have even made pseudoephedrine available only with a prescription. Mexico has also tightened its restrictions on this and other methamphetamine precursor chemicals. But manufacturers adapt to these restrictions via small- or large-scale "smurfing" operations: obtaining pseudoephedrine from multiple sources, below the legal thresholds, using multiple false identifications. Manufacturers in Mexico are also increasingly using a different production process (called P2P, from the precursor chemical phenyl-2-propanone) that does not require pseudoephedrine.

Methamphetamine production also involves a number of other easily obtained chemicals that are hazardous, such as acetone, anhydrous ammonia (fertilizer), ether, red phosphorus, and lithium. Toxicity from these chemicals can remain in the environment around a methamphetamine production laboratory long after the laboratory has been shut down, causing a wide range of damaging effects to health. Because of these dangers, the US Environmental Protection Agency has provided guidance on cleanup and remediation of methamphetamine laboratories. According to the DEA, there were 10,287 methamphetamine laboratory incidents in 2011, which include laboratories and "dumpsites" or "chemical and glassware" seizures. The vast majority of these were in the Midwest.

6.2.1 Mechanism of Action

Unlike cocaine, which is a monoamine transporter reuptake inhibitor, methamphetamine is a substrate and enters the neuron via the monoamine amine transporters. Its binding affinity at the transporters is less than that of cocaine with K_D's of >10 μM at SERT, 2.8 μM at DAT, and 660 nM at NET.[43] Once entering the neuron, a proposed mechanism for methamphetamine, as described in Chapter 2, is binding to vesicular monoamine transporters (VMAT2) that are located in the presynaptic neuron resulting in the release of dopamine.[110] In line with its binding affinity, methamphetamine is most potent at norepinephrine release ($IC_{50} = 12.3$ nM), followed by dopamine release ($IC_{50} = 24.5$ nM), and serotonin release ($IC_{50} = 736$ nM). In comparison, amphetamine is more potent with a similar profile of norepinephrine release ($IC_{50} = 7.1$ nM), followed by dopamine release ($IC_{50} = 24.8$ nM), and serotonin release ($IC_{50} = 1,765$ nM).

It is suggested that the most potent physiological action of stimulants, such as methamphetamine, which produce amphetamine-like subjective effects is the release norepinephrine. For example, the stimulants (+)-amphetamine, (−)-ephedrine, phentermine, and MDMA are 3.5-fold, 19-fold, 6.6-fold, and 4.8-fold more potent at norepinephrine release than dopamine release. This strongly suggests that, in addition to dopamine release, the release of norepinephrine contributes to the positive subjective effects produced by these substrate-type stimulants.[393]

Methamphetamine, like other amphetamines, is an agonist of the trace amine-associated receptor 1 (TAAR1). The (S) enantiomer ($EC_{50} = 1.5$ μM) is about two times as potent as the (R) enantiomer ($EC_{50} = 3.3$ μM) in a human cell line assay.[394]

6.3   MEDICATION DEVELOPMENT

Most drug strategies for the development of medications to treat stimulant addiction have focused on methods to reduce the effects of excess dopamine. These include partial antagonists at dopamine receptors, D_3 agonists, or indirect modulation by interaction with the opioid, serotonin, norepinephrine, and σ receptor systems. Xi of National Institute on Drug Abuse (NIDA) has written an excellent review of preclinical and clinical advances in the development of medication for cocaine addiction.[395]

6.3.1   Phosphodiesterase Inhibition: Ibudilast

Ibudilast (AVE411, MN-166) is a nonselective phosphodiesterase (PDE) inhibitor with some preference for PDE's 3, 4, 10, and 11 with K_i's of 1–10 μM.[396] It is of particular interest in that it targets PDE in the brain's glial cells instead of neurons. Ibudilast is prescribed in Asia to treat asthma and poststroke dizziness, and it is being tested in Eastern Europe for the treatment of multiple sclerosis and in the United States for other neurological conditions. In all of these uses, it has established a good safety profile. The main course of action is that of a anti-inflammatory agent.

Some drugs of abuse can induce neuroinflammatory processes by the activation of glial cells that then release pro-inflammatory cytokines. Methamphetamine, by its general neurotoxicity properties, is known to induce neuroinflammation resulting in changes of behavior. The research group of Beardsley has shown that infusion of ibudilast at doses of 1–10 mg/kg can attenuate self-administration of methamphetamine by rats in a dose–response-dependent manner. The attenuation observed was at a methamphetamine dose of 0.03 mg/kg per infusion. The effects of the drugs were surmounted at a higher dose of methamphetamine of 0.1 mg/kg per infusion.[397] Ibudilast has been shown to reduce methamphetamine seeking in shock- or meth-induced reinstatement studies in rodents.[398]

The mechanism of action may not be totally dependent on PDE inhibition as injection of the antibiotic minocycline at doses of 10–60 mg per day also attenuated methamphetamine self-administration. However, ibudilast was found to be about 13 times more potent than minocycline so reduction of neuroinflammation in of itself may be insufficient.[397]

An additional possible mode of action has been discovered. Ibudilast was found to inhibit the catalytic and chemotactic activity of macrophage migration inhibitor factor (MIF). MIF is stored in the cytosol of glial cells, can be released out of the cell, and appears to help regulate the immune system in response to stress or infection.[399] Secretion of MIF has been found to contribute to hyperalgesia that is a common effect of morphine use. Ibudilast binds weakly to MIF with a K_D of 55 μM and is a noncompetitive inhibitor as evidenced by the pattern of the Lineweaver-Burk plot. Solution phase 2D ^{1}H-^{15}N NMR showed that ibudilast causing small chemical shift changes of amino acids in the active site area as well as distal to the active site. An analog of ibudilast, AV1013, was cocrystallized with MIF. Analysis of the resultant structure showed that AV1013 did not bind in the active site but at a site distal to it. By inference then, we can conclude ibudilast would bind similarly. Furthermore, it was found that AV1013 bound to a site different from that of a pseudosubstrate for MIF, p-hydroxyphenylpyruvate. MIF acts as a tautomerase with substrates such as p-hydroxyphenylpyruvate. As such ibudilast can be putatively classified as an allosteric inhibitor of MIF.

It must be stressed that no relationship between MIF production and methamphetamine use has been reported. At most, it is known that use of morphine does not appear to change MIF levels in patients taking opioids for chronic pain.[400]

Ibudilast is being developed by MediciNova and is in Phase I(b) clinical trials as a treatment for methamphetamine relapse. In 2013, the company in collaboration with University of California, Los Angeles and NIDA will conduct a 2-year Phase II clinical trial to study the safety and efficacy of the drug. Part of the study will examine patients with a codiagnosis of HIV as the concurrence of HIV infection and addiction is a recognized problem.

The synthesis of ibudilast has not been published in the academic literature but a large number of patent syntheses are reported. They generally follow the basic course shown in Figure 6.1 of N-amination of 2-methyl pyridine followed by a Chichibabin-type cyclization with isobutyric anhydride to form the pyrazolo[1,5-a]pyridine ring. Electrophilic acylation on the aza-pyrrole ring with isobutyric anhydride in the same pot completes the synthesis.

Few structure–activity relationship (SAR) studies have been reported, none with respect to structural effects on methamphetamine relapse. Of the PDEs, it is not clear if one should be targeted in preference to the others. Research teams at Kyorin Pharmaceutical and Heriot-Watt University have started to publish work on developing idubilast for its anti-inflammatory and bronchodilatory activity. Selectivity for PDE4 and or PDE3 can be obtained as shown in Figure 6.2 by removal of the isopropyl keto group and functionalization of the pyridine ring.[401]

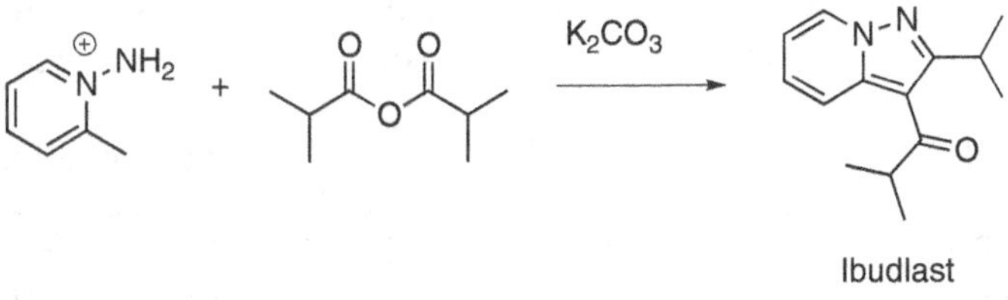

Ibudlast

FIGURE 6.1   Synthesis of ibudilast

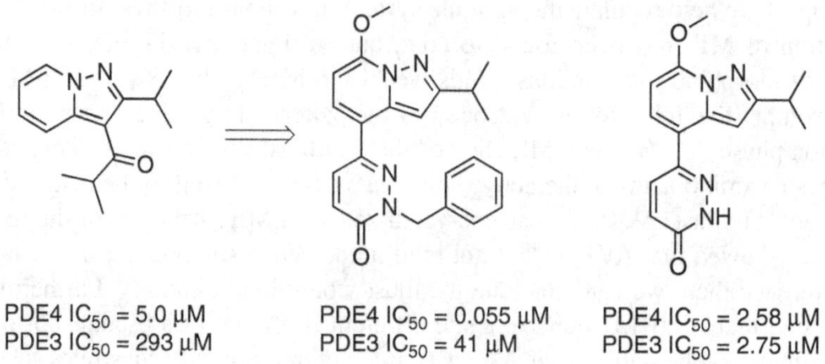

PDE4 IC_{50} = 5.0 μM   PDE4 IC_{50} = 0.055 μM   PDE4 IC_{50} = 2.58 μM
PDE3 IC_{50} = 293 μM   PDE3 IC_{50} = 41 μM   PDE3 IC_{50} = 2.75 μM

FIGURE 6.2   SAR of ibudilast

Some interesting pharmacological tools that are bioavailable are thus being made that could be of interest in exploring the role of the different PDEs in methamphetamine relapse.

An inhibitor for a second phosphoesterase target, PDE7, is being investigated for cocaine addiction. The compound OMS527 is discovered by Asubio Pharma (Daiichi Sankyo subsidiary) and is licensed to Omeros. A press release (March 19, 2013) from Omeras claim they have identified the mechanism by which PDE7 inhibition reduces dopamine release initiated by drugs of abuse but it has not been published (as of July 2013). According to the company web site, they are currently preparing an Investigational New Drug (IND) file for submission to the FDA. The structure of OMS527 does appear to have been released at the time of writing this book.

6.3.2   GABA-AT Irreversible Enzyme Inhibitors: Vigabatrin (CPP-109) and CPP-115

Vigabatrin, γ-vinyl GABA, a small polar amino acid analog of GABA, is a mechanism-based irreversible enzyme inhibitor of GABA aminotransferase (GABA-AT) that originated from the laboratories at Merrell International in France in the 1970s. Vigabatrin is approved for treatment of epileptic seizures and infantile spasms and is sold under the trade name Sabril™. The drug is currently of interest for the treatment of cocaine addiction and is the subject of several clinical trials. The mechanistic rationale is that inhibition of GABA transaminase prevents metabolism of GABA to succinic semialdehyde (4-oxobutanoic acid), the resultant increased levels of GABA suppresses the increase in dopamine release caused by cocaine. Stabilizing the homoeostatic levels of dopamine can reduce its reinforcing effects and may reduce craving.

Catalyst Pharmaceutical Partners (CPP) is investigating vigabatrin for the treatment of cocaine addiction. In 2011, CCP partnered with researchers at the University of Pennsylvania to conduct a 60 subject, double-blind, placebo-controlled, investigator-sponsored study to evaluate the use of vigabatrin (CCP-109) for the treatment of patients addicted to both cocaine and alcohol.[402] In 2010, CCP also

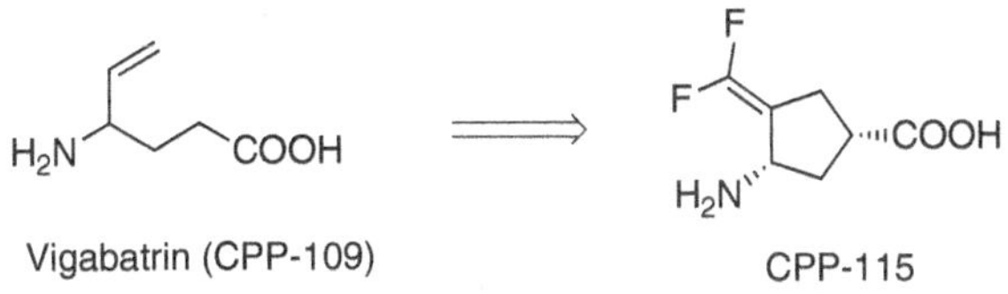

FIGURE 6.3   Structures of vigabatrin and CPP-115

partnered with NIDA and the Veterans Administration Cooperative Studies Program, to initiate a registration-directed US Phase II(b) clinical trial of CPP-109 in patients with cocaine addiction.[403]

A cyclic analog of CPP-109, CPP-115 (Figure 6.3), has gained fast track status by the FDA for the treatment of cocaine dependency. A Phase I(a) study in 55 subjects has shown that CPP-115 showed no serious or adverse events at doses of 5–500 mg (CPP press release May 22, 2012).

CPP-115 was developed in the Silverman laboratory at Northwestern University and is also an irreversible inhibitor of GABA-AT. An extensive description of the pharmacokinetics, pharmacodynamics and pharmacology of the drug has been reported.[404] A case study of how the drug was discovered is available.[405] The drug shows excellent selectivity with no inhibition of GABA transporters at 1 mM, no inhibition of GABA binding to $GABA_{A,B,C}$ receptors at 100 μM, and at 10 μM no significant interaction in a screening assay of 111 targets. The pharmacokinetic data in animals are very good, and the drug has been shown using positron emission tomography (PET) and microdialysis experiments to reduce dopamine levels in rodents upon exposure to cocaine. In behavioral studies, CPP-115 is not rewarding and blocks cocaine-induced conditioned place preference. In all, at much lower doses than required for vigabatrin.

CPP-115 was first synthesized as shown in Figure 6.4 starting with a protected azabicycloheptanedione.[406] Due to the need to prepare the starting azabicycloheptanedione in several steps as shown in Figure 6.5 and subsequent use of the highly pyrophoric and dangerous t-butyl lithium, this is most likely not the process used on a manufacturing scale. The azabicyclo compound is prepared as a racemic mixture by a cycloaddition reaction between cyclopentadiene and chlorosulfonyl isocyanate.[407] The key rearrangement step of the racemic (shown with the (−) enantiomer) azabicycloheptenone proceeds through a bromonium ion that is reminiscent of the norbornyl carbocation rearrangment.[408]

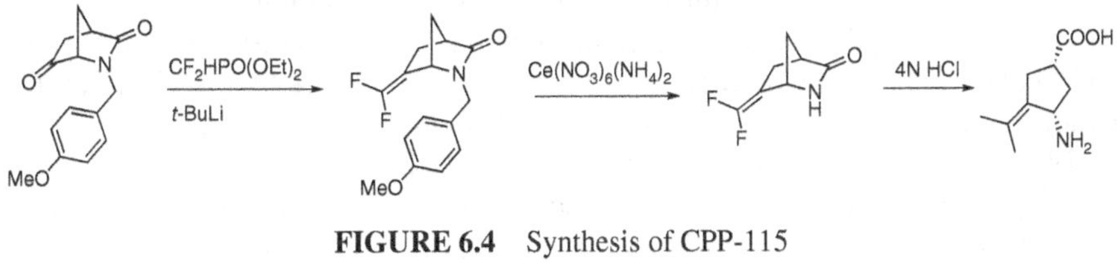

FIGURE 6.4   Synthesis of CPP-115

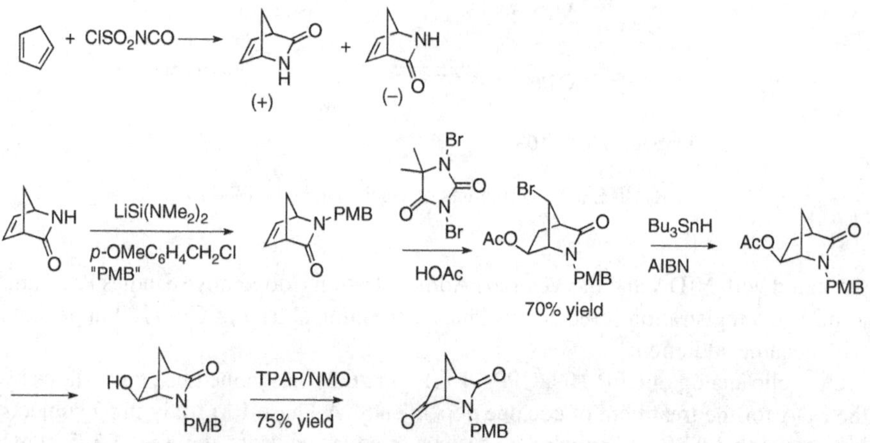

FIGURE 6.5 Synthesis of azabicycloheptanedione

What is the mechanism of action of CPP-115? GABA-AT converts the amino group of GABA into an aldehyde via an imine exchange of GABA with the coenzyme pyridoxal 5′-phosphate. Pyridoxal 5′-phosphate is covalently linked to GABA-AT by an imine bond with Lys-329. The amino group of GABA attacks this imine linkage, forming a new imine bond to pyridoxal 5′-phosphate, breaking the covalent connection of GABA-AT and pyridoxal 5′-phosphate. GABA, now being in the active site of the enzyme, undergoes an imine tautomerization and imine hydrolysis to form the metabolite succinic semialdehyde.

Both CPP-115 and vigabatrin inhibit this process by covalent bond formation with Lys-329. Silverman has discussed this in detail but briefly, upon imine formation of either CPP-115 or vigabatrin with pyridoxal 5′-phosphate, conjugation of the vinyl group with the imine develops allowing 1,4-conjugate addition of the amino group of Lys-329. The presence of the electron-withdrawing gem difluoro group in CPP-115 increases electrophilicity of the carbon atom leading to a 187 times increase in the inactivation of GABA-AT by CPP-115 versus vigabatrin.

6.3.3 Disulfiram—Multimodal Enzyme Inhibitor

The main therapeutic use of disulfiram is for the treatment of alcoholism so the drug will be discussed in more detail in Chapter 7, but some limited clinical trials have been conducted on its use with respect to cocaine use. The trials reported to date have provided promising, yet mixed results on whether disulfiram can be useful for the treatment of cocaine abuse. A common complication is comorbid use of cocaine and alcohol so there may be additional therapeutic possibilities for disulfiram for the treatment of concurrent cocaine and alcohol addicton.

Two of the most recent trials will be discussed. A study at Yale University investigated the effects of disulfiram and a Twelve-Step Facilitation (TSF) program separately and in combination on cocaine-dependent individuals maintained on methadone.[409] More specifically, they conducted a randomized, placebo-controlled, double-blind (for medication condition), factorial (2 × 2) trial with four treatment conditions: disulfiram plus TSF, disulfiram plus standard counseling only, placebo plus TSF, and placebo plus standard counseling in the context of a community-based methadone maintenance program. Participants ($N = 112$) received either disulfiram (250 mg per day) or placebo in conjunction with daily methadone maintenance. They concluded that while TSF was associated with modest reductions in cocaine use, disulfiram was effective only in individuals without concurrent alcohol abuse.

In a second study at the Michael E. DeBakey Veterans Affairs Medical Center, 17 nontreatment-seeking, cocaine-dependent volunteers were treated with 250 mg per day of sulfiram for 3 days, and after a 2-week period, the same individuals were treated with a placebo.[410] The effects of each phase on the reinforcing properties of cocaine were then determined. A dosage to subject weight relationship was observed. When disulfiram dosage was adjusted for weight in mg disulfiram/kg body weight, the number of cocaine infusions decreased relative to disulfiram dose (mg/kg). The researchers concluded that a disulfiram dose of approximately 4 mg/kg most effectively reduced the reinforcing effects of cocaine relative to money and this dose may be an optimal target dose in the clinic.

A mechanism of action on how disulfiram promote cocaine abstinence has been suggested to be by the inhibition of dopamine β-hydroxylase with a reduction in the brains norepinephrine/dopamine ratio. This modulation of neurotransmitter levels would correct a hypodopamineric state that has been proposed to exist in cocaine addicts. Microdialysis experiments in rats have shown that disulfiram reduced extracellular norepinephrine in areas of the reward system while increasing dopamine levels. The authors propose that disulfiram, by inhibiting norepinephrine synthesis and release, causes a lack of norepinephrine at α_2-adrenergic autoreceptors, removing the major inhibitory control on noradrenergic activity. The lack of norepinephrine would result in dopamine release from adrenergic terminals.[411]

Disulfiram is known to increase levels of cocaine in the plasma and reduce the clearance of cocaine.[412] In human volunteers, cocaine doses of 0.25 and 0.5 mg/kg combined with 62.5 and 250 mg doses of disulfiram resulted in an about twofold increase in cocaine plasma levels and about a twofold decrease in cocaine clearance (Cl in $1 \ min^{-1} \ kg^{-1}$). This appears counterintuitive to how disulfiram can reduce cocaine consumption. One proposal is the increased levels of cocaine may reduce the subjective effects of cocaine such that upon relapse harmful activities like binge use is reduced. It is interesting to speculate that in this sense disulfiram could be producing the functional effects of a partial agonist.

The microdialysis study mentioned above shows a complex interaction of cocaine and disulfiram. Administration of cocaine increased levels of dopamine and norepinephrine in the medial prefrontal cortex and nucleus accumbens. When disulfiram was given, the amount of cocaine-induced norepinephrine was reduced in both the medial prefrontal cortex and nucleus accumbens, but the levels of dopamine induced by cocaine *increased* in the medial prefrontal cortex and *no* effect on dopamine levels was seen in the nucleus accumbens. The effects of disulfiram on the CNS are clearly complex, region-specific, and require more study.

The mechanism of enzyme inhibition by disulfiram is presumably due to chelation of the two Cu atoms in the active site of dopamine β-hydroxylase by the sulfur atoms in diethyldithiocarbamate, a well-known metabolite of disulfiram. A 10-μM concentration of disulfiram is known to completely inhibit dopamine β-hydroxylase.[413] Very little investigation appears to have been conducted on the exact mechanism of how diethyldithiocarbamate inhibits dopamine β-hydroxylase. Additional inhibition of carboxylesterases and of cholinesterase may contribute by reducing the metabolism of cocaine.

6.3.4 Dopamine-β-hydroxylase Inhibition: Nepicastat

Nepicastat (SYN117) was developed by Roche Bioscience as an alternative therapeutic approach to the use of β-blockers for the treatment of heart disease. Nepicastat is a potent competitive inhibitor of dopamine-β-hydroxylase with an $IC_{50} = 9$ nM. The S-enantiomer, shown below in the synthesis, is about three times more potent than the R-enantiomer.[414] Dopamine is oxidized to norepinephrine by dopamine-β-hydroxylase located in vesicles in presynaptic noradgenergic neurons. The theory is that by reducing the metabolism of dopamine, synaptic levels of dopamine will increase leading to a reduction in craving. The decreased levels of norepinephrine may also reduce feelings of stress thus helping to prevent relapse.

Dopamine-β-hydroxylase exists as a tetramer and contains two Cu^{2+} atoms in the active site that are coordinated to the imidazole groups of histidines. The Cu^{2+} atoms are responsible for the oxidizing activity of the enzyme. The mechanism of oxidation requires that the Cu^{2+} atoms be first reduced to Cu^+ by ascorbate. The enzyme has yet to be crystallized but a homology model is available.[415] Using this model, docking experiments suggest that nepicastat binds to the active site by, presumably, forming a salt bridge between its protonated primary amine and the carboxylate salt of Glu265 in the enzyme. This same interaction was modeled for dopamine.

Model studies with nepicastat in rats have shown that it can reduce the reinforcing properties of cocaine and can attenuate relapse-like behavior produced by cocaine, formerly cocaine-paired cues, and physiological and pharmacological stressors. There was no significant effect on food seeking.[416]

The Finnish company Biotie is developing nepicastat for the treatment of cocaine dependency and also for post-traumatic stress disorder. In collaboration with NIDA, a Phase II trial is scheduled to start in 2013. If the clinical trials are successful, utilizing dopamine-β-hydroxylase as a therapeutic target could be a very interesting approach. Being an enzyme, it could be much easier to develop selective drugs for it than receptors.

A synthesis for nepicastat does not appear to have been published but methods for doing so are disclosed in US patent 5538988. The following synthetic scheme shown in Figure 6.6 may be applicable.

The sole chiral center is fixed by using an enantioselective reduction in the ketone with lithium aluminium hydride and a chiral ligand. The key cyclization step is based on a Marckwald-like heterocyclization reaction. This same heterocyclization method was used in the synthesis of the antibiotic etomidate.[417]

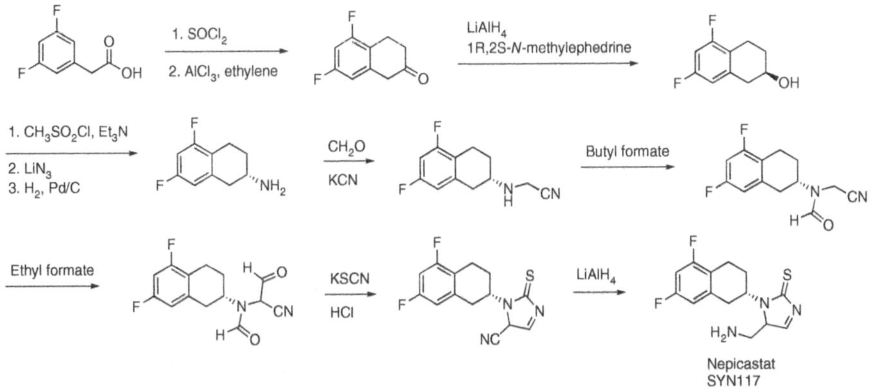

FIGURE 6.6   Synthesis of nepicastat

6.3.5   Aldehyde Dehydrogenase 2 Inhibition: CVT-10216

Another method by which levels of dopamine may be modulated is by inhibition of the enzyme aldehyde dehydrogenase 2 (ALDH2). ALDH2 is responsible for the oxidation of DOPAL (3,4-dihydroxyphenylacetaldehyde) to DOPAC (3,4-dihydroxyphenylacetic acid) as outlined in Figure 6.7.

A characteristic feature of increasing concentrations of DOPAL is the Picket–Spengler reaction of DOPAL with dopamine to form tetrahydropapaveroline (THP) via intramolecular electrophilic aromatic substitution of the imine intermediate. THP itself has been implicated in modulating dopamine. This was indeed found to be true when the ALDH2 inhibitor CVT-10216 (Figure 6.8) suppressed cocaine self-administration and prevented cocaine- or cue-induced reinstatement in rats. It was determined that the behavioral effects were due to *decreased levels of dopamine* and that the metabolite THP was actually the responsible drug that was causing the reductions in dopamine.[418] A key point in this study was they were investigating the

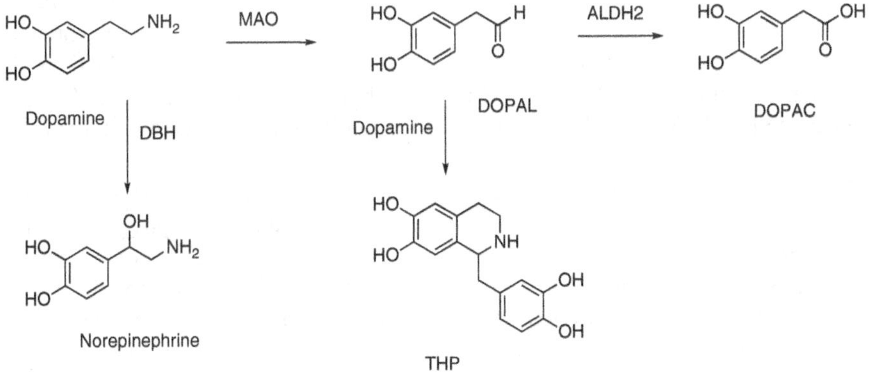

FIGURE 6.7   Oxidation of dopamine by aldehyde dehydrogenase 2: CVT-10216

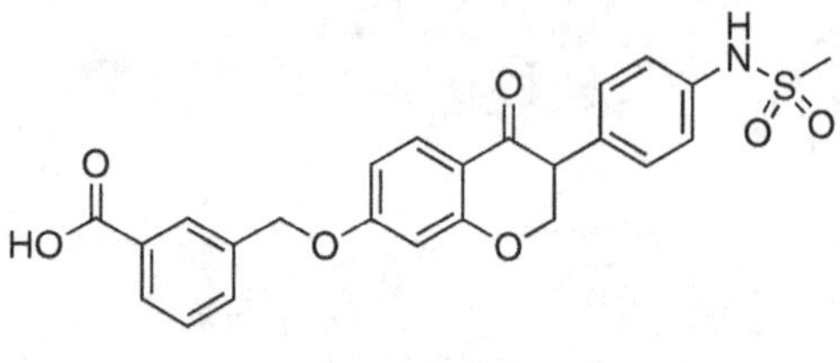

CVT-10216

FIGURE 6.8   Structure of CVT-10216

effects of CVT-10216 in the *presence* of cocaine with CVT-10216 attenuating the increases in dopamine caused by cocaine.

CVT-10216 was discovered to decrease dopamine levels by preventing the axonal synthesis of dopamine. THP was also found to be an inhibitor of tyrosine hydroxylase, the enzyme that oxidizes tyrosine to dopamine by the introduction of a phenolic –OH group ortho to the OH on tyrosine. The compound does not appear to have been examined in a clinical trial.

From this work, several interesting new approaches arise. One is the inhibition of aldehyde dehydrogenase, the second is the inhibition of tyrosine hydoxylase using the tetrahydroisoquinoline chemotype. Both enzymes have important homostatic roles so if inhibition would lead to adverse side effects remains to be seen.

6.3.6   Cholinergic System M$_1$/M$_4$ Agonists: VU 0357017 and Xanomeline

Several compounds that bind to both the M$_1$ and M$_4$ receptors and act as agonists have shown potential in attenuating cocaine discrimination and self-administration in mice.[419]

The use of knockout mice has proven particularly informative, and in this model, the presence of both receptors is not mandatory for a cocaine "effect." Both M$_1$$^{-/-}$ and M$_4$$^{-/-}$ mice could be trained to discriminate 10 mg/kg cocaine from saline.[420] In fact, neither receptor is required as double knockout M$_1$$^{-/-}M_4$$^{-/-}$ mice could also discriminate cocaine from saline.[177] This simply shows that cocaine can interact with other receptors to cause an effect. The M$_1$ selective agonist VU0357017 could attenuate cocaine responding in wild type and M$_4$$^{-/-}$ mice but not in M$_1$$^{-/-}$ mice, indicating that VU0357017 affects its attenuating effect through the M$_1$ receptor.

VU0357017   It has proven difficult to produce subtype selective agonists for the muscarinic receptors. The Vanderbilt University drug discovery program though by using a functional high-throughput cell-based screen and subsequent diversity-oriented synthesis discovered a series of highly selective M$_1$ allosteric agonists with EC$_{50}$'s ranging from 150 to 500 nM. These compounds had no detectable agonist activity at other mAChR subtypes and have favorable ancillary pharmacological profiles when evaluated in a large panel of GPCRs and other potential CNS targets (<50% inhibition at 10 μM).

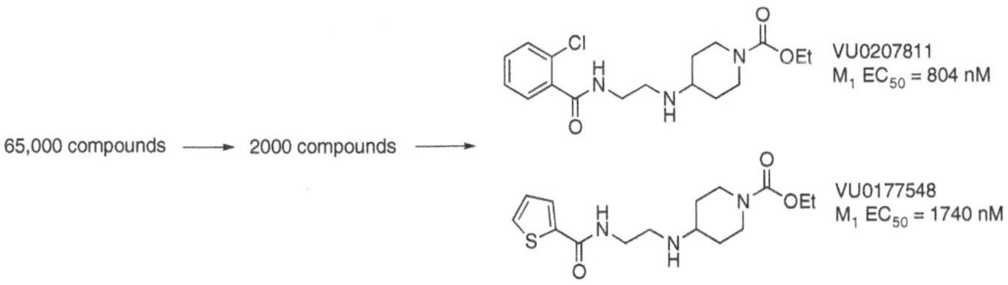

65,000 compounds ⟶ 2000 compounds ⟶

VU0207811
M_1 EC_{50} = 804 nM

VU0177548
M_1 EC_{50} = 1740 nM

FIGURE 6.9 Development of M_1 selective ligands

Lead generation was initiated by screening the 65,000 member compound library from the NIH supported Molecular Libraries Production Center Network to give 2000 primary agonist hits. To confirm these hits, they were tested in duplicate using a M_1 assay and a M_4 assay. Compounds showing initial selectivity for M_1 over M_4 were tested in triplicate in a 10-point concentration-response series against the muscarinic panel (M_1-M_5). From this, two leads with >50-fold selectivity for M_1 over M_4 with Ec_{50}'s of 804 nM and 1740 nM arose. An outline of the scheme is shown in Figure 6.9.

Subsequent lead optimization led to VU0357017 with an EC_{50} at M_1 of 198 nM. The mono-HCl salt was soluble in water at >25 mg/mL and with a log BB of 1.46 entered the rat brain giving a 4:1 brain/plasma ratio of 4339 ng/mL in the brain to 1054 ng/mL in the blood. At a dose of 10 mg/kg delivered intraperitoneal, VU0357017 was able to reverse cognitive deficits induced by the orthosteric mAChR antagonist scopolamine in a contextual fear conditioning behavioral assay.

The observed SAR of this series shows that the presence of a amide and secondary amine N–H group, a hydrophobic aryl group, and the carbamate groups are critical for M_1 activity (Figure 6.10).[421]

Xanomeline Xanomeline was in a Phase III clinical trial for the treatment of cognitive dysfunction in Alzheimer's disease when it had to be removed due mainly

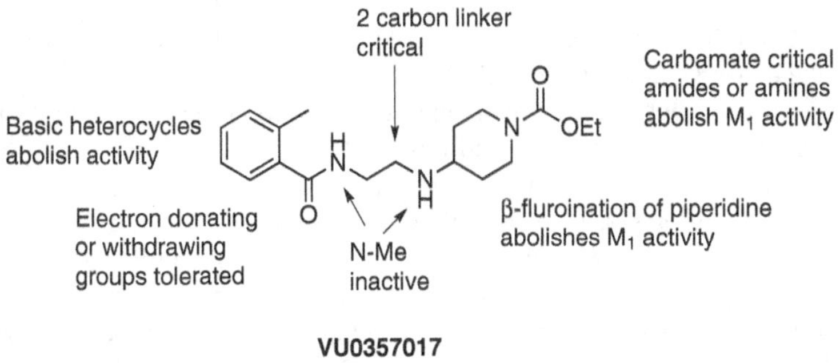

VU0357017

FIGURE 6.10 SAR of M_1 agonists (Adapted with permission from Lebois et al.[421] Copyright © 2009 American Chemical Society)

TABLE 6.1 Binding Affinities of Xanomeline at Cloned Neurotransmitter Receptors

Receptor	K_i (nM)
hM_1	79
hM_2	125
hM_3	39
hM_4	20
hM_5	40
$h5\text{-}HT_{1A}$	63
$h5\text{-}HT_{1B}$	50
$h5\text{-}HT_{1D}$	6
$h5\text{-}HT_{1E}$	2511
$h5\text{-}HT_{1F}$	316
$h5\text{-}HT_{2A}$	125
$h5\text{-}HT_{2B}$	20
$h5\text{-}HT_{2C}$	40
$h5\text{-}HT_6$	1259
$h5\text{-}HT_7$	125
hD_2	1000
hD_3	398
$h\alpha\text{-}1B$	1585

to adverse gastrointestinal effects.[422] Xanomeline is a multi-target drug with strong binding to all five muscarinic and several serotonin receptors as shown in Table 6.1.[423]

Although it is has moderate binding selectivity to muscarinic receptors, it has a remarkable selective agonist functional *in vitro* activity profile with EC_{50}'s at $M_1 = 10$ nM, $M_3 = 25$ μM, $M_5 = 2$ μM and $IC_{50} = 25$ nM at M_4 (agonist activity was assessed by measuring the inhibition of cAMP formation stimulated by forskolin in CHO cells expressing M4 receptors). Note the lack of correlation between binding affinity and functional activity.

Xanomeline can be classified as a selective M_1/M_4 agonist. This profile is curious, as the M_1 receptor is $G\alpha_q$ coupled increasing intracellular Ca^{2+} concentrations and activating inwardly rectifying potassium channels, while M_4 is $G\alpha_{i/o}$ coupled and inhibits the conversion of ATP to cAMP.

In genetic knockout studies, xanomeline was able to produce rightward shifts in the cocaine dose-effect curves in wild type and in both $M_1{}^{-/-}$ and $M_4{}^{-/-}$ mutant mice. This illustrates that the drug affects its pharmacological activity through both the M_1 and M_4 receptors.[420] Xanomeline also attenuates cocaine discrimination and abolished cocaine self-administration.[177] The drug has little effect on food-maintained behavior indicating that it is not a systemic depressant or stimulant.

The drug is receiving interest as a potential pharmacotherapeutic agent for schizophrenia. While studying the effects on prepulse inhibition in $M_1{}^{-/-}$ and $M_4{}^{-/-}$ mutant mice Thomsen found that it displays antipsychotic-like effects mainly through the M_4 receptor.[424]

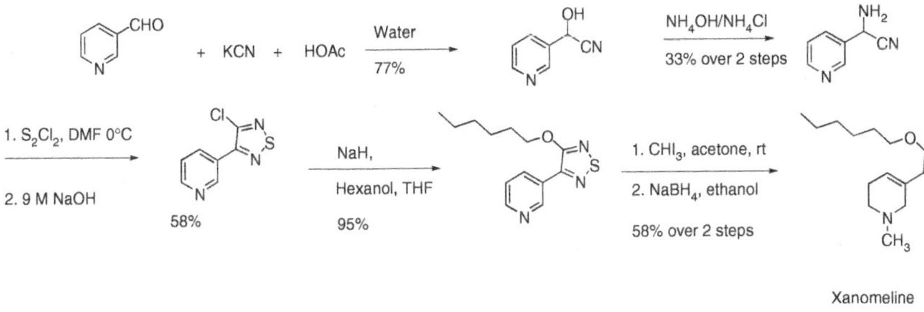

FIGURE 6.11 Synthesis of Xanomeline

The drug is readily synthesized by the scheme shown in Figure 6.11. A Strecker reaction furnishes the pyridine amino nitrile that is reacted with sulfur monochloride to form the thiadiazole ring. Nucleophilic aromatic substitution of the heterocycle 3-chloride atom with hexanol, N-methylation, and reduction in the resultant pyridinium ring completes the sequence. The Strecker reaction can be carried out in two steps as shown or in a one-pot method (KCN, NH_4OH, NH_4Cl, water) in low yield.

Xanomeline is derived from the natural product arecoline that is found in the areca nut. The areca nut is used to produce betel quid that has been chewed since antiquity for its production of euphoria, a sense of well-being, generation of heightened alertness, to combat against hunger and for increased stamina. Betal quid is a concoction consisting of the nut of the palm tree *Areca catechu*, quicklime and some type of psychoactive leaf of the plants that is rolled together in the betel leaf. Betal chewing is said to be the fourth most commonly used drug in the world after tobacco, alcohol, and caffeine.[425,426]

Arecoline is a mild stimulant and muscarinic partial agonist with binding selectivity for the M_2 ($EC_{50} = 25$ nM) and M_4 ($EC_{50} = 100$ nM) receptors though it also binds strongly to the M_3 ($EC_{50} = 316$ nM) and M_5 ($EC_{50} = 630$ nM) receptors with poor binding to the M_1 receptor ($EC_{50} = 3$ μM).[427] As seen in Figure 6.12, xanomeline is composed of the arecoline piperidine ring system with the ester of arecoline being replaced by a 1,2,5-thiadiazole bioisotere.

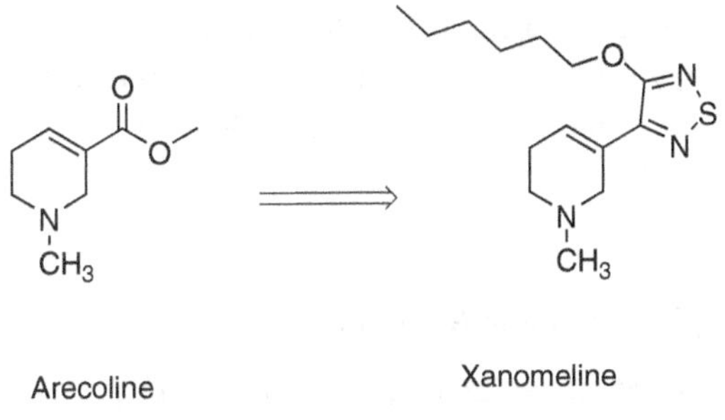

Arecoline   Xanomeline

FIGURE 6.12 Xanomeline from arecoline

Few SAR studies of xanomeline have been reported. The original report on the synthesis and pharmacological properties of xanomeline investigated variations at position 3 (hexyloxyether) of the 1,2,5-thiadiazole ring.[428] The binding and functional assays used whole tissues with nonselective radioligands so it was not possible to get a good idea of selectivity, however the following conclusions can be reached. Oxygen and sulfur can be interchanged though the presence of a sulfur atom appears to increase M_2 functional activity. Replacement of oxygen with nitrogen reduced binding affinity. Comparing the presence or absence of a heteroatom (O, S vs. H, alkyl), it is clear that oxygen or sulfur at position 3 is important for strong binding. There was an interesting $M_{1,4}$ binding affinity and functional activity relationship with chain length. A U-shaped curve of binding affinity versus chain length for $-O(CH_2)_nCH_3$ developed with maximum binding for O-pentyl. Functional activity (% maximum activity) versus chain length gave an inverse U-shaped curve with maximal functional activity about equal for O-pentyl and O-hexyl.

Looking at changes in the heterocyclic ring, if we focus on $M_{1,4}$ receptors, replacement of the sulfur atom with oxygen (1,2,5-thiadiazole to 1,2,5-oxadiazole) in the heterocyclic ring reduced binding to the $M_{1,4}$ receptors approximately 40-fold with a complete loss of functional activity. Modified Neglect of Diatomic Overlap (MNDO) level calculations suggest that the thiadiazole will be a better hydrogen bond acceptor unit than the oxadiazole.[429] The thiadiazole ring is more polarized with the sulfur atom bearing a partial positive atomic charge (+0.561) and the nitrogens negatively polarized (−0.328 to −0.399). The oxadiazole ring oxygen only has a partial positive atomic charge of +0.0658 and the nitrogens of −0.0595 to −0.123.

A curious finding during the research on xanomeline was that it appears to bind to the M_1 receptor in a wash-resistant manner when the receptor is overexpressed in CHO cells.[430] Persistent binding and agonist functional activity was detected even after extensive washing of the cells; the drug though could be displaced by the antagonist atropine. Further studies suggested that xanomeline was binding to an allosteric site on the receptor. When the hexyloxy chain is shortened to butoxy or less the wash-resistant phenomenon disappears.[431] To further study the importance of the alkyl chain twenty compounds were prepared where the length, polarity, and charge of the chain was modified.[432] The binding affinities were measured only at the M_1 receptor. Relative to xanomeline (O-n-hexyl) introduction of polarity or charge in the chain reduced binding affinity at least 10-fold with the exception of the ester O-$(CH_2)_4$COOEt with IC_{50} = 59 nM versus 81 nM for xanomeline. The results did not present an unambiguous solution as to the mechanism of wash resistance.

Fusion of heterocyclic rings to the tetrahydropyridine ring has no substantial effect on binding affinities to all five muscarinic receptors and can actually switch functional activity from an agonist to an antagonist.[427]

6.3.7 Metabotropic Glutamate Receptors

Modulation of the metabotropic glutamate receptors has proven an effective strategy on controlling cocaine addiction. Evidence pointing to the significance of the metabotropic glutamate receptors during cocaine withdrawal are; there is decreased $mGlu_{2/3}$ signaling, metabolic activity in prefrontal cortex is reduced and the firing rate of glutaminergic afferents from the prefrontal cortex leading to the nucleus accumbens are reduced, and there is a decreased synaptic glutamate release.

Of the different metabotropic glutamate receptors $mGlu_2$-positive allosteric modulators and $mGlu_5$-negative allosteric modulators have shown promise in the treatment of cocaine dependence and pain while $mGlu_5$-positive allosteric modulators are being examined for the treatment of schizophrenia. The medicinal chemistry of $mGlu_5$-positive allosteric modulators[433] and $mGlu_5$-negitive allosteric modulators[434] has been reviewed. Researchers at the Vanderbilt Program in Drug Discovery program have been especially active in developing glutaminergic active compounds.

As an interesting aside, it has been found that in fragile X syndrome, a common form of mental retardation, the lack of FMRP (fragile X mental retardation protein) potentiates a $mGlu_5$-mediated signaling pathway involving the endocannabinoid 2-arachidonoylglycerol (2-AG).[435] There were reduced levels of 2-AG due to enhanced activity of its metabolic enzyme diacylglycerol lipase.

6.3.8 Metabotropic Glutamate Receptor Subtype 2 Positive Allosteric Modulators (mGlu$_2$ PAM)

Several studies using different chemotypes have shown that modulation of the $mGlu_2$ receptor could be useful for the treatment of cocaine addiction.

LY354740   Studies of constrained glutamic acid analogs by Eli Lilly for the treatment of anxiety lead to the discovery of the Group 2 mGlu ($mGlu_2$ and $mGlu_3$) selective agonist (+)-LY354740 (Figure 6.13). The compound was orally bioavailable and possessed anxiolytic activity in the plus maze model of anxiety (ED_{50} = 0.5 mg/kg).[436] Further pharmacological characterization (+)-LY354740 was shown to attenuate naloxone-precipitated symptoms in morphine withdrawal in morphine-dependent mice and attenuate symptoms of nicotine withdrawal in rats.[437,438] A through pharmacological study in rodents showed LY354740 was also effective in attenuating symptoms of anxiety without the side effects commonly associated with diazepam.[439] The importance of the metabotropic glutamate receptors in cocaine addiction was demonstrated when it was found that the nonselective $mGlu_{2/3}$

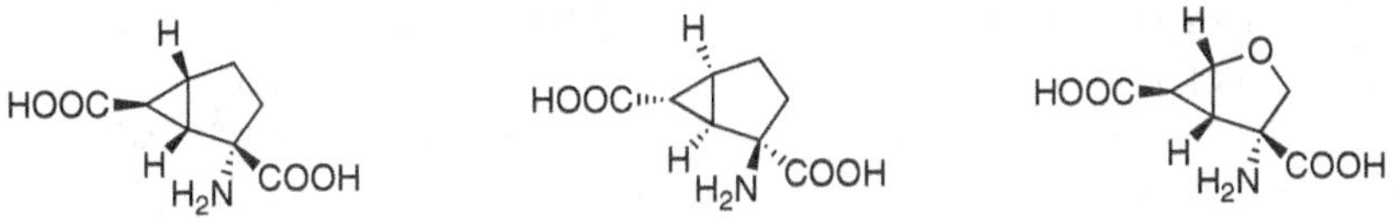

(+) LY354740
$mGlu_{2/3}$ IC_{50} = 180 nM
$mGlu_{2/3}$ EC_{50} = 55 nM

(−) LY366563
$mGlu_{2/3}$ IC_{50} > 100 μM
$mGlu_{2/3}$ EC_{50} > 300 μM

LY379268
$mGlu_2$ EC_{50} = 2.7 nM
$mGlu_3$ EC_{50} = 4.6 nM

FIGURE 6.13   Constrained glutamic acid mGlu$_{2/3}$ PAMs

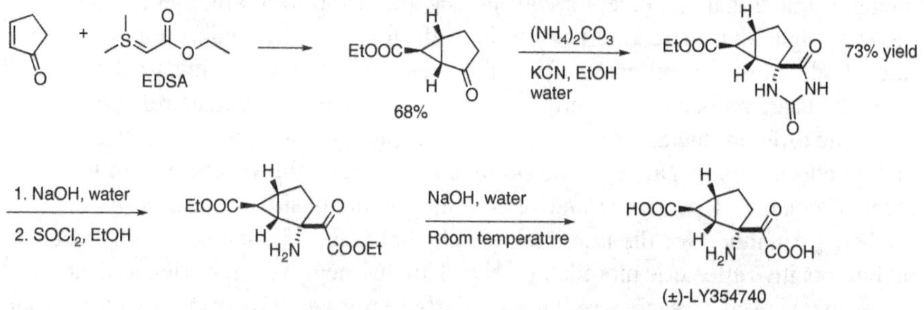

FIGURE 6.14   Synthesis of LY354740

agonist LY379268 (Figure 6.13) attenuated cocaine self-administration in rodents[205] and squirrel monkeys.[440] Side effects observed were reduced responding for food and food-seeking behavior and were thought to be due to mGlu$_3$ binding. Imre has written a detailed review covering preclinical properties of LY379268.[441]

LY354740 was originally prepared as a racemate and separated into enantiomers by resolution of diastereomers produced with (R)(S)-1-phenylethylamine. The [3.1.0] bicyclic core is made by the reaction of the sulfur ylid ethyl (dimethyl-sulfuranylidene)acetate (EDSA) with cyclopentanone. The cyclopropanation reaction proceeds to give predominately the *cis* diastereomer (shown below) in a 69:31 ratio. The diastereoselectivity of the reaction could be enhanced to 97:3 by performing the cycloaddition at room temperature either with isolated EDSA or with EDSA generated *in situ* from ethyl (dimethylsulfonium)acetate bromide using 1,8-diazabicyclo[5.4.0]undec-7-ene as the base. Formation of the hydantoin produced an 87:13 ratio of diastereomers with the major one shown in Figure 6.14. Hydrolysis of the two esters gave the final compound as a racemic mixture.

Alternatively, the two enantiomers could be prepared by resolution of the hydantoin followed by the same sequence as above. The structure of (+)-LY354740 was confirmed by X-ray crystallography anaylsis.[436] A 12-step enantioselective synthesis starting with the optically active (1S,2R)-1-amino-2-hydroxycyclopentanecarboxylic acid has been reported.[442]

Removal of the two methylene groups of the cyclopentane ring of LY354740 and introduction of a hydrophobic aryl group generated a series of cyclopropyl analogs that were found to be potent antagonists of mGlu$_{2/3}$. One of these (labeled carboxycyclopropyl methoxyphenethyl glycine (CMG) in Figure 6.15) had good binding affinity for mGlu$_2$ (IC$_{50}$ = 430 nM) and for mGlu$_3$ (IC$_{50}$ = 160 nM).[443] The methoxy group was recognized as an entry point for the introduction of a radiolabel so from this a ^{11}C labeled PET ligand has been developed.[444] By converting the polar CMG into the dimethyl ester prodrug carboxycyclopropyl methoxyphenethyl glycine dimethyl ester (CMGDE) (Figure 6.15), good brain bioavailability was achieved. It would be expected that at physiological pH CMG would exist in a highly ionized form with an overall negative charge making passive diffusion through the blood–brain barrier difficult.

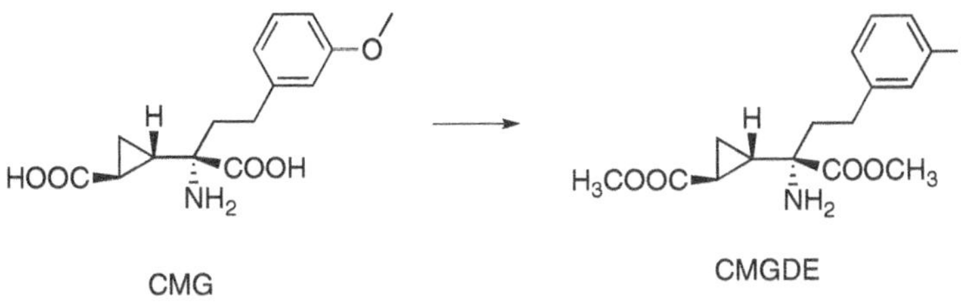

FIGURE 6.15   Conversion of CMG to CMGDE

Both [^{11}C]-CMG and [^{11}C]-CMDGE were synthesized. MicroPET imaging studies in rats showed that [^{11}C]-CMG did not enter the brain but that 2.5–3.5% of the injected dose per cm^3 of [^{11}C]-CMDGE was observed in the brain 2 minutes after intravenous injection. Injection of a mGlu$_{2/3}$ antagonist resulted in a 20–30% decrease in the [^{11}C]-CMDGE signal. The PET ligand appeared to be widely distributed though the rat brain with most of the ligand signal lost after 40 minutes. This was considered to be due to a combination of drug elimination and decay of the [^{11}C] radioisotope with its short half-life of 20 minutes. Synthesis of [^{11}C]-CMDGE is a two-step process requiring about 70 minutes (Figure 6.16).

BINA analogs   To develop a selective mGlu$_2$ agonist, biphenyl-indanone (BINA) was reported in 2005 by researchers at Merck as part of a program to discover compounds for schizophrenia.[445] BINA originated from a tetrazole lead that was discovered through library screening using a Ca^{2+} Fluorescent Imaging Plate Reader (FLIPR) functional assay. The tetrazole was a weak mGlu$_2$ potentiator with poor pharmacokinetics. It was shown to be a positive allosteric modulator as it was inactive in the absence of glutamate. Other leads discovered were a series of biphenyl acetophenones. Although these biphenyl acetophenones possessed good functional activity at mGlu$_2$ they were not selective for mGlu$_2$ versus mGlu$_3$ and had a poor brain/plasma ratio (B/P = 0.1%). Combining these two compounds as shown in Figure 6.17 in a subsequent SAR study that involved the synthesis and evaluation of 24 compounds led to BINA that has a good brain/plasma ratio of 0.75 (*ex vivo* analysis).

Building on this work, researchers at Sanford-Burnham Medical Research Institute with collaborators at Vanderbilt and UC San Diego developed a heterocyclic BINA analog (Figure 6.18) in an SAR study involving 14 compounds.[446]

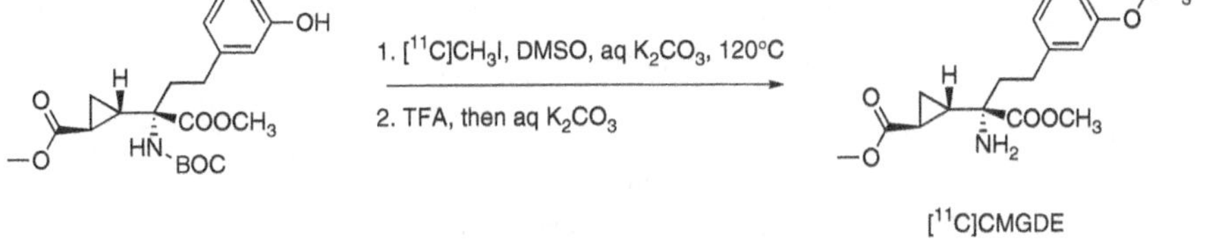

FIGURE 6.16   Synthesis of [^{11}C]-CMGDE

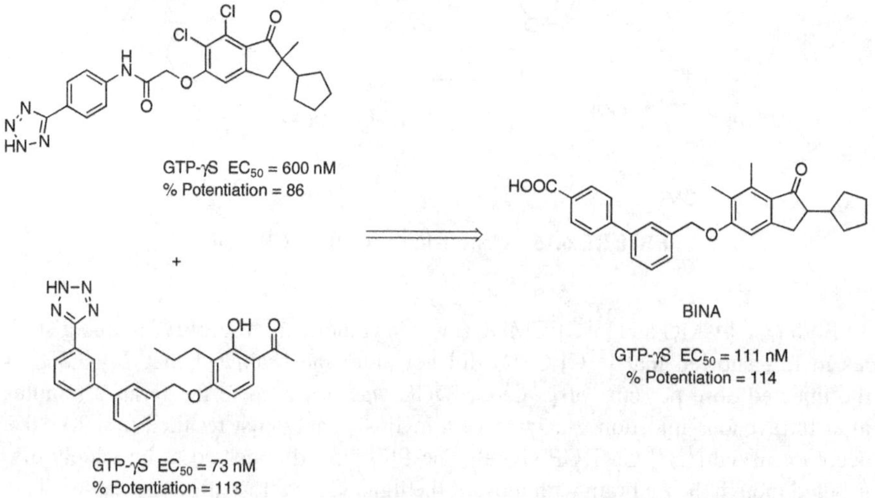

GTP-γS EC_{50} = 600 nM
% Potentiation = 86

+

GTP-γS EC_{50} = 73 nM
% Potentiation = 113

BINA

GTP-γS EC_{50} = 111 nM
% Potentiation = 114

FIGURE 6.17   Discovery of BINA

The compound is about twofold more potent than BINA (EC_{50} = 170 nM vs. 380 nM), shows excellent oral bioavailability (F = 86%, brain/plasma ratio = 4.80), and at an intravenous dosage of 20 mg/kg in rats reduced cocaine self-administration without affecting food responding. It was hypothesized that the greater $mGlu_2$ selectivity over $mGlu_3$, and/or action as a positive allosteric modulator, was responsible for the lack of effect on food responding.

Beyond the work described above, other groups are exploring positive allosteric modulators of $mGlu_2$ for use as antipsychotic and anxiolytic drugs. Some representative structures are shown in Figure 6.19.[447,448] From this work, many new chemotypes are being discovered that have not yet been examined in animal models of addiction.

6.3.9   Metabotropic Glutamate Receptor Subtype 5 Negative Allosteric Modulators ($mGlu_5$ NAM)

Both agonism and antagonism of the $mGlu_5$ receptor produces therapeutically useful behavioral changes. Modulation of $mGlu_5$ activity could be useful in the treatment

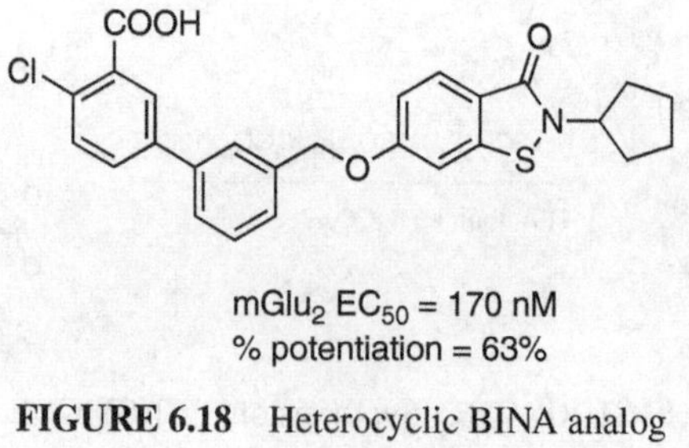

$mGlu_2$ EC_{50} = 170 nM
% potentiation = 63%

FIGURE 6.18   Heterocyclic BINA analog

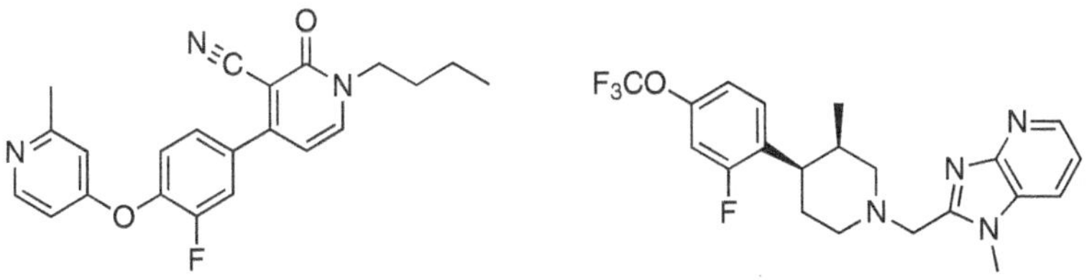

Janssen/Addex
$mGlu_2$ EC_{50} = 361 nM
E_{max} = 233%

Pfizer
$mGlu_2$ EC_{50} = 35 nM

FIGURE 6.19 Representative $mGlu_2$ PAMs

of schizophrenia, Parkinson's disease, anxiety, and drug addiction. For example, the bioavailable $mGlu_5$ PAM 3-cyano-*N*-(1,3-diphenyl-1H-pyrazol-5-yl)benzamide reversed amphetamine-induced deficits in prepulse inhibition in rats. A commonly accepted model of schizophrenia involves deficits in prepulse inhibition, that is, the reduced ability to "filter" sensory perceptions.[449]

Fenobam Fenobam was developed in the 1970s at McNeil Laboratories as a non-benzodiazepine drug for the treatment of anxiety. It underwent several clinical trials in the 1980s where it was reported to have a generally good safety profile but the drug was never carried forward. In one report, symptoms of detachment "... *feeling of unreality as in a dream-like state...*" were observed at doses of 200–300 mg per day were noted.[450] At that time, its mechanism of action was unknown. Fenobam was "rediscovered" by a lead discovery program at Hoffman-La Roche. They were searching for compounds that could modulate $mGlu_5$ activity using a functional high-throughput screening FLIPR assay. At this time, fenobam was found to be an $mGlu_5$ antagonist that bound to an allosteric site on the receptor. As will be discussed, the site of interaction was found to be the same as 2-methyl-6-(phenylethynyl)-pyridine (MPEP). The pharmacology of the compound was explored in great detail. Fenobam binds to $mGlu_5$ receptor with a K_d = 31 nM and a B_{max} = 2951 fmol/mg protein. The radiolabeled compound was displaced from the $mGlu_5$-binding site by other $mGlu_5$ antagonists with K_i's of 6–41 nM. The compound was *not* displaced by glutamate showing that the fenobam binding site is distinct from the glutamate-binding site.

The compound was determined to be an antagonist. It inhibited the quisqualate-evoked intracellular calcium response mediated by human $mGlu_5$ receptor with an IC_{50} = 58 nM. Quisqualic acid is a potent agonist for the Group 1 mGlu receptors ($mGlu_1$ and $mGlu_5$). Of note is that quiaqualate did not displace [^{3}H]-fenobam in the above-mentioned binding assays. Fenobam has excellent binding selectivity and functional activity. No specific binding at a concentration of 10 μM was detected in profile screening assay involving 86 GPCRs, ion channels, transporters, and enzymes. There was no functional activity detected at the $mGlu_1$, $mGlu_2$, $mGlu_4$, $mGlu_7$, and $mGlu_8$ receptors.

This is of interest as fenobam has recently been examined by several research groups for its effects on cocaine and methamphetamine use in rodents. Fenobam is an attractive drug due to its clinical history, if it proves to be beneficial in human clinical trials of addiction it would help validate the mGlu$_5$ receptor as a target of interest.

In a report from researchers at NIDA, the sulfate salt of fenobam was investigated for its effects on cocaine-seeking and cocaine-taking behavior in rats. Oral administration of fenobam sulfate at doses of 30 or 60 mg/kg was found to inhibit intravenous cocaine self-administration, cocaine-induced reinstatement of drug-seeking behavior, and cocaine-associated cue-induced cocaine-seeking behavior in rats. Fenobam sulfate also inhibited sucrose self-administration and sucrose-induced reinstatement of sucrose-seeking behavior, but had no effect on locomotion.[451] Lack of locomotor inhibition indicates that the drug is not affecting behavioral changes through some type of general sedation. The reduction in food-taking behavior is somewhat disconcerting. The authors state though that this is also seen with other mGlu$_5$ antagonists so perhaps it is a general physiological effect. They emphasis however that the effects on drug taking were more pronounced than on food taking.

The authors propose a mechanism of action where blockage of postsynaptic mGlu$_5$ receptors located in the VTA and nucleus accumbens by fenobam would prevent glutamate from binding, thus blocking its excitory effects. As we discussed in Chapter 2, there is an increase in synaptic glutamate levels during stimulant addiction. In this case, one would assume that fenobam, upon binding to an allosteric site on the receptor, causes a conformational change in the receptor that prevents glutamate from binding.

A research group at the University of Arizona investigated the effect of fenobam on methamphetamine-seeking behavior in rats. Rats were first trained to self-administer methamphetamine and then subjected to an extinction procedure. Fenobam at doses of 5–15 mg/kg was able to produce significant attenuation of reinstatement of methamphetamine self-administration when the rats were injected with a priming does of methamphetamine and when subjected to a methamphetamine-seeking cue (light and sound), both in a dose-dependent manner. Unfortunately, there was also a likewise attenuation of food-seeking behavior at the same doses.[452]

The Arizona group also examined the effect of fenobam on the reward system as a whole. In a study reminiscent of Olds and Milner, electrodes were implanted in the medial forebrain bundles of the brain and mGlu$_5$-positive allosteric and mGlu$_5$-negative allosteric modulators were administered. The threshold of electricity required for intracranial self-stimulation was then recorded by varying the intensity (μA) of the current at a fixed frequency of 100 Hz. In the absence of any drug, a current of 90 μA was required for self-stimulation. At a dose of 30 mg/kg fenobam, the intensity of the current required to initiate intracranial self-stimulation was increased by 68%.[453] This represents a decrease in the function of the brain reward system. This appears to be in line with the antagonistic effect of fenobam inhibiting binding of the excitory neurotransmitter glutamate, which in turn would decrease neuronal firing.

The authors propose that the ability of fenobam to decrease self-administration of cocaine and methamphetamine may be due to fenobam decreasing the baseline

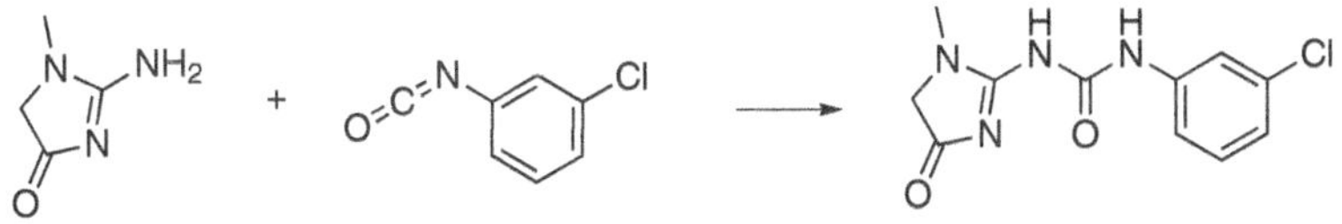

FIGURE 6.20   Synthesis of fenobam

activity of the brains reward system. This we can interpret as fenobam reducing the reinforcing and motivational effects of cocaine and methamphetamine. They caution though that the increase in the intracranial self-stimulation threshold could also be interpreted as fenobam inducing in the animal an overall negative affective state, that is, the animal is in a "depressive state."

Finally, the presence or lack of the receptor was examined in mGlu$_5$ gene knock-out mice. It was found the mGlu$_5$ receptor was not important in terms of modulating methamphetamine self-administration and conditioned place preference. There was also no effect on food responding. It was found though the time required to extinguish responding to methamphetamine was increased in the mice missing the mGlu$_5$ receptor. In other words, the mice lacking the mGlu$_5$ receptor would keep pressing the methamphetamine lever for a much longer time than the wild-type mice even though no methamphetamine was being delivered. This can be interpreted as a deficit in learning. This was not observed with food (sucrose) so it was a response specific to methamphetamine.[454]

In theory, an antagonist should give similar results as in the mGlu$_5$ gene knockout mice experiments. We have seen though there are some ambiguities, this is most likely a reflection of the complexity of behavioral pharmacology experiments and their interpretations. Nonetheless, it is clear though that the mGlu$_5$ receptor is important in cocaine and methamphetamine addiction.

The synthesis of fenobam shown in Figure 6.20 is straightforward; simply mixing the natural product creatinine with 3-chlorophenylisocyanate according to the original McNeil Laboratories patent US398315, produced the compound in good yield.

Several tautomeric forms are possible for creatinine so an interesting chemical question is if fenobam also exists in tautomeric forms. The following tautomeric forms shown in Figure 6.21 are possible.

The compound should maintain the urea group as a C=O bond is much stronger than a C=N bond, 724–757 kJ/mol versus 598 kJ/mol, respectively. The question then

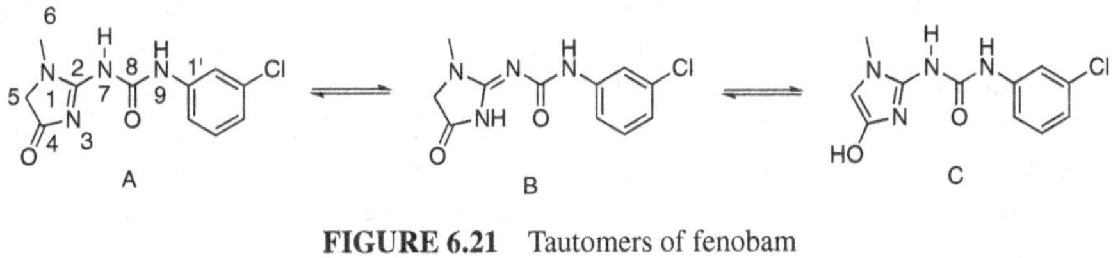

FIGURE 6.21   Tautomers of fenobam

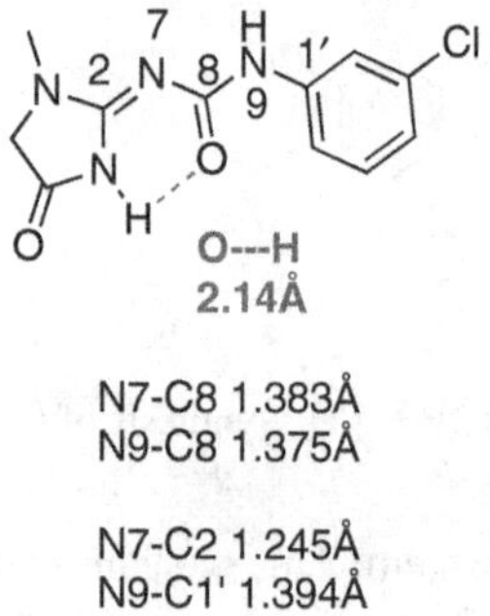

N7-C8 1.383Å
N9-C8 1.375Å

N7-C2 1.245Å
N9-C1' 1.394Å

FIGURE 6.22   Solid-state structure of fenobam

focuses on the structure 2-aminoimidazolinone ring. In the solution state in DMSO, the 1D ^{1}H- and ^{13}C-NMR spectra point toward tautomers A/B being preferred. A singlet in the ^{1}H-NMR at 4.01 ppm and a triplet at 51.1 ppm in the proton-coupled ^{13}C-NMR spectrum is assigned to the imidazolinone methylene group, which excludes tautomer C. Choosing between tautomers A and B is more ambiguous. The urea carbon is observed as a singlet so coupling with the N-H hydrogens is not observed. The imidazolinone carbonyl carbon is weakly coupled only to the methylene hydrogens ($^2J_{CH} = 5$ Hz) and C2 of the imidazolinone ring appears to be weakly coupled with the methylene hydrogens and the N-CH$_3$ hydrogens.[455] The authors conclude tautomer A being preferred based on comparison of the chemical shifts of 2-amino-1,5-dihydro-1-methyl-4H-imidazol-4-one hydrochloride in deuterated DMSO. Beyond this, it is difficult to make an unambiguous assignment of A versus B as is often the case with tautomers.

In the solid state, however, it is proposed that tautomer B is preferred over A by about 8.5 kcal/mol based on single crystal X-ray diffraction data.[456] The location of the H on the ring N was determined based on difference Fourier electron density maps. The intramolecular hydrogen bond between the urea carbonyl oxygen and the imidazolinone N-H is at a reasonable distance of 2.14 Å. The urea carbonyl C-N bond lengths are also different as summarized in Figure 6.22. The slightly shorter N9-C8 bond can be rationalized by a greater resonance interaction between N9 with C8 than N7 with C8.

The structure of fenobam is drawn in the literature as either form A or B. This may be a moot point except it could be important with regard as to how fenobam interacts with the mGlu$_5$ receptor. The solid-state tautomeric form was used in a receptor-based pharmacophore molecular modeling study from Hoffmann-La Roche.[457] The rat protein sequence was used for this study and the seven-transmembrane helices of the rmGlu$_5$ receptor were aligned toward the transmembrane helices of bovine rhodopsin. Using the homology model combined with site-directed mutagenesis, it is proposed that the binding pocket for fenobam is composed of transmembrane helices 5, 6, and 7. Docking studies propose three key hydrogen-bonding interactions, the 4-oxo carbonyl group of the imidazolidone ring with a serine hydroxyl OH group

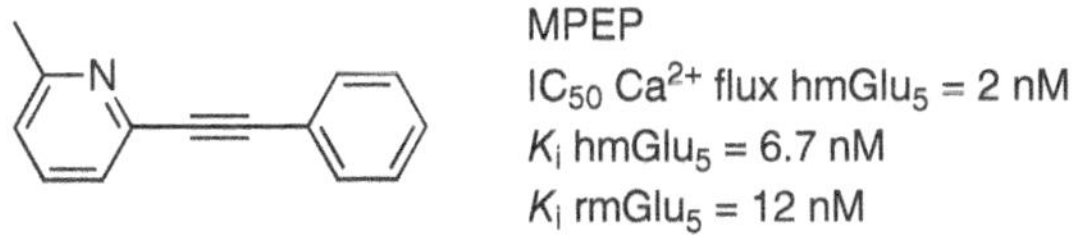

MPEP
IC_{50} Ca^{2+} flux $hmGlu_5$ = 2 nM
K_i $hmGlu_5$ = 6.7 nM
K_i $rmGlu_5$ = 12 nM

FIGURE 6.23   Structure and binding affinities of MPEP

($S657^{3.39}$), the urea carbonyl oxygen with a threonine hydroxyl group ($T780^{6.44}$), and the ring amide nitrogen with the threonine hydroxyl group.

Diarylethenes   The functional role of $mGlu_5$ receptor in drug addiction was first reported in 2001 when mutant $mGlu_5^{(-/-)}$ mice were found to not self-administer cocaine.[56] In the same study, it was also found that the selective $mGlu_5$ antagonist (or inverse agonist[458]) MPEP (Figure 6.23) dose dependently decreased cocaine self-administration in wild-type mice without affecting food responding. Furthermore, pretreatment of mice with MPEP (1, 5, and 20 mg/kg i.p.) 10 minutes prior to cocaine (15 mg/kg i.p.), D-amphetamine (2 mg/kg i.p.), nicotine (0.5 mg/kg i.p.), morphine (5 mg/kg i.p.), or ethanol (2 g/kg i.p.) dose dependently reduced the development of conditioned place preference for cocaine only.[459]

A study in squirrel monkeys showed that MPEP attenuated cocaine self-administration, cocaine-induced reinstatement of drug seeking, and the discriminative stimulus effects of cocaine without significantly affecting motor functions or operant behavior in the context of drug discrimination.[55]

The diarylethene class of compounds is synthesized by a Sonogoshira cross-coupling of a substituted alkyne and an aryl halide. The synthesis of the MPEP analog, MTEP, from the Merck labs is outlined in Figure 6.24.[460] In this study, the goal was to increase the aqueous solubility of MPEP while maintaining its potency. The methyl pyridine group of MPEP was replaced with a bioisosteric thiazole ring that initially increased binding affinity to $mGlu_5$ from a K_i of 12 to 6 nM but slightly reduced the log D to 3.3 from 3.5. Then, replacement of the phenyl group with a 3-pyridyl group brought the binding affinity back to 16 nM, but more importantly the log D was decreased from 3.5 for MPEP to 2.1 for MTEP, meaning it will be more

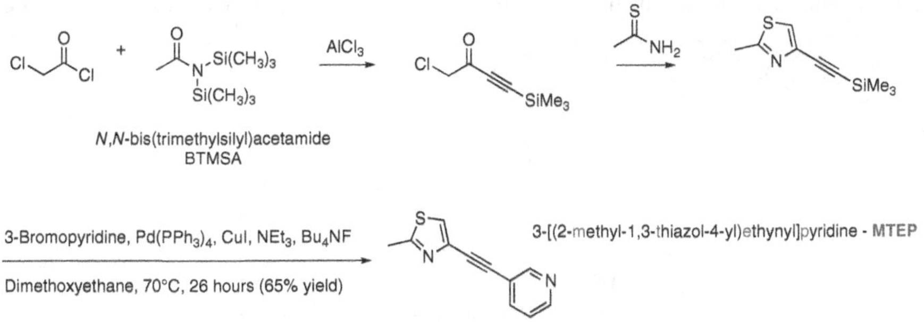

N,N-bis(trimethylsilyl)acetamide
BTMSA

3-Bromopyridine, Pd(PPh₃)₄, CuI, NEt₃, Bu₄NF
Dimethoxyethane, 70°C, 26 hours (65% yield)

3-[(2-methyl-1,3-thiazol-4-yl)ethynyl]pyridine - MTEP

FIGURE 6.24   Synthesis of MTEP

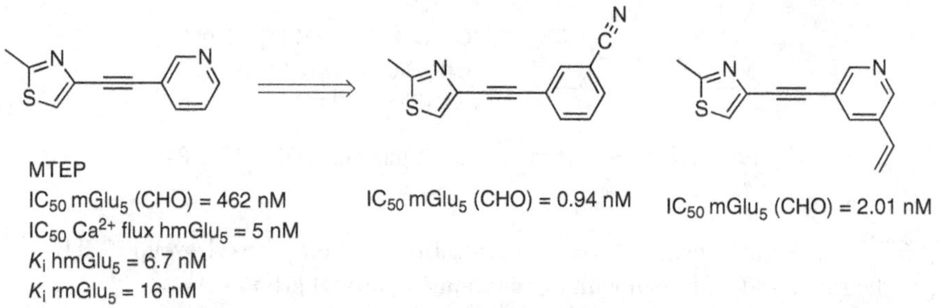

MTEP
IC_{50} mGlu$_5$ (CHO) = 462 nM
IC_{50} Ca^{2+} flux hmGlu$_5$ = 5 nM
K_i hmGlu$_5$ = 6.7 nM
K_i rmGlu$_5$ = 16 nM

IC_{50} mGlu$_5$ (CHO) = 0.94 nM IC_{50} mGlu$_5$ (CHO) = 2.01 nM

FIGURE 6.25   SAR study of MTEP

water soluble. It was indeed found that at a dose of 3 mg/kg i.p. the concentration of MTEP in the rat brain was increased to 1.4 µM relative to MPEP at 0.83 µM.

An extensive SAR study on the structurally similar MTEP involving 58 compounds discovered that the (2-methyl-1,3-thiazol-4-yl)ethynyl group is important to mGlu$_5$ antagonism.[461] The SAR study and functional activities is summarized in Figure 6.25. This work led to the discovery of the two compounds shown in Figure 6.25 that were significantly more potent than MTEP. It was also shown the MTEP attenuates cue-induced reinstatement of cocaine self-administration.

While investigating different mGlu$_5$ antagonist chemotypes using high-throughput screening and rational design the 3-cyano-5-fluorophenyl ring emerged as a common chemotype. A series of Suzuki couplings of 3-cyano-5-fluorophenylboronic acid with 6,6- and 6,5-fused heterocycles gave 27 compounds. From this a 2-methylbenzothiazole analog (Figure 6.26) was found with good functional negative allosteric modulator activity (IC_{50} = 61 nM) though it had somewhat poor metabolic stability and was highly protein bound (99.4% bound human plasma).[462] More importantly is was bioavaliable to the brain when delivered intraperitoneally showing almost a twofold higher level in the rat brain (AUC$_{0-6 h}$ = 1530 ng h/g) than in plasma.

The new benzothiazole compound was investigated in a couple of behavioral studies in mice. The first was to see whether it exhibited a similar effect as other mGlu$_5$-negative allosteric modulators in the marble-burying assay. This assay was initially developed to test new anxiolytic compounds. Mice are known to bury objects such as marbles in their bedding. It was found that low doses of anxiolytic

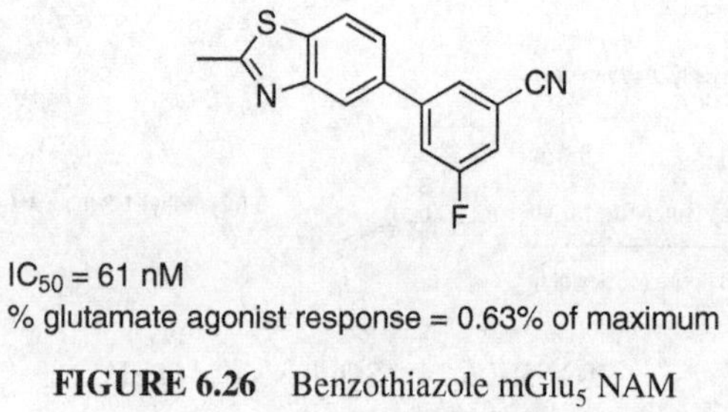

IC_{50} = 61 nM
% glutamate agonist response = 0.63% of maximum

FIGURE 6.26   Benzothiazole mGlu$_5$ NAM

benzodiazepines attenuate this behavior. Other mGlu$_5$ negative allosteric modulators such as fenobam and MPEP also attenuate this burying behavior. The benzothiazole at doses of 30 mg/kg attenuated marble-burying behavior by about 50%. This is much less potent than MTEP where a dose of 15 mg/kg practically abolished marble-burying behavior.

The second behavioral assay used the operant sensation-seeking (OSS) model. The benzothiazole, as well as MTEP, was found to dose dependently reduce the mice responding to OSS stimuli without affecting food intake.

The same docking study mentioned above for fenobam also performed docking experiments on MPEP. MPEP was found to dock onto the mGlu$_5$-binding site in a similar fashion as fenobam. Binding site amino acids tryptophan W784$^{6.48}$, phenylalanine F787$^{6.51}$, and tyrosine Y791$^{6.55}$ were found to be critical for the antagonist effect of MPEP. Only one distinct hydrogen bond of threonine T780$^{6.44}$ with the pyridyl N was observed. With the binding mode used, however, there were more hydrophobic interactions of MPEP with mGlu$_5$ than with fenobam. In all this can be used to rationalize the stronger binding of MPEP ($K_d = 3.1$ nM) than that of fenobam ($K_d = 55$ nM) at the rat mGlu$_5$ receptor.

Both MPEP and MTEP have not been used in clinical trials. The main concern is that of safety. Besides their mGlu$_5$ activity they also modulate the activity of other important systems. MPEP is a NET blocker, NMDA antagonist, a monoamine oxidase 2 enzyme inhibitor, and a mGlu$_4$ agonist.[451] MTEP is a cytochrome P450 1A1 enzyme inhibitor. The CYP1A1 metabolic enzyme is important for the metabolism of potentially harmful foreign chemicals (xenobiotics). Inhibition of this enzyme could lead to undesired side effects.

6.3.10   Metabotropic Glutamate Receptor Subtype 7 Agonist (mGlu$_7$)

AMN082   The mGlu$_7$ receptor subtype has just recently become a potential target of interest. The mGlu$_7$ receptor is coupled to G$_{i/o}$ proteins and is located presynaptically were it functions as an inhibitory autoreceptor. Activation of the receptor ultimately results in decreasing the release of glutamate into the synapse. In 2005, researchers from Novartis reported the first selective mGlu$_7$ ligand AMN082 (Figure 6.27).[463] The molecule was discovered through a high-throughput screening campaign and

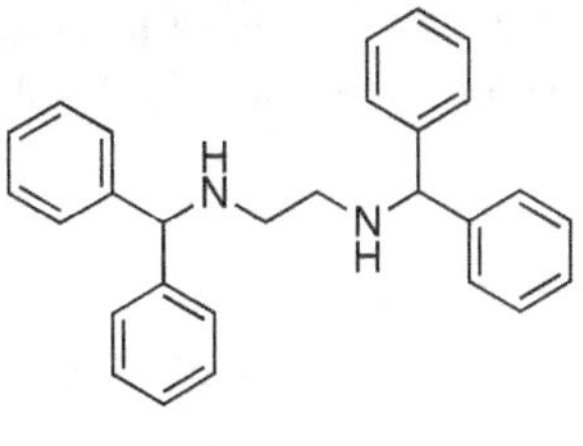

AMN082

FIGURE 6.27   Structure of AMN082

extensive pharmacology studies showed the compound behaved as an allosteric agonist of mGlu$_7$. AMN082 is orally bioavailable, passes the blood–brain barrier, and possesses good efficacy with EC$_{50}$ = 64 nM in a cAMP functional assay. The compound did not affect binding of orthosteric ligands to the glutamate-binding site nor their functional potency. As such, AMN082 should be referred to as an allosteric agonist and not as a positive allosteric modulator.

Li has just recently reviewed the pharmacology of AMN082 in animal models of drug addiction. In brief, AMN082 dose dependently attenuates the rewarding effects of cocaine, inhibits self-administration of cocaine, and inhibits reinstatement of drug-seeking behavior. *In vivo* microdialysis showed no effects on cocaine-induced extracellular dopamine levels in the nucleus accumbens but did increase extracellular GABA levels and surprisingly, glutamate levels. As an agonist acting on a presynaptic autoinhibitor receptor, glutamate levels should be decreased. A complex mechanism was proposed to explain this; increased levels of glutamate due to AMN082 will bind to presynaptic mGlu$_{2/3}$ autoreceptors and antagonize cocaine-induced increases in extracellular glutamate, thus inhibiting reinstatement and other effects. These contradictions will need to be clarified.

The role of the mGlu$_7$ receptor in addiction is just being explored so it will be interesting to see how it progresses.

6.3.11   Glutaminergic System Modulator: *N*-acetylcysteine

N-acetylcysteine (NAC; Figure 6.28) is used as a prodrug of cysteine and is sold over-the-counter as an antioxidant via its oxidation to glutathione. *N*-acetylcysteine regulates extracellular concentrations of glutamic acid by releasing glutamate from glial cells. Animal studies have shown that upon chronic cocaine use lower amounts of glutamic acid are present in the nucleus accumbens. Following extinction (withdrawal) the reduced glutamic acid levels persist. However, when the animals are exposed to cocaine-related cues a large increase in glutamic acid levels occurs. Measuring *in vivo* glutamate and dopamine levels using microdialysis, cocaine-induced reinstatement increased glutamate levels from an extinction state concentration of 20 pmol to as high as 120 pmol.[464] Of note should be that the basal level of glutamate when the rats had not been exposed to cocaine was about 50 pmol. Exposure to cocaine had decreased glutamate levels (in rats at least) 50%. This rapid flux of glutamate is what is thought to drive drug-seeking behavior.

Upon entering the brain NAC is hydrolyzed to cysteine that then dimerizes via oxidative disulfide bond formation to form cystine as shown in Figure 6.28. Cystine

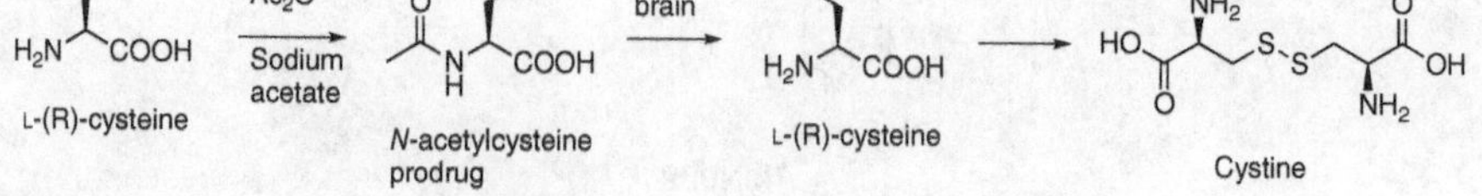

FIGURE 6.28   Formation of cystine from cysteine

is thought to regulate glutamate levels by binding to GLT1 on glia cells stimulating the exchange of glutamate for cystine. Once inside the neuron cystine is reduced to two molecules of cysteine.

Levels of glutamate in cocaine-addicted humans have been studied by ^{1}H-MRS.[465] In ^{1}H-MRS glutamate appears as a broad singlet at 2.35 ppm. The levels of glutamate in the left dorsal anterior cingulated cortex were found to be significantly higher in a cocaine-dependent group versus a control group. Upon administration of 2400 mg of NAC glutamate levels in the cocaine-dependent group were reduced to those similar to the control group. Glutamate levels in this study were expressed as a ratio relative to endogenous creatine. The higher levels of glutamate were also correlated with an increase in impulsivity that was measured by the psychological Barratt Implusiveness Scale exam.

N-acetylcysteine was studied in a double-blind clinical trial to inhibit the reaction of cocaine addicts to cues they associate with cocaine use. In the study, 15 patients were treated with NAC (600 mg every 12 hours) or placebo for 3 days. The patients then completed a cue-reactivity procedure that involved presentations of four categories of slides (cocaine, neutral, pleasant, and unpleasant). There was a statistically significant reduction in craving, desire to use, and interest for patients who had taken NAC versus placebo.[466] Craving was also reduced in a small study ($n = 4$) where cocaine-dependent human subjects were given 1200–2400 mg of NAC per day for 4 days.

Overall, NAC has shown only moderate efficacy in clinical trials. The effects of NAC appear to be concentration-dependent. It has recently been found that at low doses of NAC, the released glutamate stimulates presynaptic receptors (mGlu$_{2/3}$) on neurons; at higher doses, more glutamate is released and an additional, postsynaptic receptor (mGlu$_5$) is also stimulated.[467] The presynaptic stimulation attenuates neuronal activity in the nucleus accumbens and, in rats, reduces the tendency to respond to cocaine-associated cues. The postsynaptic stimulation has opposite effects: it intensifies neuronal activity and partly offsets the positive effect on animals' responses to drug cues.

The researchers propose that a drug or drug combination that both increases nonsynaptic extracellular glutamate to stimulate mGlu$_{2/3}$ and inhibits mGlu$_5$ may reduce the risk of relapse more effectively than NAC alone. A low dose of NAC combined with fenobam would be an example of such a combination therapy.

6.3.12 D$_3$ Partial Agonist/Antagonist

Pharmacological evidence of the importance of the D$_3$ receptor and the development of D$_3$ receptor antagonists and partial agonists for the treatment of drug addiction has been well reviewed.[88,183,468] Cocaine increases dopamine levels in the synapse by binding to DAT and preventing the reuptake of dopamine into the presynaptic neuron. The increased levels of synaptic dopamime result in prolonged binding to postsynaptic D$_3$ receptors to which dopamine binds with strong affinity (K_i of 27 nM).

While much excellent work has been done on the development of D$_3$ antagonists and partial agonists, of the thousands of compounds made and evaluated, few have

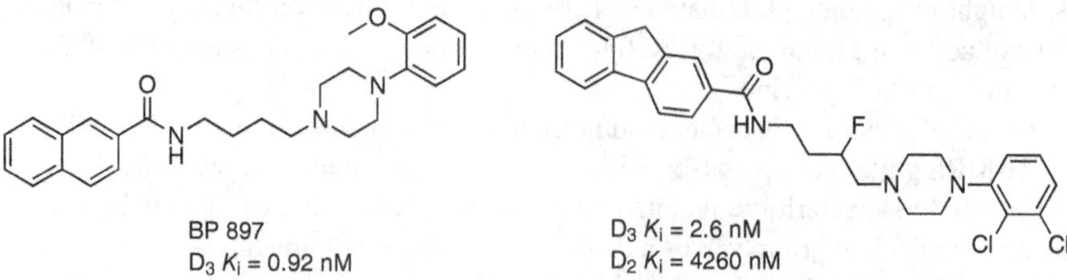

BP 897
D_3 K_i = 0.92 nM
D_2 K_i = 61 nM

D_3 K_i = 2.6 nM
D_2 K_i = 4260 nM

FIGURE 6.29   Selective D_3 ligands

advanced to clinical trials. GlaxoSmithKline has completed a Phase I study that evaluated the safety, tolerability, blood concentrations and effect following repeated oral doses of the D_3 antagonist GSK618334 in 40 healthy male and female volunteers for 21 days. No serious adverse effects were reported though at repeated doses of 10, 25, or 75 mg 73% of the subjects reported adverse effects with headaches being the most common (GlaxoSmithKline (GSK) result summary VM2010/00012/00). Oddly, 50% of the subjects taking the placebo also reported adverse effects. Only two patients taking the 75 mg dose had to be withdrawn from the study.

It should be noted that of the compounds reported in the literature, it had proven difficult to achieve desired D_2/D_3 ratios > 100. Recently, though compounds with a D_2/D_3 selectivity of over 1000 have been reported. For example, one of the first selective compounds discovered, the D_3 partial agonist BP897 (Figure 6.29), has many of the pharmacophoric features present in D_3 partial agonist and antagonists. Important pharmacophores are an aryl region containing a H-bond acceptor amide group connected via a short-chain spacer to a piperizine ring connected in turn via one of the ring nitrogen atoms to a second aryl region. Building on this and previous work, an SAR study involving 30 compounds led to several extremely selective D_3 compounds as shown in Figure 6.29.[469] Although they are highly selective, they do suffer from very high clogP values of >5, reducing their water solubility. This problem may not be unique though to just this series, random evaluation of the compounds reported in the literature reveal that many of the compounds are quite hydrophobic.

The exact neuronal site of action for these compounds is not known. Pramipexole, a D_3 selective agonist, has been shown in PET studies with primates to have greatest effects in the prefrontal and limbic cortex regions of the brain.[470] Perhaps the compounds above exert their effects in the same region.

6.3.13 D_2 Partial Agonist

The antipsychotic drug aripiprazole (Abilify) attenuates behavioral effects of D-amphetamine and is proposed to be of possible use for the treatment of stimulant abuse. Aripiprazole is considered as an atypical antipsychotic drug as it is a D_2 partial agonist and also a 5-HT$_{1A}$ partial agonist. First-generation antipsychotics (e.g., thorazine) were D_2 antagonists, while second-generation antipsychotics were

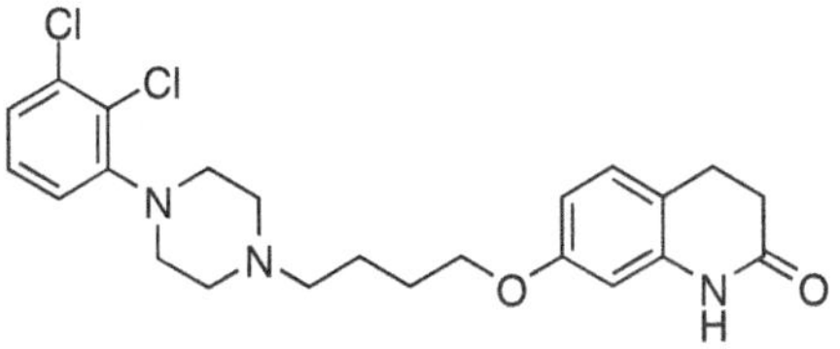

Aripiprazole (Abilify)
OPC-14597

K_i D_1 = 265 nM   5-HT$_{1A}$ = 1.7 nM
D_2 = 0.34 nM   5-HT$_{2A}$ = 3.4 nM
D_3 = 0.8 nM   5-HT$_{2C}$ = 15 nM
D_4 = 44 nM   5-HT$_3$ = 501 nM
D_5 = 95 nM
α_1 = 100 nM

FIGURE 6.30   Structure and binding profile of aripiprazole

antagonists at both the D_2 and $5HT_{2A}$ receptors. Though aripiprazole is considered to be D_2/5-HT$_{1A}$ selective partial agonist, it can be seen from the receptor-binding profile in Figure 6.30 it also has significant binding to the D_3 and 5-HT$_{2A}$ (antagonist) receptors.[471,472]

Drugs such as aripiprazole are considered dopamine "stabilizers" so it would be of interest to determine their effects on drug addiction. A double-blind, placebo-controlled trial investigating the use of aripiprazole to decrease methamphetamine use in 90 methamphetamine-dependent adults has just been completed. The patients were given aripiprazole at dosages up to 20 mg per day or placebo for 12 weeks along with substance abuse counseling with a 3-month follow-up. Unfortunately, the researchers found that aripiprazole did not reduce methamphetamine use relative to the placebo.[473]

From our discussion, it would seem that a potential strategy for medication development could be a compound with dual activity as a D_2 partial agonist and D_3 antagonist. Such compounds are being developed for antipsychotic medication. Just for the sake of discussion, a random example from the literature is the imidazolidinone in Figure 6.31 that was recently reported by GlaoxoSmithKline.[474]

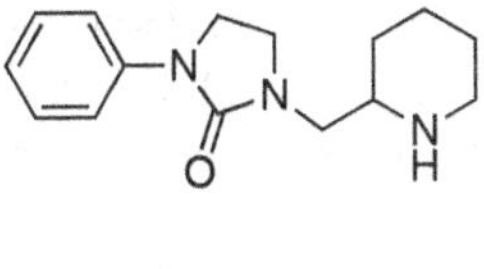

D_2 partial agonist K_i = 79 nM, EC_{50} = 5 nM
D_3 antagonist fK_i = 25 nM

FIGURE 6.31   Structure of a D_2 partial agonist/D_3 antagonist

The compound was found to have acceptable binding at the D_2 and D_3 receptors along with, more importantly, excellent selectivity. A broad profiling screen against a large number of receptors, transporters, ion channels, and enzymes showed little or modest off-target binding. The imidazolidinone has poor oral bioavailablity due to high metabolism but is bioavalilable by subcutaneous administration and enters into the brain (rat). A 3.0 mg/kg dose by s.c. administration gave a brain/blood ratio 5.5 with C_{max} (brain) = 667 ng/g and C_{max} (blood) = 142 ng/mL. This compound does not appear to have been examined in an addiction assay, but it would be interesting to see how a selective D_2 partial agonist/D_3 antagonist would affect the behavior of rodents.

6.3.14   DAT Reuptake Inhibition

An attractive strategy inline with a partial agonist approach is to develop compounds that could compete with cocaine binding to the DAT. Compounds that could compete with cocaine binding (i.e., K_d drug $< K_d$ cocaine) but with reduced efficacy by acting as a partial agonist or antagonists could in principle reduce the effects of cocaine.

Prototypic DAT blockers that have received extensive investigation are GBR 12909 (vanoxerine) and benztropine (Cogentin). Excellent reviews on the pharmacology and medicinal chemistry of GBR 12909, benztropine, and their analogs are available.[475,476]

GBR-12909 was developed at Gist-Brocades (acquired by Koninklijke DSM N.V.) in the Netherlands in 1980. It consists of a central piperizine core connected to a benzhydrol group and an aryl group by an alkyl linker. As we saw with aripiprazole, the piperizine core is a common motif with many dopamine receptor ligands.[477] The compound has good selectivity at the DAT receptor relative to the other monoamine transporters with K_i's of 12, 105, and 497 nM at DAT, SERT, and NET, respectively.[476] NIDA had actually started a Phase I clinical trial in cocaine-experienced users, but it had to be abandoned when the appearance of QTc interval prolongation symptoms occurred.

Benztropine is an older drug that was first reported by Merck in the 1952 patent US 2595405. As was the case at that time, the benzhydrol group was based on antihistimine drugs that were being developed. The structure of benztropine contains both the cocaine tropane structure and the benzhydrol ether of GBR12909 (Figure 6.32).

Cogentin has been used in the treatment of Parkinson's disease but is less favored today. Newman and Kulkarni have written an excellent review on the SAR of benztropine as referenced above so we will not discuss that in detail here. Briefly, to obtain maximal binding to DAT and selectivity relative to SERT and NET, the benzhydrol group should be in an axial position with F or Cl atoms at the meta or para positions. Even though it binds well to DAT, benztropine does not have any abuse liability, which makes it unique.

The binding profile of benztropine is quite different than that of cocaine. Benztropine binds at DAT with a $K_i = 118$ nM versus 187 nM for cocaine. No appreciable binding at NET or SERT is found for benztropine, whereas cocaine binds equally well at SERT with a $K_i = 175$ nM. Benztropine binds strongly to the M_1 and M_2 receptors

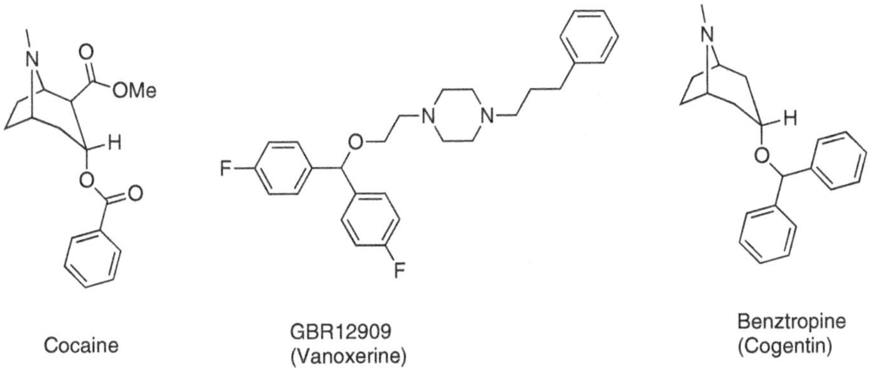

FIGURE 6.32 Structures of DAT reuptake inhibitors GBR12909 and benztropine

with K_i's of 0.6 and 2.6 nM respectively.[478] Bentropine is less potent than cocaine in terms of DAT blockage with IC_{50}'s of 403 nM versus 236 nM, respectively.

Binding of benztropine is thought to compete directly with cocaine and dopamine. Docking into molecular models based on the structure of the bacterial homolog LeuT combined with site-directed mutagenesis supports a benztropine-binding site that overlaps, but is not identical, with the cocaine- and dopamine-binding pocket. The binding pocket it considered to be deeply buried in the transporter between TM3 and TM10. Key interactions for benzotropine with DAT were hydrophobic interactions with a valine and alanine and a H-bond formed between the serine OH group and the oxygen of benztropine.[479]

In the one clinical trial published to date, doses of 1–4 mg of benzotropine did not have any affect on cocaine use in recreational users of cocaine.[480] On the other hand, no adverse effects due to benztropine were observed. No other clinical trials have been registered on ClinicalTrails.gov.

6.3.15 DAT Reuptake Inhibitor and σ-Receptor Antagonist: Rimcazole

The σ receptors appear to be a unique set of receptors. Evidence of the σ receptor was first reported in 1976 at which time it was considered a member of the opioid family. As time progressed, it was clear that it was not and now appears to be in a separate class; for example, it has cell-signaling functions but is not a GPCR. There are two known subtypes, σ_1 and σ_2, but many aspects of their structure, biochemistry, and pharmacology are still being explored. Reviews of the discovery of the σ receptor and of their biology and function with relation to addiction and CNS disorders are available.[481–483]

The σ_1 receptor has been implicated in stimulant addiction. Agonists have been found to be reinforcing and can be blocked by σ_1 antagonists in a dose-dependent manner. Rats can be trained to self-administer σ_1 agonists and the agonists can substitute for cocaine.[484] The agonists further have a synergistic effect on cocaine self-administration (leftward shift of dose-effect curve). However, σ_1 antagonists do not affect cocaine self-administration though they attenuate other behavioral effects such as cocaine-induced convulsions and acute locomotor effects. The latter effects have been extensively reviewed.[485] Cocaine has low micromolar binding affinity for σ_1 (2–8 μM)[486] and the lack of effect on antagonizing the self-administration of cocaine suggests that the primary addictive effects of cocaine do not involve direct agonism of σ_1 receptor.

In the early 1980s, Wellcome Research Laboratories studied the carbazole piper-izine drug Rimcazole as a possible therapeutic agent for schizophrenia. Clinical trials did not prove fruitful, but it did show that the compound had some interesting properties.

Pharmacological studies revealed that it was unlike other antipsychotic drugs at that time as it was a weak antagonists of dopamine receptors, did not effect dopamine synthesis and metabolism, and had very poor affinity to other receptors associated with psychosis. Further studies discovered that it had moderate binding affinity to the σ receptor (IC_{50} = 500 nM), interacted with the σ receptor as an antagonist,[487] and had good binding affinity to DAT (K_i = 103 nM) and inhibited the reuptake of dopamine (IC_{50} = 4.2 μM).[488] A review of the pharmacology of rimcazole is available.[489]

What is of interest to us is that over time it became clear that rimcazole attenuated many of the behavioral effects of cocaine. As the binding affinity to DAT is greater than to σ_1 receptor, the development of dual DAT/σ_1 inhibitors became of interest to medicinal chemists. A rather extensive literature of ligand SAR studies for the σ receptors is developing; the majority is directed to the treatment of pain. A few are directed specifically at addiction.[486,490–493]

To illustrate these studies, let us look at one involving self-administration of cocaine.[494] In this interesting study, it was determined that rimcazole and two analogs (Figure 6.33) were effective in reducing cocaine self-administration in rats and that rimcazole itself is not self-administered.

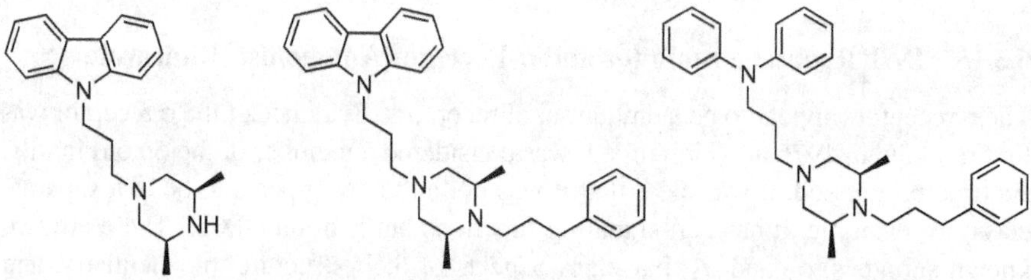

	Rimcazole		
σ_1	908	97	104
σ_2	302	183	145
DAT	224	61	263
SERT	825	219	44
NET	2160	3640	2490

FIGURE 6.33   Binding affinities of DAT/σ rimcazole ligands (K_i, nM)

Introduction of the hydrophobic propyl phenyl group to the secondary piperizine nitrogen lead to a dramatic increase in binding to the σ_1 receptor with a twofold enhancement of σ_2 binding. This modification also enhanced binding to DAT and SERT. Opening of the carbazole ring to make the more flexible diphenyl amine had a small affect on σ binding, while it decreased DAT binding about fourfold and increased SERT binding to such an extent that the new compound was now showing selective binding to SERT.

The attenuation of cocaine self-administration is due to action at both DAT and σ receptor, when a selective DAT or selective σ receptor antagonist where given separately, no attenuation in self-administration of cocaine was observed. When they were combined (same dose as separate), however, attenuation of cocaine self-administration was again observed.

As the authors note, rimcazole itself was abandoned in clinical trials due to adverse side effects; however, the strategy of developing dual DAT reuptake inhibitor/σ antagonists with reduced side effects for the treatment of cocaine abuse appears sound.

6.3.16 σ_1 Agonist: SA4503

The role of σ agonists in addiction appears very complex. For example, two separate studies investigated the effects of the σ_1 selective agonist SA4503.

Santen Pharmaceuticals in Japan first reported SA4503 in 1996 as part of a program to develop compounds that could control central cholinergic function. The drug was reported to possess excellent binding affinity and selectivity to σ_1 receptor (IC_{50} = 17 nM) versus σ_2 (IC_{50} = 1800 nM) with little affinity for 36 other receptors, ion channels, and second messenger systems even at 1 nM concentration. The drug is bioavailable to the brain and was effective in reducing scopolamine-induced memory impairment in rats as measured in a passive avoidance task behavioral assay.[495,496] Recently, binding affinities for SA4503 at the σ_1 and σ_2 receptors were reported as K_i = 5 and 63 nM, respectively, in guinea pig brain homogenates.[497] The difference in measured binding affinities was ascribed to the methods used to obtain σ_2-binding affinities.

SA4503 labeled with ^{11}C has been used for PET studies of σ_1 receptors in monkeys and humans. In a healthy human volunteer, a dose of 560 MBq corresponding to 9.7 nmol (15.1 mCi/9.7 nmol) with specific activity >10 TBq/mmol (270 Ci/mmol) was administered. Imaging showed that there was high uptake in the brain, heart, liver, and kidneys. There was relatively low radioactivity in the thorax, extremities, and lower abdomen.[498] Further PET studies using [^{11}C]-SA4503 show that σ_1 receptors are concentrated in brain areas of the limbic system, including areas involved in motor function, sensory perception, and endocrine function.[499]

Researchers in Japan found that SA4503 can attenuate methamphetamine, cocaine, and morphine-induced conditioned place preference suggesting that SA4503 inhibits the rewarding effects of the drugs.[500] Meanwhile, researchers in Missouri found that SA4503 in drug discrimination tests does not substitute for methamphetamine in rats trained to discriminate methamphetamine from saline.[501] Injection of SA4503 into the rats did not cause an increase in the percent responding to the methamphetamine

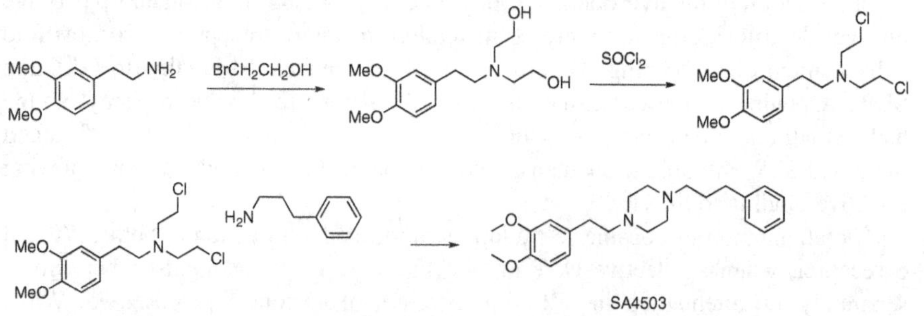

FIGURE 6.34 Synthesis of SA4503

lever. Pretreatment of the rats with 0.3–1.0 mg/kg SA4503 followed by injection of the drug of abuse either potentiated or has no effect on the percent responding for methamphetamine, cocaine, or D-amphetamine. In this case, it appears the σ_1 agonist potentiates the stimulus properties of methamphetamine. More work needs to be done to determine whether a selective σ_1 agonist can be useful for the treatment of addiction, especially in light of its potential reinforcing effects mentioned above.

The original synthesis of SA4503 shown in Figure 6.34 began with the bis alkylation of 3,4-dimethoxyphenethylamine using bromoethanol. Treatment of the primary alcohol with thionyl chloride gave the dichloride that served as a bis-electrophile. Formation of the piperizine ring and completion of the synthesis was achieved with 3-phenylpropylamine.

A classical QSAR study discovered that σ_1 binding correlated best with an electronic term S_π that was described as representing superdelocalizability over the highest occupied π orbitals.[502]

6.3.17 Mirtazapine

Mirtazapine (Remeron, ORG 3770) is a generic drug that was developed by Organon (Merck) for use as an antidepressant. Mitrazapine has a broad binding profile acting as an antagonist at the 5-HT_{2A}, 5-HT_{2C}, 5-HT_3 and α_{2A} receptors and is also a κ-opioid agonist. The compound binds very poorly to DAT and SERT with K_D's > 10 μM and has modest binding to NET with a $K_D = 4.6$ μM.[43]

Mirtazapine is a prototypic noradrenergic and specific serotonergic antidepressant, and it is postulated that mitrazapine enhances central noradrenergic and serotonergic activity by acting as an antagonist at presynaptic α_2-adrenergic inhibitory autoreceptors and heteroreceptors. Adrenergic receptors located in areas of the mesocorticolimbic system are involved in drug reward, drug craving, and drug-seeking behavior so modulation of these receptors could have a direct effect on addiction.

Mirtazapine was investigated in a double-blind, randomized, controlled, 12-week trial versus placebo (NCT00497081) to determine whether it could reduce methamphetamine use in men who have sex with men (MSM) and were active methamphetamine users. This trial was designed to address the high level of the human

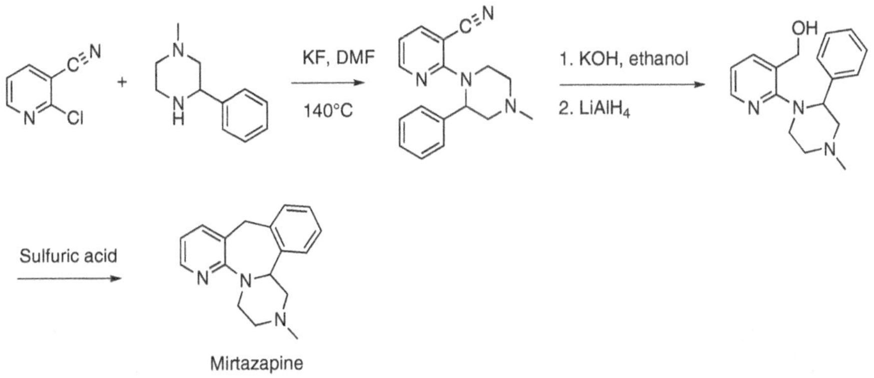

FIGURE 6.35   Synthesis of mirtazapine

immunodeficiency virus that can be associated with methamphetamine use and unprotected sex. The participants ($n = 60$) were actively using, methamphetamine-dependent, sexually active MSM being seen weekly for urine sample collection and substance use counseling. Despite a moderate medication adherence of 48%, it was determined that use of mirtazepine (15 mg per day for 1 week, then 30 mg per day), in conjunction with counseling, reduced urine positive tests from 73% to 44% along with a significant reduction in sexual risky behaviors.[503]

Taking advantage of the antidepressant effects of mirtazepine, it was investigated if it could reduce cocaine use in subjects suffering from depression. The 12-week study involved 24 subjects in which half received 45 mg mirtazepine daily and the other half placebo. While the quality of sleep was improved no reduction in cocaine use relative to placebo was observed.[504] The selective serotonin reuptake inhibitor escitalopram and mirtazapine are being investigated in a clinical trial (NCT00732901) to establish the relationship between the 5-HT$_2$R and impulsivity and cue reactivity in cocaine-dependent subjects.

The synthesis of mirtazapine (Figure 6.35) starts with a nucleophilic aromatic substitution reaction of 2-chloro-3-cyanopyridine with 2-phenyl N-methyl piperizine followed by hydrolysis of the nitrile to a carboxylic acid and subsequent reduction to a primary alcohol. A Friedel–Crafts alkylation ring closure forms the azepine ring.

The physicochemical properties of mirtazapine have been extensively studied. It was found that the seemingly simple isosteric replacement of CH in the antidepressant mianserin by N in mirtazapine (Figure 6.36) has remarkable effects on physicochemical and pharmacologically properties. For example, this simple switch reduces NET binding from $K_i = 71$ nM for mianserin to 4.6 μM for mirtazapine.

The charge distributions as indicated by NMR and calculated by semi-empirical quantum mechanics differ, not only for the changed pyridine, but also in other regions of the molecule. The 3° nitrogen atom conjugated to the changed aromatic ring is less negatively charged in mirtazapine (−0.20) than in mianserin (−0.26). Consequently, the oxidation potential of mirtazapine (0.83 V) is significantly higher than that of mianserin (0.76 V). Another result of this difference in charge distribution is

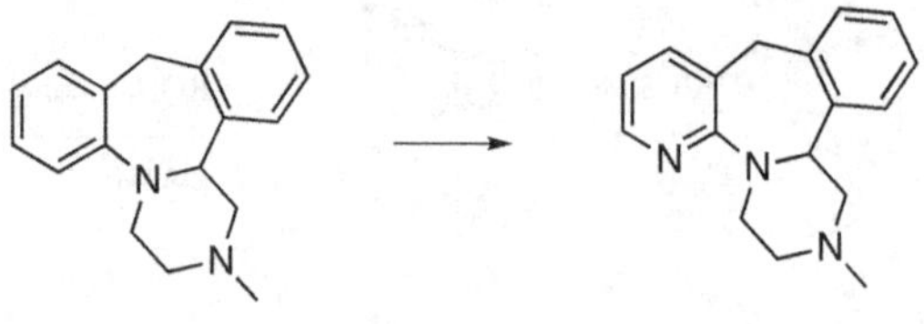

Mianserin   Mirtazapine

FIGURE 6.36 Conversion of mianserin to mirtazapine

that the (calculated) dipole-moment vectors of the compounds are oriented roughly perpendicular to each other. The experimental dipole moment of mirtazapine (2.63 D) is, moreover, three times larger than that of mianserin (0.82 D); mirtazapine is, therefore, more polar than mianserin and this is reflected in a lower retention index. Finally, the acidity of mirtazapine ($pK_a = 7.1$) is slightly higher than that of mianserin ($pK_a = 7.4$).[505]

6.3.18 5-HT$_{2A}$ Antagonist and 5-HT$_{2C}$ Agonist

Selective antagonism of the 5-HT$_{2A}$ receptor or agonism of the 5-HT$_{2C}$ receptor has been shown to suppress both cue- and cocaine-evoked reinstatement (cocaine-seeking) after a period of cocaine self-administration and extinction. To investigate whether drugs containing these pharmacological properties could act in a synergistic manner a proof-of-concept analysis was conducted using the 5-HT$_{2A}$ antagonist M100907 and the 5-HT$_{2C}$ agonist WAY163909 (Figure 6.37) in a cocaine self-administration/reinstatement assay.

Combined administration of a dose of the two drugs at concentrations where each was ineffective alone, synergistically suppressed cocaine-induced hyperactivity, inherent and cocaine-evoked impulsive action, as well as cue- and cocaine-primed reinstatement of cocaine-seeking behavior.[506]

The mechanism by which this synergism between the 5-HT$_{2A}$ receptor antagonist and 5-HT$_{2C}$ receptor agonist occurs is unknown at present. The 5-HT$_{2A}$ and 5-HT$_{2C}$ receptors share a high degree of homology, overlapping pharmacological profiles, and utilize diverse, second messenger signaling systems. The authors propose that the

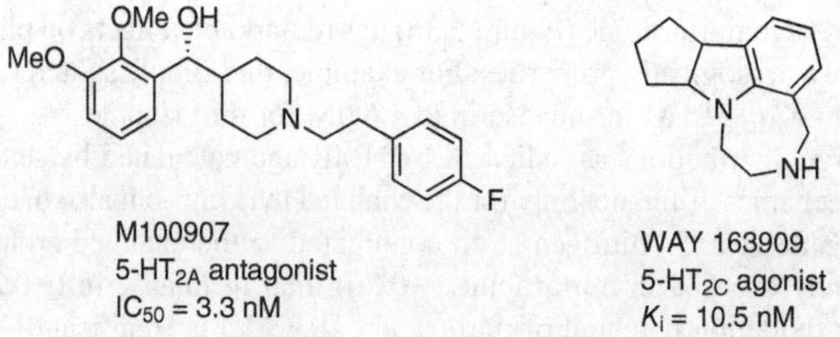

M100907
5-HT$_{2A}$ antagonist
IC$_{50}$ = 3.3 nM

WAY 163909
5-HT$_{2C}$ agonist
K_i = 10.5 nM

FIGURE 6.37 Structures of 5-HT$_{2A}$ antagonist M100907 and 5-HT$_{2C}$ agonist WAY163909

synergistic effects of M100907 and WAY163909 they discovered may be transduced through $5\text{-}HT_{2A}$ and $5\text{-}HT_{2C}$ receptor signaling in the same neurons or interacting neurons and/or separate nodes of the neural circuitry underlying these behavioral phenotypes.

6.3.19 $5\text{-}HT_{1A}$ Partial Agonist and D_3/D_4 Antagonist: Buspirone

Buspirone (Buspar®) is a clinically available, nonbenzodiazepine anxiolytic medication that acts on both serotonin and dopamine systems. Buspirone main functional activity is thought to occur by binding both presynaptically and postsynaptically to serotonin $5\text{-}HT_{1A}$ receptors ($K_i = 28$ nM) in the brain. At presynaptic receptors in the dorsal raphé, buspirone acts as a full agonist, while it acts as a partial agonist on binding to postsynaptic $5\text{-}HT_{1A}$ receptors in the hippocampus.[507] Substantial binding is also observed at the D_3 and D_4 receptors with K_i's of 98 nM and 29 nM, respectively, with moderate binding to the D_2 receptor ($K_i = 484$ nM). At both the D_3 and D_4 receptors, buspirone acts as an antagonist.[508] In addition to the $5\text{-}HT_{1A}$ and $D_{3/4}$ binding, significant binding affinity is measured at the $5\text{-}HT_{2B}$ ($K_i = 214$ nM), $5\text{-}HT_{2C}$ ($K_i = 490$ nM), $5\text{-}HT_6$ ($K_i = 398$ nM), $5\text{-}HT_7$ ($K_i = 375$ nM), α_{1A} ($K_i = 138$ nM), and σ ($IC_{50} = 263$ nM). This breadth of binding and functional activities makes buspirone a true multi-target drug.

Buspirone lacks the prominent sedative effect that is associated with more typical anxiolytics and has no significant affinity for benzodiazepine receptors and does not affect GABA binding *in vitro* or *in vivo* when tested in preclinical models. Buspirone hydrochloride tablets are rapidly absorbed in man and undergo extensive first-pass metabolism. In a radiolabeled study, unchanged buspirone in the plasma accounted for only about 1% of the radioactivity in the plasma. Following oral administration, plasma concentrations of unchanged buspirone are very low and variable between subjects. Peak plasma levels of 1–6 ng/mL have been observed 40–90 minutes after single oral doses of 20 mg. The single-dose bioavailability of unchanged buspirone when taken as a tablet is on the average about 90% of an equivalent dose of solution, but there is large variability. A multiple-dose study conducted in 15 subjects suggests that buspirone has nonlinear pharmacokinetics. Thus, dose increases and repeated dosing may lead to somewhat higher blood levels of unchanged buspirone than would be predicted from results of single-dose studies. An *in vitro* protein-binding study indicated that approximately 86% of buspirone is bound to plasma proteins.

The pharmacological effects of D_3 and D_4 antagonism on cocaine self-administration were recently examined in rhesus monkeys. Buspirone significantly reduced responding maintained by cocaine and shifted the dose–effect curve downwards. Buspirone had minimal effects on food-maintained responding. In cocaine discrimination studies, buspirone (0.1–0.32 mg/kg, intramuscular injection) did not antagonize the discriminative stimulus and rate-altering effects of cocaine in four of six monkeys.[508] These findings indicate that buspirone selectively attenuates the reinforcing effects of cocaine in a non-human primate model of cocaine self-administration, and has variable effects on cocaine discrimination.

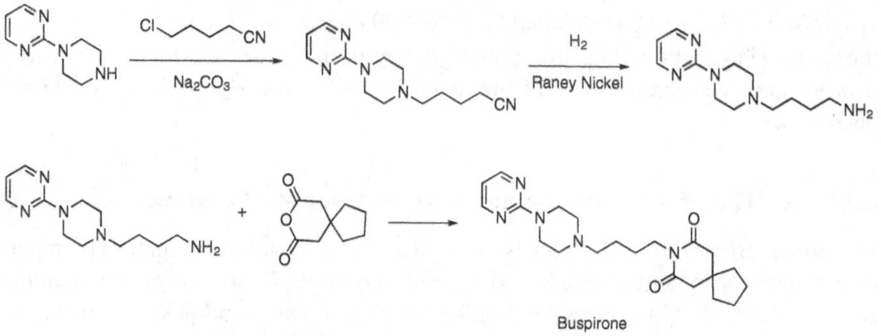

FIGURE 6.38   Synthesis of buspirone

A NIDA Clinical Trials Network study is scheduled that will use a novel two-stage process to evaluate buspirone (60 mg per day) for cocaine-relapse prevention. The study includes pilot ($N = 60$) and full-scale (estimated $N = 264$) trials. Both trials will be randomized, double-blind, and placebo-controlled and both will enroll treatment-seeking cocaine-dependent participants engaged in inpatient/residential treatment and scheduled for outpatient treatment post-discharge. All participants will receive contingency management in which incentives are given for medication adherence as evaluated by the Medication Events Monitoring System. The primary outcome measure is maximum days of continuous cocaine abstinence, as assessed by twice-weekly urine drug screens and self-report, during the 15-week outpatient treatment phase.[509]

The synthesis of buspirone is shown in Figure 6.38. It starts with N-alkylation of the piperazine pyrimidine with 5-chloropentanenitrile. Reduction in the nitrile to the amine, followed by formation of a phthalamide by condensation with the spiro-anhydride gives buspirone.

6.3.20   Antidepressants

We have seen where antidepressants have been investigated for efficacy in the treatment of drug addiction. The general rationale for the use of antidepressants is their recognized modulation of monoamine transmitters' levels. Results of small-scale clinical trials that have investigated the efficacy of commercial antidepressants as medication to treat drug addiction are equivocal, a couple of examples will be discussed.

The well-known selective serotonin reuptake inhibitor Fluoxetine (Prozac®), at 60 mg per day, was investigated in a 33-week outpatient clinical trial cocaine dependence that incorporated abstinence-contingent voucher incentives (money). The subjects ($N = 145$) were both cocaine- and opioid-dependent and were being treated with methadone. The study incorporated four conditions: fluoxetine plus voucher incentives (FV), placebo plus voucher incentives (PV), fluoxetine without vouchers (F), and placebo without vouchers (P). Fluoxtine did not prove more efficious than placebo. In fact, the placebo plus voucher incentives group had the longest treatment retention and lowest probability of cocaine use, 31%, versus a 56% probability of cocaine use for the fluoxetine plus voucher incentives group.[510]

Nefazodone, a SERT/NET reuptake inhibitor with 5-HT_{2A} antagonist activity, was examined in an 8-week double-blind, placebo-controlled trial with subjects ($n = 69$) who were cocaine-dependent and suffered from depression. Cocaine use was monitored by self-reporting and thrice-weekly urine analysis and secondary outcome measures included assessments of psychiatric functioning, cocaine craving and social functioning. The results were somewhat equivocal though nefazodone did appear to reduce craving after it had been administered for several weeks.[511]

Despite these and other studies, anecdotal clinical evidence still points toward usefulness for antidepressants. A report from physicians in Hungary that used trazodone, a serotonin antagonist and reuptake inhibitor, to treat patients with addiction and depression suggest that based on patients' reports and clinical observations, improvement of their depressive conditions and sleep problems can potentially decrease the risk of relapse of drug and alcohol dependence.[512]

6.3.21 Modafinil

Modafinil is approved for the treatment of narcolepsy under the tradename Provigil®. The drug is classified as a psychostimulant and has an abuse liability. Modafinil has a complex pharmacological profile. It binds weakly to DAT ($IC_{50} = 6.4$ μM) and NET ($IC_{50} = 35.6$ μM) but not to SERT ($IC_{50} > 500$ μM) and has been shown to occupy striatal DAT and thalamic NET sites in rhesus monkeys using PET imaging.[513] In an interesting *in vivo* 2D COSY study in rats, 600 mg/kg of modafinil administered by intraperitoneal injection was found to increase glutamate levels by 28%. Furthermore, PET studies with [^{11}C]-raclopride (D_2/D_3 selective) show that modafinil at doses of 200 mg (dosage used for treatment of narcolepsy) and 400 mg (dosage used for treatment of ADHD) elevates extracellular dopamine in humans.[514] For example, modafinil decreased [^{11}C]-raclopride binding in the nucleus accumbens by about 19% which directly reflects the increase in dopamine concentration. Of note is that these findings are similar to that reported for a 20-mg oral dose of methylphenidate (Ritalin®) in normal individuals.

Taken together it is clear that through action on DAT synaptic concentrations of dopamine can be increased leading to a potential for abuse and dependence, though at this time it appears to be less than other, more well known, psychostimulants.

Morning-dosed modafinil decreased daytime sleepiness and promoted nocturnal sleep.[515] It is suggested that drugs such as modafinil that promote healthy sleep patterns could be useful as an adjunctive medicine to help moderate withdrawal or abstinence symptoms. Compounds that enhance GABA transmission such as vigabatrin and topiramate also help normalize sleep patterns in cocaine users. In a small-scale clinical trial, modafinil, at 200–400 mg per day, attenuated smoked cocaine self-administration in frequent users.[516] Recent larger clinical trials, however, have not been promising, and it appears modafinil will not be useful as a drug to treat cocaine addiction.[517,518]

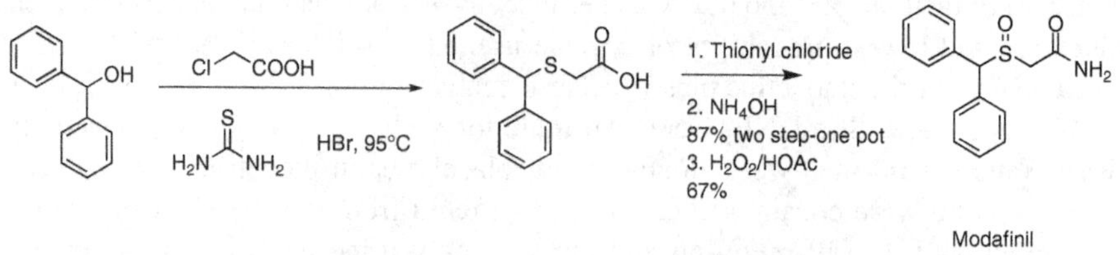

FIGURE 6.39 Synthesis of modafinil

Modafinil has proven an interesting to drug to work with perhaps in part from its easy availability and known pharmacokinetics. From a synthetic chemists' point of view, it is an attractive molecule to work with, as the synthesis is straightforward as shown in Figure 6.39 with several points of diversity readily amendable to modification. The compound is readily synthesized by reacting benzhydrol with chloroacetic acid and thiourea in the presence of hydrobromic acid to produce the diphenylmethylthio acetic acid. This is then converted to the acid chloride that reacts with ammonia to produce the amide.

The sulfur atom of sulfoxides is a chiral center so modifinil exists as a pair of enantiomers that are shown in Figure 6.40. Both enantiomers have been prepared and analyzed.[519] The enantiomers were resolved by reacting the racemic carboxylic acid with α-methylbenzylamine to form a mixture of diastereomeric salts. The diastereomers were then separated by fractional crystallization and the resolved salts converted to the free acid. The more elegant method of enantioselective oxidation was tried but unfortunately failed.

The (−)-R enantiomer binds about twice as strong to DAT than the (+)-S enantiomer with K_i's of 3.26 μM and 7.64 μM, respectively. No binding at SERT and NET was detected up to concentrations of 10 μM. Both enantiomers produce locomotor activity in mice in a dose-dependent fashion though both are less potent then cocaine. The (+)-S enantiomer appears to be slightly less potent than the (−)-R enantiomer.[520]

Modafinil has a clogP of 0.937 yet is very water insoluble at < 1 mg/mL. Few medicinal chemistry SAR studies have been reported. In one recent study, the effects of *p*-halo-substitution, sulfoxide removal, and modification of the primary amide

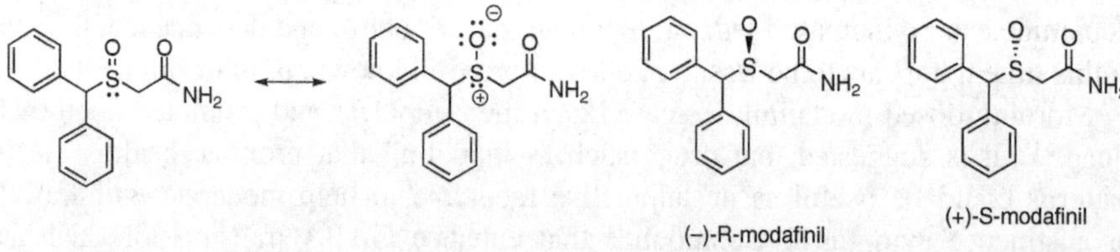

(−)-R-modafinil

(+)-S-modafinil

R

S

FIGURE 6.40 Enantiomers of modafinil

group on binding to monoamine transporters was examined.[520] Bisbromo substitution at the para position of the aryl rings increased DAT binding, while replacement of the amide by a 3° *N,N*-propylphenyl group increased binding to all three DAT, SERT, and NET receptors.

6.3.22   VMAT2 Inhibition: Lobeline

The natural product lobeline (Figure 6.41) has shown promise as a potential medication for treating methamphetamine abuse. The compound is isolated from the plant *Lobelia inflata* that is indigenous to the eastern United States and Canada. Native Americans first discovered its medicinal uses and extracts were used in the early 1900's to restore breathing after shock.[388] Its pharmacological effects are similar to nicotine so lobeline was first investigated with regard to the nAChRs. This compound is fairly promiscuous acting as an μ-opioid receptor antagonist (K_i = 740 nM),[521] a $\alpha_4\beta_2$ nAChR antagonist (K_i = 4 nM), α_7 nAChR antagonist (K_i = 6.3 μM), VMAT2 inhibitor (K_i = 2.8 μM), SERT inhibitor (K_i = 47 μM), and a DAT inhibitor (K_i = 28 μM).[522]

The first SAR study of lobeline was reported by Glennon who was interested in the importance of the keto and hydroxyl group on binding to nAChRs.[523] This was followed by extensive SAR studies by Crooks and Dwoskin.[524] The SAR studies of Crooks focused on phenyl substitution and modification of the ethylene linker. Many of the analogs prepared for the SAR study were derived by modification of lobeline. Other compounds were generally derived from 2,6-disubstituted pryridines as outlined in Figure 6.41. The 2,6 positions are modified, the methyl pyridinium salt is made and then sequentially reduced with sodium borohydride and palladium/carbon hydrogenation to generate the piperidine. As with xanomeline, presumable the hydride reduction stops at the tetrahydropyridine stage and catalytic hydrogenation is required to produce the piperidine ring. Alternatively, the pyridine ring can be exhaustively reduced to the piperidine with hydrogen and platinum oxide.

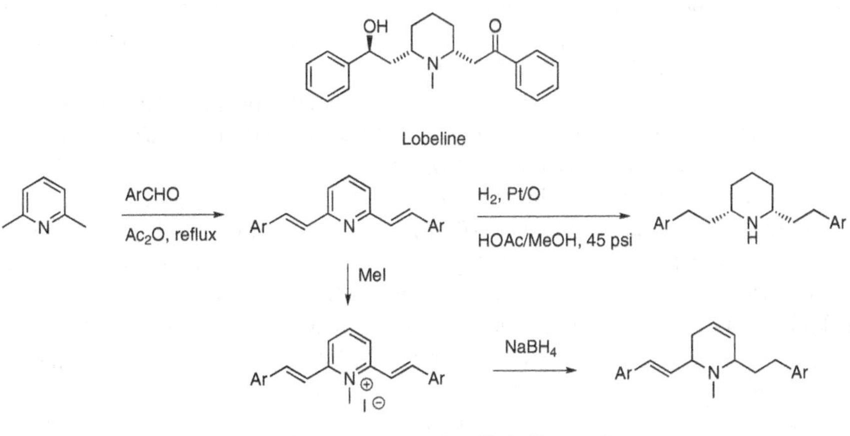

FIGURE 6.41   Synthesis of lobeline analogs

The SAR study revealed that the piperdine nitrogen and *cis* stereochemistry are critical for VMAT2 binding but the presence of the hydroxy and keto groups is not required.

Lobeline is not self-administered by rodents, it attenuates methamphetamine self-administration, and increasing the dose of methamphetamine does not surmount attenuation.[525,526] It has been reported that lobeline was in Phase I clinical trials for the treatment of methamphetamine abuse (e.g., NCT00439504), but no results have been published and it is not clear if the study took place. Lobeline has been investigated as a potential treatment for smoking cessation. The compound itself appears to have no major adverse effects but a recent Cochrane Analysis concluded it is not effective in causing smoking cessation.[527]

6.3.23   DAT/NET Inhibition and nAChR Antagonist: Bupropion

Bupropion, a DAT inhibitor, was tested for efficacy in increasing the weeks of abstinence in methamphetamine-dependent patients ($n = 151$).[528] Seventy-two patients were randomized to placebo and 79 to sustained-release bupropion (150 mg twice daily). It was found that bupropion was effective in increasing the number of weeks of abstinence in patients with low-to-moderate methamphetamine dependence. No further clinical trials for stimulant addiction are listed on clinicaltrials.gov as of February 2014. There is interest in if bupropion combined with varenicline (Chapter 8) affects alcohol consumption (NCT00580645). Bupropion is approved for the use of smoking cessation so the chemistry of bupropion will be discussed in Chapter 8.

6.3.24   GABA Uptake Inhibitor: Tiagabine

Tiagabine (Gabitril®) is a GABA uptake inhibitor that was developed by Cephalon and is used as an anticonvulsant to treat partial seizures. By blocking the GAT-1 transporter, GABA uptake into presynaptic neurons is prevented, allowing more GABA to be available for receptor binding on the surfaces of postsynaptic cells. According to the Gabitril medication guide, tiagabine also binds to the histamine H1, 5-HT$_{1B}$, benzodiazepine, and chloride channel receptors at concentrations 20–400 times those inhibiting the uptake of GABA.

The drug arose from studies in the 1980s to develop lipophilic derivatives of the GABA uptake inhibitor nipecotic acid. Nipecotic acid (piperidine-3-carboxylic acid) proved to be an efficient inhibitor of GABA transporters, but it was too polar to cross the blood–brain barrier following peripheral administration. Building on work from Smith-Kline French and Parke-Davis/Warnet-Lambert, Novo Nordisk reported the discovery of tiagabine (NNC 05-0328) in 1993.[529] Tiagabine proved to be a potent and bioavaliable anticonvulsant that inhibited GABA uptake with an IC$_{50}$ = 67 nM and inhibited chemo-induced seizures in mice with an ED$_{50}$ = 1.2 mg/kg. Tiagabine is highly selective for the GABA transporter 1 (GAT-1) with IC$_{50}$'s of > 917 μM for GAT-2, GAT-3, and GAT-4.[530]

The original synthesis is shown in Figure 6.42 starting from substituted 3-methylthiophenes.

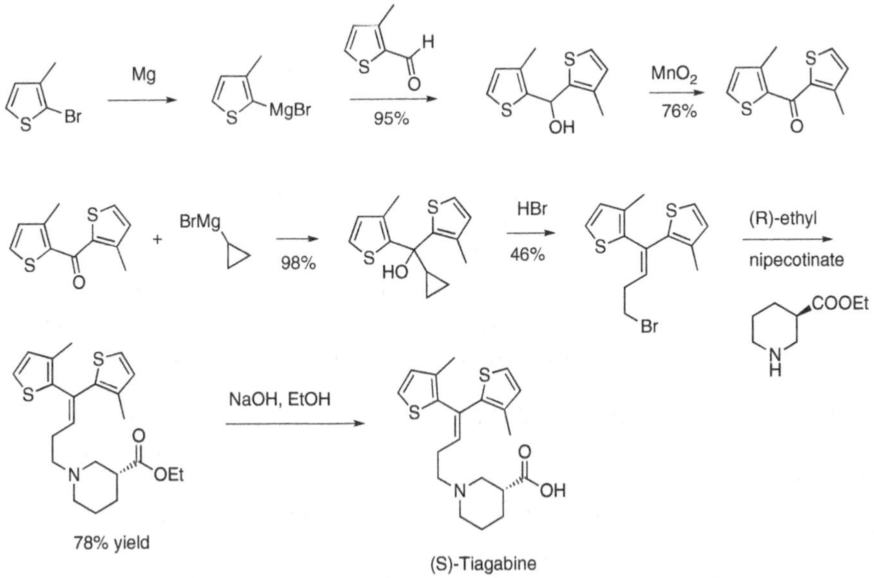

FIGURE 6.42   Synthesis of tiagabine

The results of clinical trials with tiagabine are mixed. Two small-scale (45 and 76 subjects, respectively) placebo-controlled clinical trials indicated that tiagabine at a dosage of 24 mg per day produced a moderate reduction (~30%) in cocaine use in methadone-treated cocaine addicts, while other studies demonstrated that the same doses of tiagabine neither altered the acute effects of cocaine, nor lowered cocaine use in cocaine addicts.[395,531]

6.3.25   GABA_B Agonist: Baclofen

Baclofen (Figure 6.43) is a selective $GABA_B$ receptor agonist (IC_{50} = 32 nM) that acts on the spinal cord nerves and decreases the number and severity of muscle spasms caused by multiple sclerosis or spinal cord diseases. It also relieves pain and improves muscle movement. Baclofen was first synthesized in 1962 at Ciba and

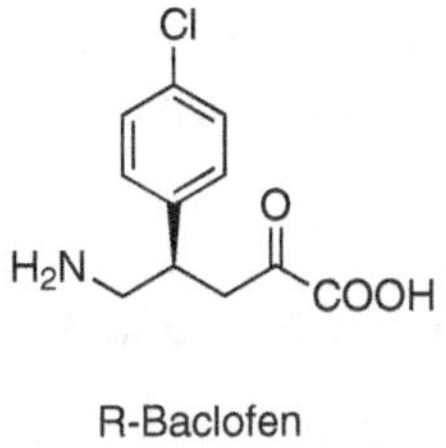

R-Baclofen

FIGURE 6.43   Baclofen

played an important role in the discovery of the $GABA_B$ receptor. During studies in tissue preparations, it was found that baclofen mimicked, in a stereoselective manner, the effect of GABA in these systems. Furthermore, ligand-binding studies provided direct evidence of distinct binding sites for baclofen on central neuronal membranes. The term $GABA_B$ was used to distinguish this site from the bicuculline-insensitive receptor, which was, in turn, designated $GABA_A$.

Modulation of GABA levels appears to be beneficial in the treatment of addiction. Elevation of endogenous GABA levels in the mesolimbic area by administration of vigabatrin, an inhibitor of GABA metabolism, or the GABA uptake inhibitor NO-711, attenuates heroin and cocaine self-administration in rats and prevents cocaine-induced increases in dopamine in this brain region. It has also been reported that gabapentin, a putative $GABA_B$ receptor agonist, reduces the craving for cocaine in dependent adults.[532]

There is substantial preclinical evidence that baclofen could be effective in the treatment of different aspects of cocaine addiction.[533] Clinical trials however have not been promising.

Subjects with cocaine dependence ($n = 35$) were given baclofen (20 mg, t.i.d) and compared with a placebo group ($n = 35$) in a 16-week double-blind screening trial.[534] The outcomes examined were study retention, cocaine use, cocaine craving, and adverse side effects. The participants were also required to attend behavioral drug counseling sessions three times a week. It was found that the group receiving baclofen had a greater retention rate, but there were no other significant differences relative to the placebo group. As is often typical in drug addiction trials there was poor retention rate, only 26% of patients receiving baclofen and 23% of the placebo group completed the study. It should be noted this is, unfortunately, not unusual. The major reason is the participant simply not showing up. There was slight increase in the number of participants reducing cocaine consumption based on drug screening of the urine for the cocaine metabolite benzoylecgonine. Of the participants taking baclofen, 64% tested negative for cocaine use while for those who did not 52% tested negative for cocaine use.

A much larger study ($n = 160$) sponsored by NIDA and Department of Veterans Affairs concentrated on nonabstinent cocaine patients in a multisite study in the United States.[535] Subjects were given up to 60 mg per day of baclofen for an 8-week treatment period. In short, no significant differences were observed in the reduction in cocaine use.

Baclofen is readily prepared from p-chlorbenzaldehyde by several methods that generally following those first presented in patent US3471548. Metal-catalyzed asymmetric syntheses have been developed; a recent one involves Pd-catalyzed allylic alkylation of cinnamyl methyl carbonate with nitromethane as the key asymmetric step.[536]

6.3.26   Combination Therapy: Baclofen and Amantadine

Amantadine (see section 5.4.2) is a multimodal drug that appears to affect its influence at monoamine transporters and NMDA receptors. Amantadine (Symmetrel®)

is approved for use in the treatment of Parkinson's disease as an agent to increase levels of dopamine in the brain by potentiating dopamine release from neurons and by blocking the reuptake of dopamine.[537]

Amanatadine has been investigated in several small clinical trials for its affect on cocaine use. The most recent published study looked at the effects of amantidine (100 mg t.i.d) in combination with baclofen (30 mg t.i.d) on cocaine use in eight male participants.[538] There appeared to be reduction in the "desire" for cocaine but no difference in the intensity of the cocaine high or the likeihood of using cocaine if given access. The combination of cocaine (up to 40 mg i.v.), amantidine, and baclofen had no significant effect on cardivascular variables. No recent clinical trials using a combination of baclofen and amantadine are reproted in Clinical Trials.gov, so this combination therapy does not appear to be of current interest.

6.3.27 GABA$_A$ Agonist and AMPA Antagonist: Topiramate

Topiramate is a positive modulator of GABA$_A$ receptors (acting at nonbenzodiazepine sites) and is used for the treatment of epilepsy. In addition, topiramate has other pharmacological actions, including antagonism of AMPA/kainate glutamate receptors, inhibition of voltage-gated sodium and calcium channels and inhibition of carbonic anhydrase.[395]

Topiramate was investigated in a 13-week study for the reduction in methamphetamine use. The study involved escalating doses from 50 mg per day to the target maintenance of 200 mg per day in weeks 6–12. The primary outcome was abstinence from methamphetamine use during weeks 6–12. Topiramate was found to be safe and well tolerated, but it did not appear to promote abstinence in methamphetamine users. It appeared however to reduce the amount taken and reduce relapse rates in those who are already abstinent.[539] The use of topiramate for the treatment of alcohol addiction looks more promising so the chemistry of the compound will be discussed in Chapter 7.

6.3.28 α$_{2A}$ Adrenergic Agonist: Clonidine

The use of clonidine has been investigated for the treatment of cocaine addiction for reasons similar to those that were mentioned for opioid addiction. A study involving 59 nontreatment-seeking cocaine users showed that clonidine at 0.1 and 0.2 mg per day orally could attenuate craving due to stress, while the 0.2 mg dosage attenuated craving due to visual cues.[540] The stress test used involved listening to an audiotape for 60 seconds describing sitting in a dentist's waiting room anticipating a painful procedure. The doses of clonidine used are similar to those used to treat high blood pressure and opioid withdrawal.

6.3.29 β-adrenergic Antagonist: Propranolol

Propranolol is a nonselective β-adrenergic antagonist (β-blocker) that was developed for the treatment of angina and hypertension. The compound has an important role in

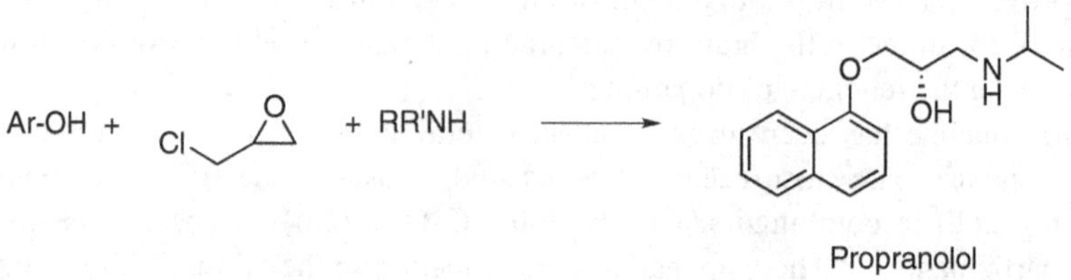

FIGURE 6.44   Construction of propranolol

the history of medicinal chemistry as the first β-blocker that was developed and generated a Nobel Prize for James Black. Propranolol belongs to the aryloxypropanolamine class of compounds from which several first-generation β-blockers were developed. The aryloxypropanolamine chemotype was a popular target of combinatorial chemistry libraries as it can be readily constructed from a three-component reaction involving a phenol, epichlorohydrin, and an amine as shown in Figure 6.44.

Propranolol has shown some benefit in reducing anxiety that is experienced by cocaine-dependent individuals undergoing withdrawal. The affects of propanolol have been suggested to reduce the somatic effects of anxiety (tremors, palpitations, etc.) versus a neural centrally active affect.[541]

The effects on cocaine abstinence are mixed.[542] A recently completed trial involving 50 patients studied the effects of propranolol on memory associated with drug cues (ClinicalTrails.gov Identifier: NCT00830362). The patients were exposed to a cocaine cue and then given 40 mg of propranolol or a placebo. Next day, they were again exposed to the cocaine cue and their craving and physiological reactivity was measured. Propranolol was able to reduce the craving of cocaine relative to the placebo. Propranolol may be useful then as an adjunctive medication. In terms of the treatment of drug addiction, antagonism of β-adrenergic receptors in general does not appear to be a useful strategy though perhaps in combination with other drugs some benefit may be derived.

6.3.30   Cannabinoid Antagonists: AM251

The rational for use of cannabinoid antagonists for the treatment of cocaine addiction has been explained by Xi as follows. The mesolimbic dopamine and the downstream nucleus accumbens GABAergic transmission have been thought to underlie cocaine reward and addiction. Growing evidence suggests that similar mechanisms may also underlie the action produced by Δ^9-THC or marijuana. It was reported that Δ^9-THC elevates extracellular DA in the nucleus accumbens. This action could be mediated by a GABAergic mechanism, that is, Δ^9-THC may initially activate CB_1 receptors located on GABAergic interneurons in the VTA and produce a decrease in GABA release, which subsequently disinhibits (or activates) VTA dopamine neurons. In addition, CB_1 receptors are also highly expressed on presynaptic glutamatergic terminals in the nucleus accumbens. Thus, activation of CB_1 receptors located on glutamatergic terminals decreases glutamate inputs onto medium-spiny GABAergic neurons in the nucleus accumbens and decrease GABA release in their projection areas, for example, the VTA. Further, CB_1 receptors are also expressed on striatal GABAergic neurons, and activation of the CB_1 receptors produces a direct inhibitory effect on medium-spiny GABAergic neurons and decreases GABA release in the ventral pallidum and the VTA. Finally, cocaine or dopamine has been shown to increase endocannabinoid release in the striatum, which subsequently increases endocannabinoid binding to CB_1 receptors located on presynaptic glutamatergic terminals and postsynaptic GABAergic neurons. Taken together, activation of CB_1 receptors located on both GABAergic and glutamatergic neurons causes an increase in nucleus accumbens dopamine and a decrease in GABA release in both the VTA and ventral pallidum. This decrease in GABAergic transmission constitutes a final common pathway underlying drug reward and addiction. Accordingly, blockade of CB_1 receptors in both the VTA and nucleus accumbens would attenuate the actions of cocaine on nucleus accumbens dopamine and ventral pallidum GABA release and therefore attenuate cocaine reward and addiction.[395]

A few studies have indicated that CB_1 antagonists may be useful for the treatment of methamphetamine addiction. In particular, the selective CB_1 antagonist AM 251 at 1 mg/kg (i.v.) reduced methamphetamine self-administration and prevented reinstatement of extinguished methamphetamine seeking in rhesus monkeys.[543] At this dosage, no effect on food responding was observed. Rats were trained for cocaine self-administration and then underwent extinction for 14 days. When the mice were exposed to cocaine-associated cues or cocaine itself, it was found that the CB_1 receptor antagonist trimonabant attenuated relapse. However it did not attenuate relapse to cocaine self-administration when the mice were exposed to stress.[544]

As a CB_1 antagonist, it would expected that AM251 would have similar side effects as rimonabant (see Chapter 9) so it has not been submitted for human clinical trials.

6.3.31   CRF$_1$-Antagonists: CP-154,526

A CRF_1-selective antagonist developed by Pfizer (CP-154,526) has been studied in some rodent models of addiction. The drug attenuated both cue-induced and methamphetamine-induced reinstatement of methamphetamine-seeking behavior in mice.[545] These results are complex though as cue-induced reinstatement was attenuated only by intracerebroventricular administration of CP-154,526, while intraperitoneal administration was sufficient to attenuate methamphetamine-induced meth-seeking behavior.

The discovery of CP-154,526 was initiated by a high-throughput screening campaign that detected the ability of compounds to inhibit binding of radiolabeled oCRF (^{125}I-oCRF) to the human CRF receptor in membranes prepared from IMR32 cells, a human neuroblastoma cell line. This effort yielded a low affinity hit (800 nM) that served as a lead for subsequent optimization. Directed synthesis of the lead then resulted in a series of novel pyrrolo[2,3-d]pyrimidines exemplified by CP-154,526. CP-154,526 bound to CRF receptors in IMR32 cells with a K_i of 2.7 nM and showed similar high affinity for cerebral cortical and pituitary sites labeled by ^{125}I-oCRF in

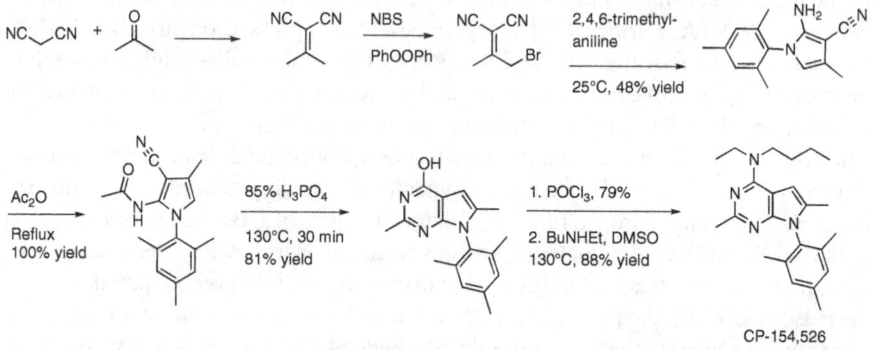

FIGURE 6.45 Synthesis of CP-154,526

multiple species. Binding assays against CRF_1 and CRF_2 singly expressed in CHO cells showed the compound is completely selective for the CRF_1 receptor with a $K_i = 2.7$ nM versus >10 µM for CFR_2.[546]

The synthesis of CP-154,526 as reported in the original SAR studies is shown below in Figure 6.45 Key steps include the Paal-Knorr synthesis of the pyrrole and the acid-catalyzed cyclization between the acetamide group and the nitrile to form the fused pyrimidine-pyrrole heterocycle core.[547]

6.3.32 Trace Amine-Associated Receptor 1

Trace amine-associated receptor 1 is a new potential GPCR target. The receptor was first reported in 2001 and binds to what are known as trace amines. Trace amines are similar in structure to the monoamine neurotransmitters (e.g., a trace amine is β-phenethylamine) but are present at much lower concentrations than the monoamine neurotransmitters. It is estimated that trace amines are present throughout the brain in the low nM range.[548] Their exact function is an open question, though studies in gene knockout mice show they have a role in the regulation of dopamine neuronal activity.[549] Their interest in addiction is that amphetamines act as agonists upon binding to TAAR1 with an $EC_{50} = 0.6$ µM for (S)-amphetamine and an $EC_{50} = 1.5$ µM for (S)-methamphetamine.[394] What exact role trace amine associated receptor 1(TARR1) has in addiction remains to be determined but it does present some potential new opportunities.[550] Development of agonist and antagonist ligands for TAAR1 seem to have perked the interest of the pharmaceutical industry. Several papers have been recently generated from Hoffmann-La Roche investigating substituted imidazole as TARR1 agonists, though the therapeutic interest appears to be for schizophrenia.[551–553]

7

MEDICATION DEVELOPMENT FOR DEPRESSANT ADDICTION

Substances of addiction that are classified as CNS depressants include alcohol, benzodiazepines, and barbiturates. The latter two are mainly prescribed for anxiety, insomnia, and generalized anxiety disorder (GAD). There is little to discuss relative to benzodiazepines and barbiturates so the majority of the discussion will be on alcohol. There does not perceive to be a need for pharmacological treatment of benzodiazepine or barbiturate addiction.

7.1 PHARMACOLOGY OF ALCOHOL ADDICTION

The abuse of alcohol (ethanol) appears to be controlled by several neurotransmitter systems. Ethanol is thought to exert its actions at the synapse and affect synaptic transmission. Ethanol is known to be a $GABA_A$ agonist where it affects its activity in concert with GABA as a positive allosteric modulator, opening the ion channel and increasing the flow of chloride ions from the extracellular fluid into the neuron. This hyperpolarizes the cell, inhibiting an action potential from developing. The exact location where ethanol binds to the $GABA_A$ ion channel is unclear.

The $GABA_A$ ion channel is located on GABAergic neurons in the ventral tegmental area of the limbic system. With acute alcohol use, *increased* chloride influx and neuron hyperpolarization occurs as described above, however, chronic alcohol use actually *decreases* chloride influx, ultimately resulting in a new homeostatic physiological state. With prolonged use of alcohol, the effects of both alcohol and GABA on the $GABA_A$ receptor are reduced.

Drug Discovery for the Treatment of Addiction: Medicinal Chemistry Strategies, First Edition. Brian S. Fulton.
© 2014 John Wiley & Sons, Inc. Published 2014 by John Wiley & Sons, Inc.

Upon alcohol withdrawal, the glutaminergic system becomes hyperpolarized; it is as if the body compensates for the lack of alcohol by going from a hypoglutaminergic state to the opposite. This is considered to be due to changes in receptor concentrations and disruption of the second messenger system caused by the presence of alcohol. This change to a hyperglutaminergic state is thought to be responsible for the sensation of craving, hence relapse. In all, there is a wide and opposing functional state of the glutaminergic system ongoing from acute (social) to chronic (addiction) use of alcohol to withdrawal.

Ethanol also binds to the glutamate-NMDA ionotropic receptor where, in concert with glutamic acid, acts as a negative allosteric modulator. The effect is a decrease in cellular calcium and sodium concentrations. Glutamate-NMDA function is inhibited by ethanol in a concentration-dependent manner over the range of 5-50 mM. This is also the concentration range that produces intoxication and that is linearly related to the intoxication potency. This suggests that ethanol induced inhibition of responses to the NMDA activation may contribute to the neural and the cognitive impairments associated with alcohol intoxication. However, the mechanism(s) of ethanol interference on NMDA receptor function remains in question.[554]

PET studies using [^{11}C]-raclopride show that acute alcohol use increases the levels of dopamine in the ventral striatum (nucleus accumbens) of young adults.[555] Perhaps further complicating treatment strategies, the correlation of alcohol use and dopamine release on subjective effects was greater in men than in women.

Complicating the direct role of alcohol in addiction is that the primary metabolite of ethanol, acetaldehyde has been implicated as a mediator of the rewarding and reinforcing effects of ethanol. Acetaldehyde is able to affect dopamine neurotransmission, increasing neuronal firing in the ventral tegmental area, thus stimulating dopamine release in the nucleus accumbens.[556,557]

Approved drugs for the treatment of alcohol addiction are disulfiram (Antabuse®), acamprosate (Campral®), and naltrexone (Vivitrol®, Depade®, Revia®). Some physicians prescribe anxiolytic drugs such as librium, valium, and phenobarbital for inpatient withdrawal treatment and oxazepam or clonazepam for outpatient withdrawal treatment.[322] Clinical information on alcohol abuse and treatment can be found in the National Institute on Alcohol Abuse and Alcoholism (NIAAA) publication: *Helping Patients Who Drink Too Much A Clinicians Guide.*

7.2 APPROVED MEDICATIONS

7.2.1 Disulfiram—Aldehyde Dehydrogenase Inhibitor

Disulfiram, (Et$_2$NCSS)$_2$, is a negative reinforcer that produces disagreeable side effects when alcohol is consumed. Its use and effectiveness is considered limited due to poor patient compliance unless given in a monitored fashion such as in a clinic. As shown in Figure 7.1, disulfiram acts as a prodrug that on reduction by glutathione reductase in the blood produces diethyldithiocarbamic acid (Et$_2$NCSSH), an irreversible inhibitor of aldehyde dehydrogenase (ALDH).

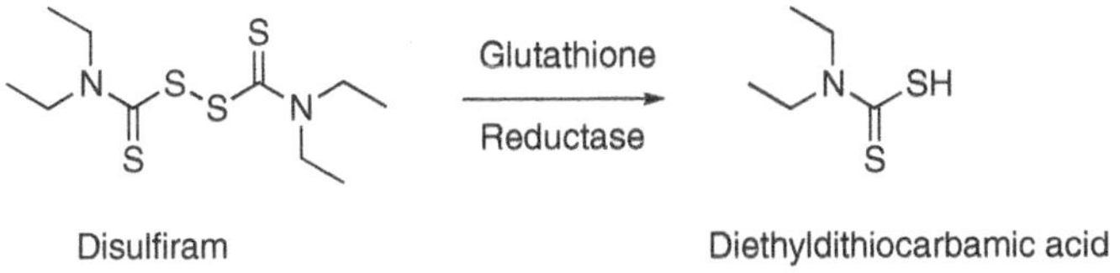

FIGURE 7.1   Enzymatic reduction in disulfiram

The normal course of ethanol metabolism is first the oxidation to acetaldehyde by alcohol dehydrogenase (ADH) followed by further oxidation to acetate by ALDH. Acetaldehyde is an extremely toxic compound that upon ethanol oxidation occurs in blood concentrations in the low picomolar range. The buildup of acetaldehyde is what is responsible for the "hangover" effect. It has been suggested that the cytotoxicity of acetaldehyde mainly results from its involvement in lipid peroxidation, generation of highly reactive free radicals, inactivation of various enzymes, or by irreversible binding to proteins and other cells constituents, with impairment of cell membrane functions.

Two isoforms of ALDH exist. ALDH1 is a mitochondrial enzyme, with K_m's of approximately 1 pM for acetaldehyde and 100 pM for its required cofactor nicotinamide adenine dinucleotide (NAD). ALDH2 is a cytosolic soluble enzyme with a K_m's of approximately 100 pM for acetaldehyde and 10 pM for NAD. Disulfiram, or strictly speaking its metabolites, preferentially inhibit ALDH1 *in vivo* by an irreversible mechanism. Acetaldehyde rapidly accumulates intracellularly, because its formation is much faster in the preceding ADH catalyzed rate-limiting step than its oxidation to acetate by ALDH.[558]

In vitro, the inactivation of mitochondrial ALDH is pH-dependent with maximum inhibition at pH 9-10. At pH 7.5, the pseudo-first-order rate constant for disulfiram inhibition of ALDH is $k_{app} = 5000$ min^{-1}.[559]

The compound is readily synthesized as shown in Figure 7.2 by reacting diethylamine with carbon disulfide to make a thiocarbamate that undergoes oxidative dimerization to form disulfiram.

Despite all the sulfur atoms, the compound is described as a light yellow odorless crystalline powder. Disulfiram has limited solubility in water (0.3 mg/mL).

As mentioned in Section 6.3.3, the reduction product diethyldithiocarbamic acid forms nicely colored complexes with metals, mainly copper, upon which blue-tinted complexes are formed. At one time, these were the basis of colorimetric assays.[560]

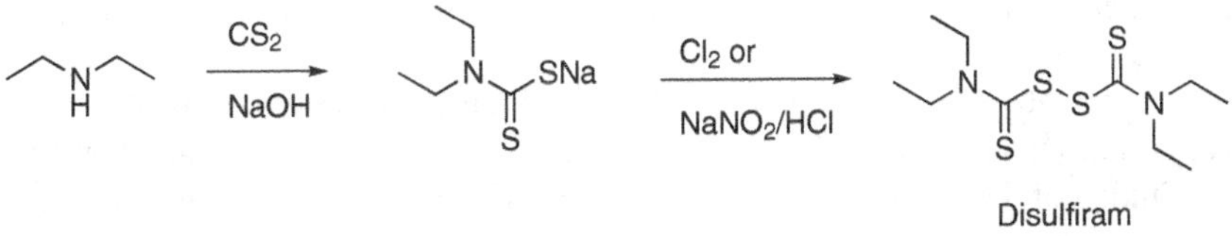

Disulfiram

FIGURE 7.2   Synthesis of disulfiram

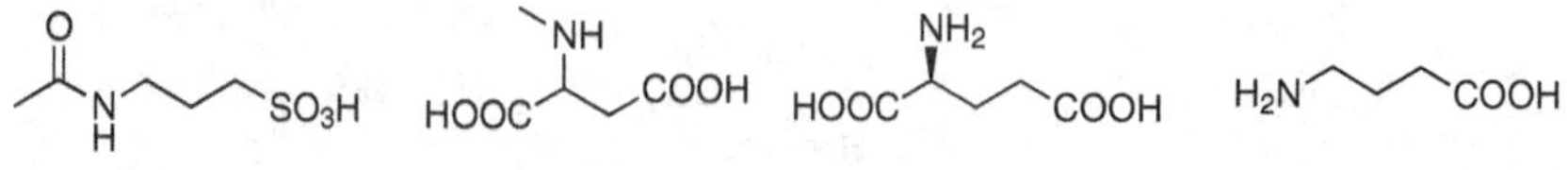

Acamprosate   NMDA   Glutamic acid   GABA

MR: 39.26 (cm^3/mol)   MR: 30.74 (cm^3/mol)   MR: 30.33 (cm^3/mol)   MR: 24.53 (cm^3/mol)
tPSA: 83.47   tPSA: 86.63   tPSA: 100.62   tPSA: 63.32
CLogP: −2.465   CLogP: −1.6556   CLogP: −2.6938   CLogP: −2.771

FIGURE 7.3   Structures of acamprosate and GABA analogs

7.2.2   Acamprosate—NMDA Antagonist and GABA$_A$ Agonist

Acamprosate (Cambral®) was approved by the FDA in 2004 for use in the maintenance of alcohol abstinence (relapse prevention). Its mechanism of action is unclear.
It is proposed to help modulate the glutaminergic and GABAergic systems by acting
as an NMDA antagonist and GABA$_A$ agonist.[561] This profile is similar to ethanol
itself. It substitutes for ethanol in humans and attenuates cue-induced craving in
alcohol-dependent patients. It acts to modulate or "smooth" the profound fluctuations of the glutaminergic system during alcohol abuse. There is increasing evidence
though that part of the activity of acamprosate may be due to its action as an mGlu$_5$
antagonist.[562] This may question if acamprosate is actually an NMDA antagonist. As
can be seen in Figure 7.3, it is a small molecule with structural and physicochemical
similarity to GABA-like neurotransmitters.

Acamprosate is not addictive and appears to have no potential for abuse; patients
maintained on the drug have developed no known tolerance for or dependence on
it. It also carries little overdose risk. Even at overdoses up to 56 g (a normal daily
dose is 2 g), acamprosate was generally well tolerated by patients. Acamprosate
is not metabolized by the liver and is excreted primarily from the kidneys. Since
acamprosate is not metabolized in the liver, individuals with liver disease, unlike
disulfiram and naltrexone, can use it. The recommended dosage of Campral is two
333-mg tablets three times a day, with or without food. It is recommended that
treatment with acamprosate should be initiated as soon as possible after alcohol
withdrawal and should be maintained if the patient relapses. Treatment duration at
this dosage ranged from 3 to 12 months in clinical trials and should be maintained
for 1 year. As always for treatment of addiction, individuals taking acamprosate
should be expected to participate fully in a treatment program's activities, including
attending 12-step or mutual-help group meetings. In addition, they may need ongoing
motivational counseling specifically geared to helping them comply with a drug
regimen.[563]

Acamprosate is readily prepared as shown in Figure 7.4 by converting the terminal
hydroxyl group of 3-amino-1-propanol into the chloro group using thionyl chloride.
Substitution of the chloro group using sodium sulfite by the Strecker Synthesis gives
3-aminopropanesulfonic acid. Acetylation of the amine with acetic acid and calcium
hydroxide produces acamprosate as the calcium salt, $Ca^{2+}(^-OSO_2(CH_2)_3NHAc)_2$.

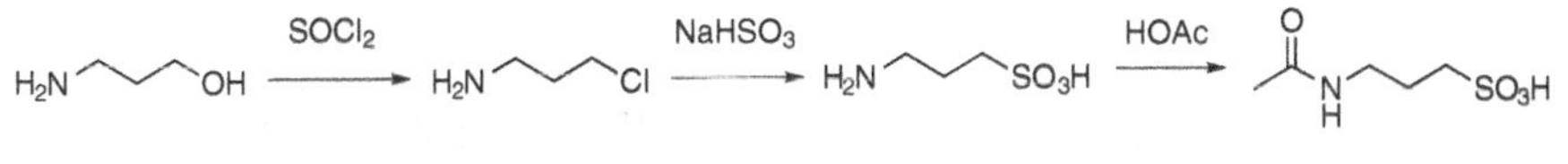

FIGURE 7.4 Synthesis of acamprosate

7.2.3   Naltrexone—Opioid Antagonist

Under homeostatic conditions, GABA released from GABA neurons binds to GABA receptors on dopamine neurons located in the ventral tegmental area. GABA binding inhibits the firing of dopamine neurons and the release of dopamine into the nucleus accumbens. Alcohol causes the release of endogenous peptide opioids that can bind to μ-opioid receptors located on the GABA neurons in the VTA. The binding of these agonist peptides (as while as morphine) to the μ-opioid receptor on the GABA neurons results in the blockade of GABA release, hence disinhibition of DA release, increasing the levels of dopamine in the nucleus accumbens. Antagonism of the μ-opioid receptor will prevent this sequence of events ultimately decreasing the levels of dopamine by allowing GABA to inhibit dopamine release. Alcohol has profound effects on the opioid system, administration of alcohol will increase the levels of β-endorphin (μ-opioid endogenous peptide) in the nucleus accumbens in rats bred to be alcohol-preferring while not in alcohol-avoiding rats.[564] Chronic ethanol exposure tends to down-regulate positive aspects of the opioid reward system and up-regulate negative aspects.

A genetic component has also been uncovered. The gene OPRM1 codes for the μ-opioid receptor. An allele exists where the more common asparagine Asn40 (A allele) amino acid is replaced by an aspartate Asp40 (G allele) amino acid. The G allele binds β-endorphin three times more strongly than the A allele. Male heavy drinkers that possessed the G allele had significant more craving for alcohol when exposed to a visual cue (beer).[565] It is suggested that those with a G allele may be at risk for alcohol-related problems by having a predisposition to strong cravings to drink alcohol.

The first clinical trials investigating whether naltrexone could be used to treat alcoholism occurred in 1983 at Philadelphia, VA, Medical Center and were further expanded and reported in 1992, simultaneously with a report from Yale University.[566] The FDA approved the opioid antagonist naltrexone (Revia®) in 1994 for the treatment of alcoholism. Naltrexone binds strongly to the μ-opioid receptor and the κ-opioid receptor with lesser binding to the δ-opioid receptor. More recently Vivitrol™ was approved in 2006 as an injectable extended-release formulation of naltrexone. The prescribed dosage of oral naltrexone is 50 mg per day. An imaging PET study investigated the effects of 50 mg naltrexone on the occupancy of the μ-opioid and δ-opioid receptors in recently abstinent alcohol-dependent subjects.[567] The level of μ-opioid receptor occupancy was measured using the displacement of [¹¹C]-carfentanil from the receptors while the displacement of [¹¹C]-methyl naltrindole was used to measure the occupancy of the δ receptors. The results were striking. Naltrexone clearly displaced all the [¹¹C]-carfentanil from the μ-opioid receptors while at the δ-opioid receptor 21% (mean) of [¹¹C]-methyl naltrindole was displaced.

Mean (^{11}C)carfentanil BP images in 21 alcoholics

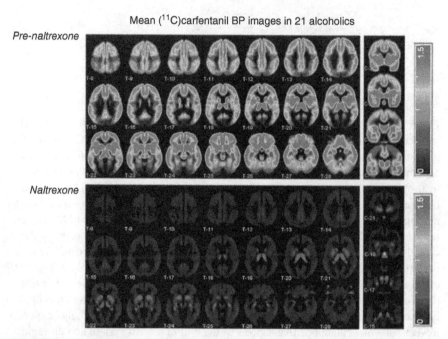

FIGURE 7.5 Binding of naltrexone in human brain. Mean images of the distribution of μ-opioid receptors in the brain of 21 alcoholics after IV administration of [^{11}C]-carfentanil during scans conducted pre-naltrexone treatment (top panel) and during naltrexone treatment (bottom panel). Images in e-book edition are color-coded according to the scale shown (0–1.5) so that highest concentrations of the radiotracer are represented by red and lowest concentrations by black/purple (Reprinted with permission from Weerts et al.[567] Copyright © 2008 Nature Publishing Group)

Images from this study are shown in Figure 7.5, in part to illustrate what is observed during the imaging studies. We can see that before the naltrexone is administered, many parts of the brain contain the radiolabel indicated by the light-colored regions. Naltrexone is delivered with practically complete disappearance of the radiolabel signal. This occurs by naltrexone binding to the opioid receptors, surmounting the binding of [^{11}C]-carfentanil and displacing it from the receptors. The study clearly shows that at the therapeutic dose of 50 mg per day, naltrexone is strongly interacting with opioid receptors in the brain. For naltrexone to be successful though complete abstinence from alcohol is required.

7.2.4   Nalmefene—Opioid Antagonist

Nalmefene (Selincro^{TM}) was first reported in 1975 during an investigation of generating new opioid antagonists from naltrexone and naloxone.[568] Nalmefene is readily

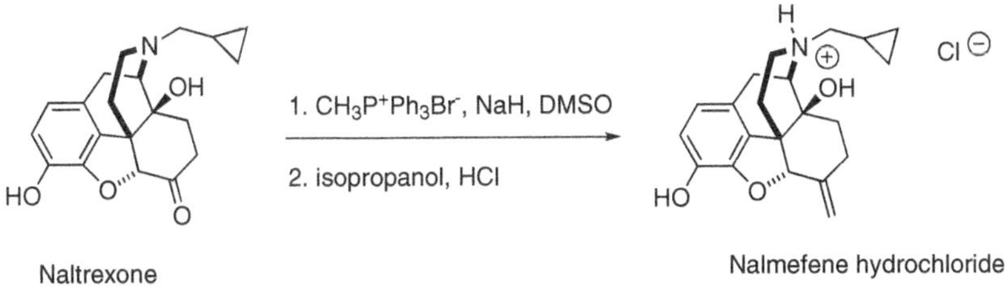

FIGURE 7.6   Synthesis of nalmefene from naltrexone

made from naltrexone by a Wittig reaction of the 6-keto group with methyl triphenylphosphonium bromide (Figure 7.6).

The binding affinities of naltrexone and nalmefene for the opioid receptors are very similar though naltrexone is somewhat more selective for the μ receptor ($K_i = 0.30$ nM) versus the δ receptor ($K_i = 16$ nM) and κ receptor ($K_i = 0.81$ nM).[569] The most striking difference is in their functional activity as nalmefene is a fairly potent partial agonist at the δ receptor ($EC_{50} = 31.6$ nM) compared with naltrexone ($EC_{50} = 230$ nM). Both naltrexone and nalmefene can be defined as μ-opioid receptor antagonists ($EC_{50} > 10,000$ nM) and δ/κ-opioid receptor partial agonists.

Nalmefene is being developed by a Biotie Therapies/Lundbeck collaboration and it received European marketing authorization in February 2013 for use in the reduction in alcohol consumption in adult patients with alcohol dependence who have a high level of alcohol consumption.

In a most remarkable Phase III trial involving 718 individuals, it was found that oral nalmefene at 20 mg per day caused a 50% decrease in the number of days of heavy drinking during the 28-week study. Even after 1 year of treatment, a 60% reduction in alcohol consumption was observed.[570]

This could proof an important first step in helping an alcoholic to reduce their liability on alcohol. It could allow a person to enter a treatment program without having to achieve total abstinence. In a sense, then, this strategy is similar to that of methadone for the clinical treatment of opioid addiction. This would seem of great promise but it has been pointed out that the FDA has not approved, and/or will not grant approval of a drug whose primary indication is the reduction of alcohol consumption.[571]

At this stage, we have an opioid antagonist and a mixed antagonist/partial agonist at the opioid receptors. The importance of each opioid receptor targets can be examined by investigating selective ligands. For example, the κ-opioid antagonist nor-BNI attenuates self-administration of alcohol in alcohol-dependent rats.[572]

7.3   MEDICATION DEVELOPMENT

7.3.1   Morphinans: Samidorphan

A potential complication of naltrexone and nalmefene are their biologically active metabolites. Naltrexone is reduced to 6-β-naltrexol on first-pass metabolism through

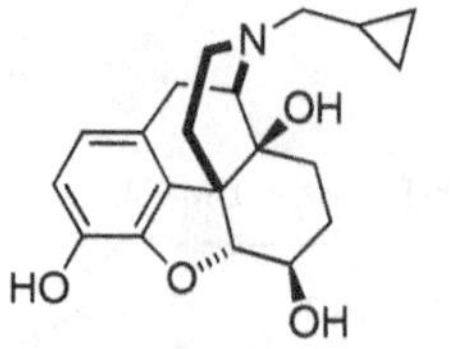

6-β-naltrexol

μ-opioid receptor K_i = 2.12 nM
δ-opioid receptor K_i = 213 nM
κ-opioid receptor K_i = 7.42 nM

FIGURE 7.7   Structure and opioid binding affinities of 6-β-naltrexol

the liver. The metabolite 6-β-naltrexol (Figure 7.7) itself is biologically active with binding to the μ- and κ-opioid receptors and *in vivo* acts as an antagonist, except when the animal is in an opioid-dependant state, then it acts as an inverse agonist.[573,574]

Nalmefene has the potential of the 6-methylene group being oxidized to, for example, an epoxide that could react with protein nucleophiles leading to toxicity effects. The corresponding 6-amino analog of naltrexol, naltrexamine, also is an antagonist and served as the starting point for the development of the widely used irreversible antagonist β-FNA (β-funaltrexamine).[575] To determine whether a more metabolically stable μ-opioid antagonist still possessing alcohol-cessation effects could be produced, a series of naltrexamine amides were prepared and their pharmacological properties with relation to alcohol consumption determined.[569] A series of 26 compounds were prepared and analyzed. From this, a *p*-chlorobenzamide analog (Figure 7.8) proved to have strong binding with partial agonism at all three opioid receptors. The compound showed an excellent *in vitro* metabolic stability profile and was the most effective at inhibiting alcohol self-administration (ED_{50} = 0.5 mg/kg). In all it shows that a strategy of partial agonism at all three opioid receptors could be a new approach for the treatment of alcoholism.

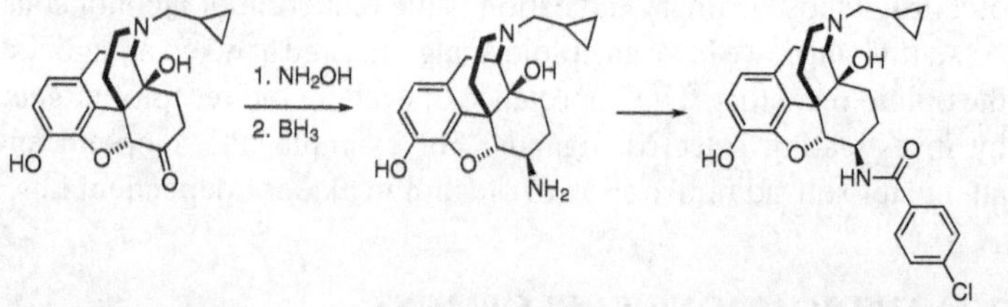

	K_i (nM)	EC_{50} (nM)	EC_{max}		K_i (nM)	EC_{50} (nM)	EC_{max}
μ-opioid receptor	0.3	<10,000			0.05	4	46
δ-opioid receptor	16	230	20		1.4	22	29
κ-opioid receptor	0.8	2.1	5		0.4	7	31

FIGURE 7.8   SAR of naltrexamine amides

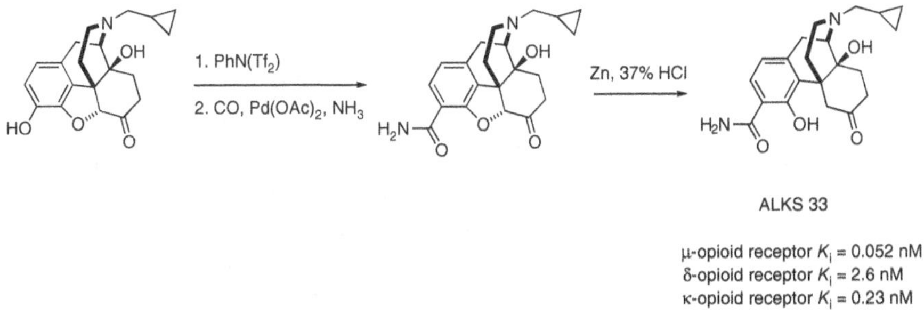

ALKS 33

μ-opioid receptor K_i = 0.052 nM
δ-opioid receptor K_i = 2.6 nM
κ-opioid receptor K_i = 0.23 nM

FIGURE 7.9   Synthesis and opioid binding affinities of ALKS 33

ALKS 33 (Samidorphan)   Alkermes is developing the naltrexone analog ALKS 33 that was first synthesized in the Wentland Laboratory at the University of Rochester. The compound is made from naltrexone as shown in Figure 7.9 by conversion of the phenolic OH group into a carboxamide followed by reductive cleavage of the furan ring.[576,577] The compound has exceptional binding affinity to the μ-opioid receptor where it acts as an antagonist.

Alkermes finished a Phase II study of 400 alcohol-dependent patients in 2010. The results have not been published but a press release (August 12, 2010) by Alkermes states that "the preliminary results of the study showed that once daily administration of ALKS 33 was generally well tolerated at all three dose levels. While the difference in complete abstinence from heavy drinking between treatment groups did not reach statistical significance, patients treated with all three doses of ALKS 33 demonstrated a significant reduction in heavy drinking days in a dose-dependent manner compared to placebo. Patients on the highest dose of ALKS 33 showed the greatest relative reduction in heavy drinking days of 41% compared to placebo." No further information has been released from Alkermes on the efficacy of ALKS 33 in addiction though the clinical effects observed mirror that of nalmefene where a reduction in drinking was observed.

Alkermes does have two other drugs in Phase II clinical trials that have ALKS 33 as an active ingredient. ALKS 3831 is a proprietary drug compound for the treatment of schizophrenia that is a combination of ALKS 33, a potent opioid modulator, and the established antipsychotic agent, olanzapine while ALKS 5461 is a combination of ALKS 33 and buprenorphine and is designed to be a nonaddictive opioid modulator for use in major depression.

7.3.2   CB₁ Antagonist: Rimonabant

The CB_1 antagonist rimonabant has been investigated in several clinical trials for use in alcohol addiction. Unfortunately, no positive effects have yet emerged. Rimonabant was investigated in a 258-patient clinical trial for the treatment of alcohol relapse. At a dosage of 20 mg per day for 12 weeks, rimonabant had no significant effect on preventing relapse.[578]

In a second study, the same dosage was used in a double-blind study of 49 nontreatment-seeking heavy alcohol drinkers for 2 weeks. No effect on alcohol consumption was observed.[579]

A Phase II study (NCT00075205) has been completed where healthy normal volunteers between 21 and 40 years of age were studied who consume between 20 and 40 alcoholic drinks per week, drink at least 4 days a week, and were not seeking treatment for alcoholism. No data have been released from this study.

The chemistry of Rimonabant will be discussed in Chapter 9.

7.3.3   GABA$_B$ Agonist: Baclofen

GABA$_B$ agonists have been shown to attenuate many features of addiction. It is proposed that the mechanism of action is to reduce dopamine levels by binding to inhibitory GABA$_B$ receptors on dopaminergic neurons in the ventral tegmental area. Ethanol is known to increase the levels of dopamine in the nucleus accumbens, GABA$_B$ agonism would help to attenuate dopamine levels. It is proposed that baclofen can improve alcohol-associated cravings and anxiety that in turn can lead to relapse. A synopsis of clinical trials conducted using baclofen for the treatment of alcohol addiction has been published.[580] The authors concluded that there is not enough evidence to support the use of baclofen as a first-line treatment option, except for those alcohol-dependent patients with moderate-to-severe liver cirrhosis in whom other pharmacological treatments are not safe or practical. However, a drug containing a component of GABA$_B$ agonist activity would appear to be useful for the treatment of addiction.

7.3.4   GABA$_A$ Agonist and AMPA Antagonist: Topiramate

As discussed in Chapter 6, topiramate (McN-4853, Topamax) can decrease the levels of dopamine by its agonist effects on GABA$_A$ ion channels by increasing GABA levels and its antagonist effects on the AMPA ion channel by decreasing glutamate levels. Compared with other drugs, we have discussed it is quite novel as it is derived from β-D-fructopyranose. The synthesis is shown in Figure 7.10

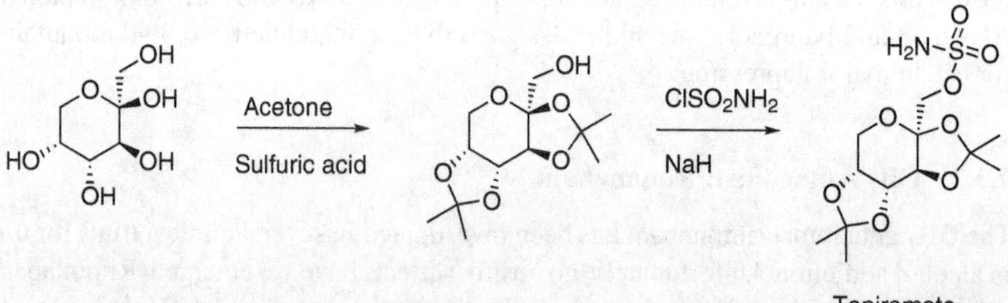

FIGURE 7.10   Synthesis of topiramate

The compound was originally synthesized and reported in 1987 by researchers at McNeil Laboratories. In an SAR study involving 26 compounds, it was found that the anticonvulsant activity was favored by an unsubstituted sulfamate group and was dependent on the overall lipophilicity.[581] Solution 360 MHz NMR and X-ray crystallography revealed that topiramate assumes a twist form in both solution and the solid state. This is interesting as it places the O attached to the anomeric center in more of an equatorial position instead of an axial position as would be predicted by the anomeric effect. Johnson and Johnson introduced topiramate as medication for the treatment of epilepsy in 1994.

The mechanism of action for the prevention of convulsions involves, in addition to the other effects mentioned above, the reduction in the firing rates of neurons by increasing the inactivation period of sodium channels. It is orally available (80%), is rapidly absorbed, and has an elimination half-life of 19-23 hours.[582]

In an interesting variation to normal protocol, building on the ability of topiramate to attenuate dopamine levels, Johnson investigated topiramate in several clinical trials for its effects on alcohol dependence despite the lack of extensive preclinical data. In one of their largest trials, they studied the effects of up to 300 mg per day topiramate in 371 alcohol-dependent persons for 14 weeks (placebo, double-blind) on their physical health, obsessional thoughts and compulsions about using alcohol, and psychosocial well-being.[583]

No significant adverse effects were observed. Patients receiving topiramate showed an improvement in overall health (e.g., decreased BMI, blood pressure) and social well-being. For example, there were less obsessional thoughts and compulsions about using alcohol. The drug did not reduce the symptoms of withdrawal. In all it is suggested that topiramate could be effective in reducing drinking. It has been noted that during the early stages of use when the dosage of topiramate is being determined, there are adverse effects that cause patients to stop usage of the drug. As time progresses though the adverse effects become much less.

To further illustrate the potential use of topiramate, two recent clinical trials that focused on the reduction in drinking and craving will be very briefly addressed. A Spanish group conducted a clinical trial lasting 6 months and involving 182 alcohol-dependent patients compared the effects of 200 mg per day topiramate versus 50 mg per day naltrexone.[584] Overall topiramate was more effective than naltrexone in reducing alcohol intake and cravings for alcohol.

A Greek clinical trial investigated the effects of up to a 75 mg per day dose of topiramate to see whether the reduced dosage might result in better patient compliance. The patient group was divided as those receiving topiramate plus psychotherapeutic treatment ($n = 30$) and those receiving psychotherapy alone ($n = 60$). Following a 10-day inpatient detoxification, the patients were monitored for 4-6 weeks on an inpatient basis followed by an outpatient 4-month basis. In general, positive outcomes were observed for both groups but those taking topiramate were more positive. There was still a high relapse rate though the relapse rate for the topiramate/psychotherapy group (67%) was less than for the psychotherapy alone group (85%). The time to relapse was also increased for 4-10 weeks.

These studies suggest that topiramate alone can be beneficial but it appears that it needs to be used in combination with another drug.

The Veterans Health Administration (VHA) published a major analysis on different medications that were used at VA facilities for alcohol addiction in 2013. Using national VHA administrative data in a retrospective cohort study, they examined time trends in topiramate use from fiscal years (FY) 2009–2012, and predictors of topiramate prescription in 375,777 patients identified with alcohol-use disorders treated in 141 VHA facilities in FY 2011. They found that among VHA patients with alcohol-use disorders, rates of topiramate prescription have increased from 0.99% in FY 2009 to 1.95% in FY 2012, although substantial variation across facilities exists. Predictors of topiramate prescription were female sex, young age, alcohol dependence diagnoses, and engagement in both mental-health and addiction specialty care, and psychiatric comorbidity. The rate of topiramate prescription was higher for patients diagnosed with alcohol dependence than for those diagnosed with alcohol abuse.

Topiramate prescription rates were also higher for patients who received care in both VHA substance-use disorder and mental-health specialty treatment settings in the same year. Further, patients with psychiatric comorbidities were much more likely to have a prescription for topiramate (83% of patients with a prescription for topiramate had at least one psychiatric comorbidity). Topiramate has mood stabilizing properties and is efficacious for treating several psychiatric disorders such as bipolar disorder, borderline personality disorder, and post-traumatic stress disorder. As alcohol-use disorders are often comorbid with psychiatric disorders, topiramate may be viewed by providers as a way to address multiple disorders with one medication.[585]

7.3.5   CRF₁ Antagonists: Pexacerfont and GSK561679

Bristol-Myers Squibb developed the CRF1 antagonist pexacerfont (BMS562086) for the treatment of mood disorders. The development program was terminated in 2008 after lack of efficacy in Phase II trials. There is now interest in the drug for the treatment of alcohol addiction.

Pexacerfont (Figure 7.11) is a potent and selective CRF_1 antagonist that is orally bioavailable and brain penetrant with an IC_{50} of 7.2 nM. It is specific for CRF_1 and has more than 1000-fold less affinity for CRF_2, and > 100-fold less affinity for

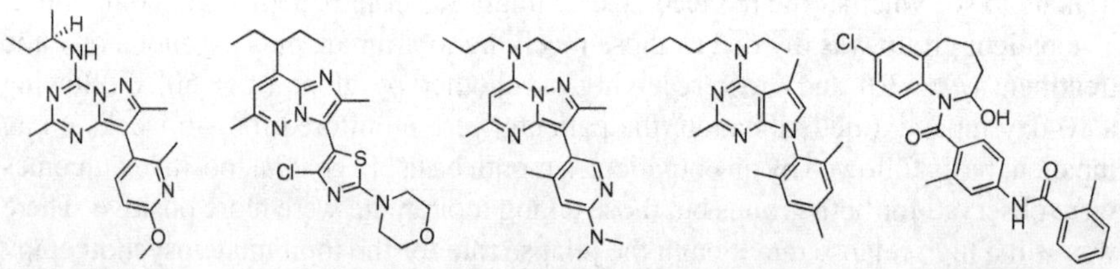

| Pexacerfont BMS 562086 | MTIP Eli Lilly | R121919 [195055-03-9] | CP254526 [157286-86-7] | SSR125543A Tolvaptan [150683-30-3] |

FIGURE 7.11   Structures of CRF_1 antagonists

the CRF-binding protein. Pexacerfont was active in the defensive withdrawal and elevated plus maze models of anxiety in rats (1-10 mg/kg, p.o.). The lowest effective dose of pexacerfont in the defensive withdrawal model resulted in ~50% occupancy of brain CRF_1 receptors. Oral doses as high as 30 mg/kg had no effect on mild stress-induced elevations of plasma corticosterone and doses as high as 100 mg/kg did not lead to overt side effects such as ataxia or alteration of locomotor activity (from NIH National Center for Advancing Translational Sciences). Researchers at BMS have published a detailed report on its discovery, development, and pharmacology.[586]

Clinical trials were conducted to determine its efficacy in the treatment of major depressive disorder (MDD) and GAD, but unfortunately no improvement in efficacy relative to placebo was detected.[587] National Institute on Alcohol Abuse and Alcoholism is planning on conducting a Phase II trial of pexacerfont (100-300 mg orally once per day) as a treatment for anxiety-related alcohol craving (NTC01227980).

Eli Lily with the NIAAA has conducted preclinical studies on the CFR_1 antagonist 3-(4-chloro-2-morpholin-4-yl-thiazol-5-yl)-8-(1-ethylpropyl)-2,6-dimethyl-imidazo[1,2-b]pyridazine (MTIP). The drug arose from a directed SAR program to obtain an oral, CNS accessible compound with high CRF_1 antagonist potency. A description of the medicinal chemistry has not been published but a detailed evaluation of the pharmacokinetic and pharmacodynamic properties of the drug is available.[588] Preclinical aspects of MTIP in rodents were compared with three other commonly used CFR_1 antagonists, R12119, SSR125543A, and CP154526 (Figure 7.11). SSR125543A is a selective vasopressin V2-receptor antagonist and CFR_1 antagonist that are approved for the treatment of clinically significant hypervolemic hyponatremia (abnormally increased volume of blood and reduction in sodium ion concentration in the plasma) and euvolemic hyponatremia (reduction in sodium ion concentration in the plasma without plasma volume expansion). However, significant liver damage has been reported and a warning on its use was issued in January 2013.

All four compounds have similar binding affinity for the $hCRF_1$ receptor of $K_i = 0.22-0.44$ nM with no significant binding at the CRF_2 receptor. MTIP showed great selectivity in a binding profile panel of 74 receptors and ion channels. MTIP at a concentration of 10 μM exhibited no inhibition of >50% while R121919 produced >50% inhibition at the hA_3 receptor, α2 adrenoceptor, AT_1, hCGRP, κ-opioid receptor, σ, Na^+ channel, and Cl^- channel. SSR125543A produced inhibition at 10 μM in the hA_3, peripheral benzodiazepine, and the Na^+ channel. One problem with CRF_1 antagonists is their limited bioavailability, MTIP surmounted this issue with an oral bioavailability of 91% versus 18%, 24%, and 70% for CP154526, SSR125543A, and R121919, respectively. In studies with alcohol-dependent rats MTIP reduced reinstatement of alcohol seeking and self-administration in a dose-dependent fashion.[588]

In all a CRF_1 antagonist with good potency, selectivity, bioavailability, and efficacy in controlling alcohol consumption and relapse was developed. No further work with this compound appears to have been done, or at least published, so its current status is unknown.

CRF_1 antagonists of the pyrazolo[1,5-a]pyrimidine core often suffer from poor pharmacokinetics, especially it was thought, a large volume of distribution and high

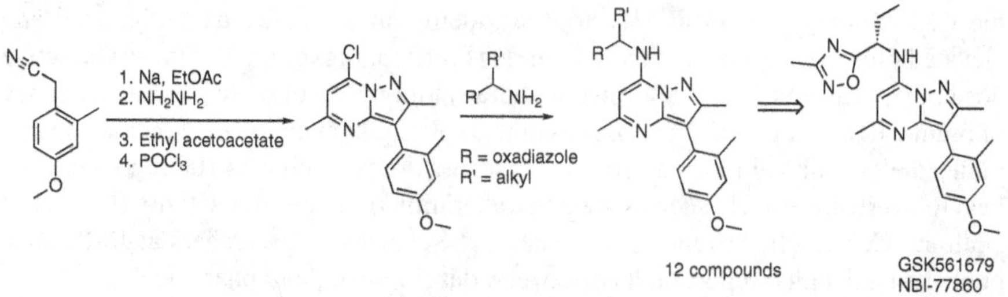

FIGURE 7.12   SAR development for NBI-77860/GSK561679

clearance. To address this, the Neurocrine/GSK groups decided to attach a heteroaromatic group onto the N-alkyl side chain at the C7 position to make the compound more hydrophilic. It was hoped that binding affinity and functional activity would be maintained while decreasing the volume of distribution (V_{dss}) and clearance. A series of 12 compounds (Figure 7.12) was prepared and from this the S enantiomer of GSK561679 had the best balance of CRF_1 binding affinity (6.3 nM), hydrophilicity (log $D_{7.4}$ = 3.8), good oral bioavailability (66%, rat), good entry into the brain (1 hour brain/plasma ratio = 1.6), and met their standards of volume of distribution (V_{dss} = 7.5 L/kg) and clearance (14 mL/min/kg).[589]

GSK561679 was evaluated in a Phase II clinical for the treatment of MDD, but no benefit was noticed relative to placebo. The drug is entering an NIAAA-sponsored study though to explore its use in the ability to reduce alcohol craving in recently detoxified alcohol-dependent women in response to stress or alcohol-associated stimuli (NCT01187511).

7.3.6   Neurokinin 1 Receptor Antagonists: LY686071

The tachykinin peptides are widely distributed in both the central and peripheral nervous systems. These peptides exert a number of biological effects through actions at the tachykinin receptor subtypes neurokinin 1 (NK_1), neurokinin 2 (NK_2), and neurokinin 3 (NK_3). The endogenous peptides are substance P (SP), neurokinin A (NKA; previously known as substance K, neurokinin a, neuromedin L), neurokinin B (NKB; previously known as neurokinin b, neuromedin K), neuropeptide K and neuropeptide g (N-terminally extended forms of neurokinin A).[42]

The NK_1 receptor is known to play a role in the modulation of stress and anxiety and some NK_1 antagonists have been investigated. The antagonist LY686017 (Figure 7.13) was examined in a small trial for efficacy in reducing cravings for alcohol in 25 alcohol-dependent patients.[590] The drug was promising in that at a dose of 50 mg per day spontaneous cravings for alcohol and responses to stress and alcohol cues were reduced.

LY686071 contains a triazole core surrounded by two pyridyl rings and an aryl ring, and was first reported in the patent literature by Eli Lily (WO 03/091226 A1) in 2003.

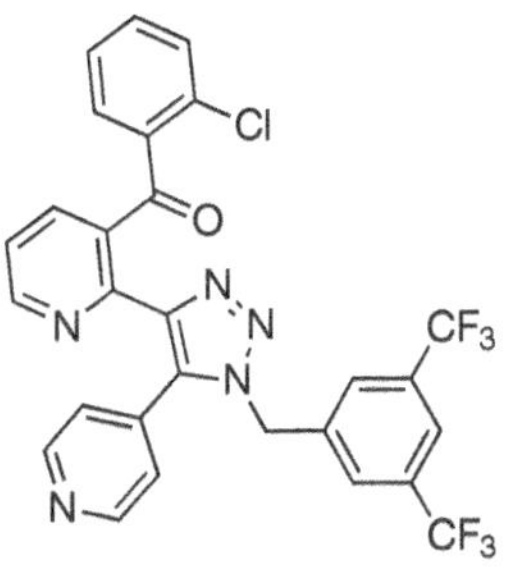

LY686017

FIGURE 7.13   Structure of LY686017

7.3.7   PPAR-γ Agonists: Pioglitazone

PPARs (peroxisome proliferators-activated receptor) are ligand-activated transcription factors that are a component of the nuclear receptor family. There are three PPAR isoforms, $PPAR_\alpha$, $PPAR_\beta$, and $PPAR_\gamma$. In general, they are activated by fatty acids but no single endogenous ligand seems to exist that binds to all three. As many endogenous cannabinoid ligands are fatty acids, there are several endogenous CB_1 agonists that also act as weak PPAR agonists, for example, anandamide binds to $PPAR_\alpha$ and to $PPAR_\beta$ with EC_{50}'s of 10-30 μM compared with $K_i = 32$ nM at CB_1. The psychoactive component of marijuana, Δ^9-THC, a CB_1 agonist with a K_i of 10 nM, binds to $PPAR_\alpha$ and to $PPAR_\beta$ with EC_{50}'s of about 0.3 μM. The synthetic CB_1 antagonists rimonabant and AM251 act as agonists on $PPAR_\alpha$ and to $PPAR_\beta$ with EC_{50}'s of about 10 μM.[591] The PPAR system appears to play a role in modulation of pain but its role in addiction is unclear.

Omeros Corporation has reported on the use of PPARγ agonists for the treatment of alcohol addiction in US Patent 8426439. According to the patent, in a European pilot study involving 12 patients, their PPAR-agonist OMS405 at 30 mg per day showed favorable results in stopping alcohol consumption relative to naltrexone or counseling.

Little public information is available of the nature of their compounds but from the patent it appears the compounds are of the thiazolidinedione heterocyclic class that is similar to the drug Actos (pioglitazone) from Takeda, which contains a thiazolidinedione ring. Pioglitazone was approved in 1999 for the treatment of non-insulin-dependent diabetes mellitus. The drug is used in a composition that can also include therapeutic agents such as opioid antagonists, opioid partial agonists, antidepressants, antiepileptics, antiemetics, CRF-1 receptor antagonists, 5-HT_3 antagonist, 5-$HT_{2A/2C}$ antagonists, or CB_1 antagonists.

Several syntheses of pioglitazone are reported in the patent literature. As an example, a fairly direct approach is shown in Figure 7.14.

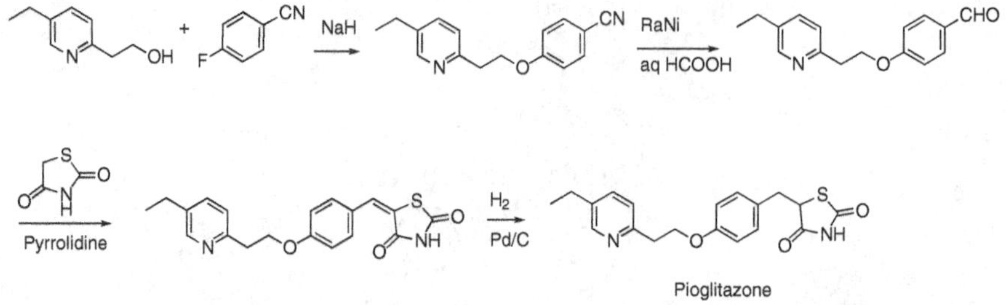

FIGURE 7.14   Synthesis of pioglitazone

The first step involves a nucleophilic aromatic substitution reaction followed by a Knoevenagel condensation benzaldehyde with 2,4-thiazolidinedione to yield a benzylidenethiazolidinedione. The double bond is then reduced to give pioglitazone. The final step has undergone an optimization exercise using a reaction response surface analysis.[592] The dependence of reaction yield (%) and impurity level (%) on the reaction conditions was examined using five independent variables (process parameters): time, amount of a catalyst, temperature, amount of the reducing agent (sodium borohydride), and substrate purification. The optimal conditions were a time of 1-2 hours at 40°C using a reducing agent composed of sodium borohydride with a cobalt chloride-dimethylglyoxime (CoCl$_2$-DMG) catalyst. Yields of 94–95% were obtained with impurity levels of 0.16–0.36%.

7.4 BENZODIAZEPINES

Chemist–drug name recognitions tend to be quite rare, however, one of the most famous is Leo Sternbach and benzodiazepines. Leo Sternbach of Roche first discovered the benzodiazepine chemotype in 1957.[593] At this time, binding assays were not used, the compounds synthesized in the laboratory went directly into animal models of sedation. This work resulted in the development of Librium, the first nonsedating drug for the treatment of anxiety.

Benzodiazepines bind to the GABA$_A$ chloride ion channel where they potentiate chloride influx. Benzodiazepines act as positive allosteric modulators, potentiating the effects of GABA binding to the GABA$_A$ ion channel. The influx of negatively charged chloride ions hyperpolarizes the neuron inhibiting an action potential. Benzodiazepines bind to sites on the interface of the α and γ subunits while GABA binds to two sites on the αβ subunits. The benzodiazepine binding site is thought to be composed of a combination of α1, 2, 3, or 5 subunits and γ2 or γ3 subunits. Studies in transgenic mice suggest that the α1 site plays a role in amnesia and sedation, the α2 and α3 sites in the anxiolytic effects of benzodiazepines, while the α5 subunit may play a role in memory.[594]

The addictive properties of benzodiazepines are thought to be due to benzodiazepines increasing the firing of dopamine neurons of the ventral tegmental area through the positive modulation of $\alpha 1$-containing $GABA_A$ receptors in nearby $GABA_A$ interneurons.[595] In other words, binding of benzodiazepines to the $\alpha 1$ subunit of $GABA_A$ receptors in the GABA interneurons *decreases* the levels of GABA, therefore, the lower amounts of GABA binding to $GABA_A$ receptors located on dopamine neurons in the VTA *increases* the release of dopamine into the nucleus accumbens.

Benzodiazepines also bind to the $GABA_A$ receptor on the dopamine neurons where the $GABA_A$ receptors contain the $\alpha 3$ subunit instead of the $\alpha 1$ subunit. The overall disinhibition effect of benzodiazepines though is greatest through the binding to $GABA_A$ receptors in the interneurons of the VTA.

Benzodiazepines are Drug Enforcement Agency Schedule IV drugs and used orally. Their use produces positive reward feeling of reduced pain and anxiety, general well-being, and lowered inhibitions. They do not produce euphoria. The abuse liability of classic benzodiazepines can be considered to be high. They produce dependence, tolerance, and have withdrawal symptoms. In 2003, approximately 1.9 million individuals in the United States used benzodiazepines for nonmedical use. They do have valid medical use as several drugs, for example, alprazolam (Xanax), are prescribed for anxiety, panic disorder, and social phobia.

To illustrate the complexities that arise in the field of drug addiction, it has been suggested that the withdrawal symptoms of benzodiazepines are very similar in scope and severity to that of serotonin selective reuptake inhibitors (SSRI) used to treat depression.[596] As such the authors suggest it is incorrect to refer to the withdrawal symptoms of benzodiazepines as due to dependence and the DSM-IV definitions need to be changed. Two commentaries to the study state that though the withdrawal aspect may be correct, one does not search out illegal sources of SSRIs in order to increase the dosage of SSRIs as can occur with benzodiazepines.[597,598]

This type of argument may seem rather academic yet it is important from the viewpoint if time and effort should be spent to develop medication to treat the various stages of benzodiazepine addiction. Definitions may dictate if a condition is considered a bona fide medical disorder, hence, worth the effort (i.e., funded). At this time, the consensus seems to be: no, it is not important.

7.5   BARBITURATES

Barbiturates are dangerous drugs due to their narrow therapeutic index (ratio of therapeutic and toxic doses) of 20. The toxic dose ranges from 2 to 10 g. Common barbiturates are thiopental, secobarbital, and phenobarbital. Pentobarbital and phenobarbital are sometimes used to treat addiction to sedative-hypnotic drugs.

They have the same mechanism of action as benzodiazepines by binding to the $GABA_A$ ion channel.

8

MEDICATION DEVELOPMENT FOR NICOTINE ADDICTION

Tobacco is one of the oldest drugs known to humans. Tobacco comes from the plant *Nicotiana tabacum* that is indigenous to the temperate climate regions of North and South America. The leaves are removed and dried and are then normally chewed or smoked. The harmful effects of tobacco are widely known and are due to the psychoactive component nicotine and to the tars and other chemicals present in the leaves. Nicotine is highly addictive and as Mark Twain elegantly expressed, the addiction can be very difficult to overcome. Current medications used for the treatment of nicotine addiction are nicotine replacement therapy (NRT), bupropion (Wellbutrin®, Zyban®), and varenicline (Chantix®).

8.1 PHARMACOLOGY OF NICOTINE ADDICTION

Nicotine binds to neuronal nicotinic acetylcholine receptors (nAChR) located in the reward system. As discussed in Chapter 2, the nAChR ion channel consists of pentameric units, for example, the $\alpha 7$ nAChR consists of five $\alpha 7$ subunits with each subunit containing a binding site for acetylcholine. Genetic knockout studies in mice indicate that the β_2, α_4, α_6, and α_7 subunits play a role in the positive reinforcing properties of nicotine. Loss of the α_5 subunit results in an *increased* level of nicotine self-administration in mice suggesting that this subunit plays a role in reducing the positive reinforcing effects of nicotine. To what extent the concentration and combination of subunits have in the reinforcing effects of nicotine are not yet known.

Drug Discovery for the Treatment of Addiction: Medicinal Chemistry Strategies, First Edition. Brian S. Fulton.
© 2014 John Wiley & Sons, Inc. Published 2014 by John Wiley & Sons, Inc.

The $\alpha 7$ nAChR appears to play a role in cognition. It is well known that gating deficits apparent in schizophrenia are transiently improved or normalized by smoking. Genetic evidence and lower concentrations of $\alpha 7$ nAChR in schizophrenic patients show the importance of the receptor in cognition.[599]

Nicotine elicits its addictive effects by two neuronal mechanisms: a direct mechanism and an indirect mechanism.[600] In the direct mechanism, nicotine binds to nAChR located on dopamine neurons in the VTA, this results in the release of dopamine in the nucleus accumbens, prefrontal cortex, and the amygdala. The indirect mechanism involves modulation of dopamine levels by nicotine binding to nAChR located on glutamate and GABAergic neurons in the VTA. Nicotine binding to α_7-containing nAChR located on the presynaptic terminus of glutaminergic neurons results in glutamate being released in the synapse. The synaptic glutamate binds to mGlu$_5$ and NMDA receptors located on dopaminergic neurons, firing an action potential, causing the release of dopamine into the reward system.

The opposite occurs with nicotine binding to $\alpha_4 \beta_2$-containing nAChR located on presynaptic GABAergic neurons. In this case, an increase in levels of GABA in the VTA results in GABA binding to GABA$_A$ receptors located on dopaminergic neurons, inhibiting the firing of the neuron, causing dopamine levels to decrease in the nucleus accumbens.

There are two main strategies for nicotine addiction medication discovery. One is to directly modulate the binding of nicotine to the nAChRs. The second is to modulate the effects of the neurotransmitters dopamine, glutamate, and GABA that are released upon binding of nicotine to nAChRs. Remember in the latter we are referring to nicotine binding to nAChRs located on dopaminergic neurons, glutaminergic neurons, or GABAergic neurons.

Nicotine increases the levels of the excitatory neurotransmitter glutamate. One strategy then would be to discover drugs that would block binding of glutamate to postsynaptic receptors, that is, an antagonist. This strategy is complex, as well as appealing, as there are at least 11 potential receptors for glutamic acid; the ionotropic ion channels NMDA, AMPA, and kainate and the eight metabotropic glutamate receptors. An interesting historical note is that the term "metabotropic" appears to have been first used by Eccles and McGeer in 1979 to classify the actions of neurotransmitters with receptors.[601] It was used to describe all receptors that interact with a neurotransmitter and do not involve opening of an ion gate. Metabotropic was applied to the newly discovered glutamate receptor due to its linked hydrolysis (metabolism) of inositol phospholipids.[602]

With a wide range of potential targets, the opportunity exists to discover a selective antagonist that might attenuate relapse without other side effects. Two potential targets are mGlu$_{2/3}$ or mGlu$_5$. In animals, compounds that decrease glutamatergic neurotransmission, such as postsynaptic NMDA receptors antagonists, excitatory postsynaptic mGlu$_5$ antagonists, or inhibitory presynaptic mGlu2 and mGlu3 agonists, decreased nicotine self-administration or reinstatement of nicotine-seeking behavior. While the mGlu receptors continue to be of interest, there is little evidence that the use of NMDA antagonists such as memantine can play a beneficial role in the treatment of nicotine or other drug addictions.[603]

8.2   APPROVED MEDICATIONS

8.2.1   Varenicline-nAChR Partial Agonist

Varenicline (Chantix®) was developed at Pfizer and approved by the FDA for the treatment of nicotine addiction in 2006. The drug is a $\alpha_4\beta_2$ nicotinic acetylcholine receptor partial agonist. An excellent first-hand case history of the development of Chantix® is available.[604] Reading of the account is highly recommended for those interested in the methodology of drug discovery and development.

When the Pfizer scientists in 1993 set out to develop a drug for the treatment of nicotine addiction, their goal was to develop a compound acting as a partial agonist that would blunt the craving and withdrawal symptoms of nicotine but without the abuse liability. Evidence that this could be achieved came from NRT where nicotine supplied via a transdermal patch was found to reduce craving and symptoms of withdrawal. A combination therapy of an nAChR agonist (nicotine) and an nAChR antagonist (mecamylamine) was shown to result in higher quitting rates than nicotine patches alone. A partial agonist approach using buprenorphine for the treatment of opioid addiction was also showing success at that time. A partial agonist would be expected to show properties of both agonist and antagonistic properties. The theoretical rationale for the use of a partial agonist to treat nicotine is illustrated in Figure 8.1.[605]

Assuming a sufficient amount of the drug can enter the brain and have a strong binding affinity to the nAChRs, when a person is not smoking a partial agonist can have a mild nicotine-like effect and relieve aspects of craving. Increasing the dose (e.g., during abuse) would cause no further effect. The antagonist efficacy of the drug could meanwhile be reducing the reinforcement effects of nicotine. This presumes again that the drug can enter the brain in sufficient quantities to efficiently compete with nicotine in its binding to nAChRs.

Using a combination of *in vivo* drug discrimination, self-administration, and *ex vivo* dopamine release assays, various nicotinic agents were tested leading to the discovery of the plant-derived $\alpha_4\beta_2$ partial agonist cytosine. This compound served as a lead scaffold upon which an intensive SAR study was conducted. One interesting aspect of the studies was that with this scaffold there was not a direct correlation between receptor-binding affinity and receptor functional activity. In a brilliant flash of intuition, it was recognized that cytosine shares significant similarity with the A, B, D ring system of morphine as illustrated in Figure 8.2.

Morphing of the cytosine and benzomorphan scaffolds and fusion of a pyrazine ring onto the benzene ring led to varenicline as shown in Figure 8.3.

Of structural interest, it appears that the N-H group could be within the shielding cone of the aromatic ring. No significant anisotropic effect is apparent though as the N-H chemical shift of 1.82 ppm[11] (free base, 400 MHz, CDCl$_3$) is well within the region typically seen for alkyl N-H groups.

How effective is varenicline in treating nicotine addiction? A normal dosing schedule is 2 mg per day (1 mg twice a day) for a 12-week period. Results from the Phase III clinical trials showed that the 4-week abstinence rates were 44% for smokers treated

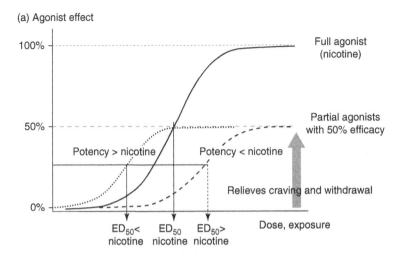

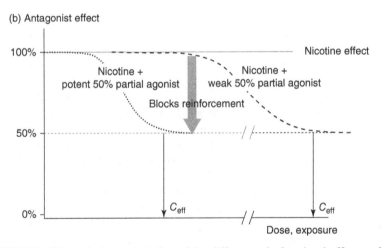

FIGURE 8.1   Theoretical representation of the differences in functional efficacy of a full agonist (nicotine) and a weak or potent partial agonist, alone and combined with nicotine. (a) Partial agonists have a smaller maximal effect than does the agonist nicotine (i.e., increasing their concentration or exposure does not further increase their effect). When not smoking, the partial agonist has a mild nicotine-like effect and can relieve craving and withdrawal, dependent on the relative functional efficacy versus nicotine, the receptor-binding affinity, and free levels in the brain. (b) To exert an antagonist effect (i.e., to block the reinforcing effect of nicotine when smoking), a partial agonist should have high binding affinity and sufficiently free levels in the brain (C_{eff}) to inhibit the effect of nicotine completely, resulting in the same maximal effect as the partial agonist alone. Partial agonists with poor binding affinity and/or low brain penetration have insufficient C_{eff} to be efficacious as an antagonist *in vivo* (Reprinted with permission from Rollema et al.[605] Copyright © 2007 Elsevier Ltd.)

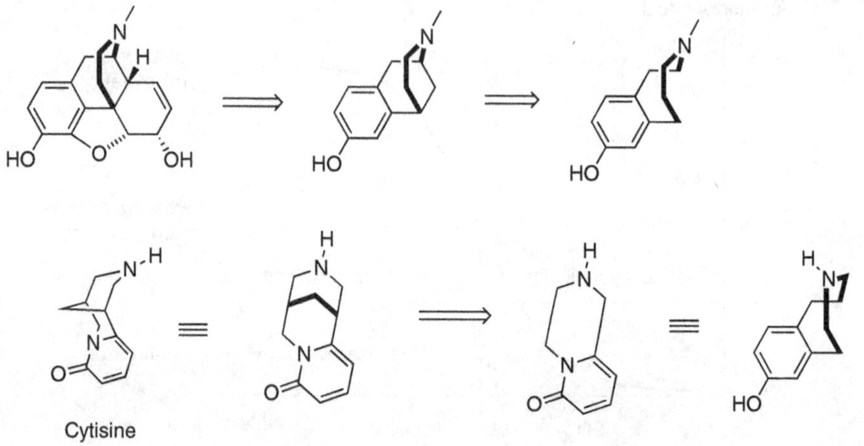

FIGURE 8.2   Structural similarities of morphine and cytisine

with varenicline, 30% for those treated with bupropion, and 18% for the placebo group. After a 1-year period only 22% of varenicline treated, 15% of bupropion treated, and 9% of the placebo group were still abstinent. It has been proposed that a much longer treatment period may be required.

The FDA in 2009 required boxed warnings to be placed on product labeling warning of reports of changes in behavior such as hostility, agitation, depressed mood, and suicidal thoughts or actions. Zyban (see section 8.2.2) was given the same requirement. In 2011, the FDA reported follow-up studies that proved somewhat inconclusive in the risk of neuropsychiatric adverse events associated with varenicline.[606] It was stated that Pfizer is conducting a large safety clinical trial of Chantix to assess neuropsychiatric adverse events, and results from this study are expected in 2017.

The normal treatment regime is dosage of the patient starting 7 days before the quitting date followed by continued dosage up to 12 weeks post-quit date. Varenicline has reduced cue-provoked craving (2 mg per day) over a 15-day study period.[607]

The $\alpha_4\beta_2$ nicotinic acetylcholine receptor is an interesting target for other CNS disorders. AZD1446 is in clinical trials for attention-deficit hyperactivity disorder (ADHD) and Alzheimer's disease. Varenicline was investigated to determine its effects on the reinforcing and discriminative effects of cocaine in rhesus monkeys.

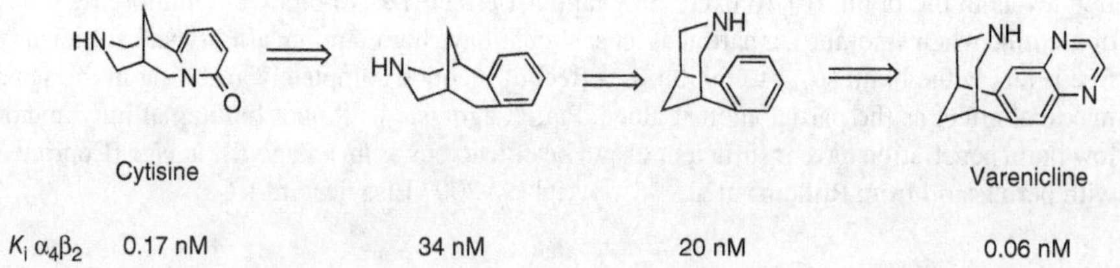

Cytisine			Varenicline
$K_i\ \alpha_4\beta_2$ 0.17 nM	34 nM	20 nM	0.06 nM

FIGURE 8.3   Development of varenicline from cytosine

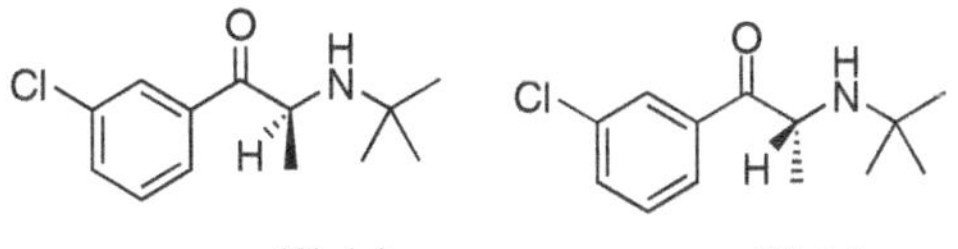

FIGURE 8.4   Enantiomeric forms of bupropion

The researchers found that though varenicline would have a low abuse liability, it potentiated the reinforcing effects of cocaine.[608]

8.2.2   Bupropion-DAT/NET Inhibition and nAChR Antagonist

Bupropion (Zyban®) was first synthesized by Burroughs Research in 1966 and was approved for the use as an antidepressant in 1989. The pharmacological properties of bupropion were first reported in 1977.[609] Bupropion hydrochloride has been widely used as a potent central nervous system stimulant for the marketed treatment of depression (Wellbutrin®) and as a smoking cessation aid (Zyban®).

Bupropion is a multi-target drug that is a DAT (IC_{50} = 2.1 µM) and NET (IC_{50} = 6.7 µM)[610] reuptake inhibitor and a micromolar antagonist at $\alpha_3\beta_4$, $\alpha_4\beta_2$, $\alpha_4\beta_4$, and α_1 nAChRs.[611] The strength of binding to the transporters favors DAT (K_d = 520 nM) followed by SERT (K_d = 9.1 µM) with very poor binding to NET (K_d = 52 µM).[43] During nicotine withdrawal, there is an abnormal decrease in dopamine concentration; bupropion is used to increase dopamine levels to homeostatic levels. Racemic bupropion is used for the treatment of depression as while as the treatment of nicotine addiction. The structures of each enantiomer are shown in Figure 8.4

There is little significant difference in the dopamine and norepinephrine reuptake inhibition of the S and R forms of bupropion. Bupropion undergoes extensive metabolism with a major metabolite arising from hydroxylation of a methyl group on the *t*-butyl amine. One study using human liver microsomes revealed that the cytochrome P450 enzyme CYP2B6 mediates the hydroxylation of bupropion.[612] Bupropion binds moderately well to CYP2B6 with a K_m = 85 µM. It was also shown that common second generation antidepressants inhibited the metabolic hydroxylation of bupropion in the low to high µM range (e.g., IC_{50} = 1.6 µM for paroxetine to 59 µm for fluoxetine) suggesting the possibility of drug-drug interactions. The resultant hydroxylated metabolite can cyclize onto the ketone group forming a hemiacetal. Both the initial hydroxy methyl metabolite and the hemiacetals are thought to contribute to the antidepressant and smoking cessation efficacy of bupropion.[613] An SAR study of 23 analogs of the hemiketal showed some promise as potential smoking cessation agents. Some of the more active compounds also possessed SERT reuptake inhibition.[614] The (R)-(−) enantiomer was reported to completely racemize at pH 7.4 with a half-life of 44 minutes.[610]

Another metabolite of bupropion is where the keto group is reduced to an alcohol generating a second chiral center.

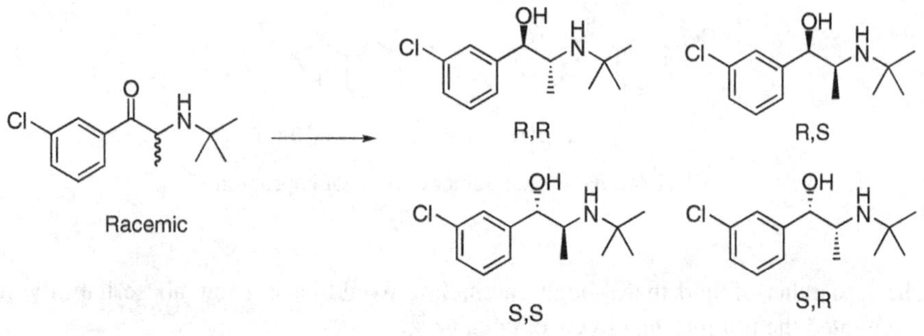

FIGURE 8.5 Hydroxyl metabolites of bupropion

Of the four metabolites shown in Figure 8.5, the 2S,3S compound proves to be the most potent metabolite acting as an antagonist at the $\alpha_4\beta_2$ nAChR with an $IC_{50} = 3.3\,\mu M$ relative to racemic bupropion of $IC_{50} = 12\,\mu M$. The 2S,3S metabolite is similar in potency or more potent than ($\pm$) bupropion in blocking catecholamine reuptake. At the DAT receptor bupropion inhibited the reuptake of $[^3H]$-DA with an $IC_{50} = 550\,nM$ versus an IC_{50} of 790 nM for the 2S,3S metabolite. At the NET receptor bupropion and the (2S,3S) metabolite inhibited reuptake of $[^3H]$-NE with IC_{50}'s of 1900 and 520 nM, respectively. In addition, the 2S,3S compound was 7–10 times more potent than bupropion in blocking nicotine-induced antinociception in mice. A follow-up study showed that bupropion and the 2S,3S-hydroxymetabolite reversed affective and somatic withdrawal signs in nicotine-dependent mice and partially substituted for nicotine in drug discrimination tests.[615] This data show that the 2S,3S metabolite shows potential as a new smoking cessation agent. Several patent applications have been filed on the use and synthesis of the 2S,3S compound.

($\pm$) Bupropion is readily synthesized from 3-chloro propiophenone as shown in Figure 8.6 and is in fact the subject of an undergraduate organic chemistry laboratory procedure.[616] An asymmetric synthesis of either enantiomer uses a Sharpless asymmetric dihydroxylation of the silyl enol ether of 3-chloro propiophenone as the key step.[617]

The concept of addiction has been generalized to non-drug situations such as gambling, sex, obesity. The idea is that in all these activities the reward system is involved. An interesting application of this involves the drug Contrave® that is being developed by Orexigen Therapeutics to treat obesity. Contrave® is a mixture

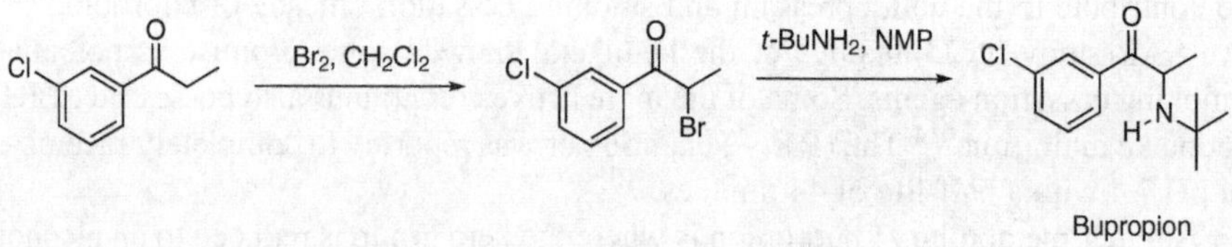

FIGURE 8.6 Synthesis of bupropion

of naltrexone and bupropion. Mercer has summarized the proposed mechanism of action as follows.[618]

Bupropion has been shown to stimulate hypothalamic pro-opiomelanocortin (POMC) neurons that release α-melanocyte-stimulating hormone (α-MSH). The resultant binding of α-MSH to melanocortin 4 receptors initiates a cascade of actions, which results in reduced energy intake and increased energy expenditure. Bupropion also causes POMC neurons to simultaneously release β-endorphin, an endogenous agonist of the μ-opioid receptor. Binding of β-endorphin to μ-opioid receptors located on POMC neurons mediates a negative feedback loop on POMC neurons, leading to a decrease in the release of α-MSH. Blocking this inhibitory feedback loop with naltrexone is thought to facilitate a more potent and longer-lasting activation of POMC neurons, thereby amplifying effects on energy balance. As a result, co-administration of bupropion and naltrexone produces a substantially greater effect on the POMC firing rate than either compound administered alone, suggesting that the drugs act synergistically.

8.3 MEDICATION DEVELOPMENT

Despite the availability of medication for the treatment of nicotine addiction, it is still difficult to overcome the addiction, so other pharmacological approaches are being investigated. Beyond nAChR partial agonists, these include CB_1 antagonists, D_3 antagonists, MAO-B inhibitors, glycine receptor antagonists, and the opioid antagonists: nalmefene and naltrexone.[619]

8.3.1 CB_1 Antagonist: Rimonabant

A more detailed description of the CB_1 antagonist rimonabant (Acomplia™) will be given in Chapter 9. There was a great interest in rimonabant for use in treating nicotine addiction. If successful, it would have opened a new therapeutic target area. Rimonabant was investigated in the Phase III STRATUS-US clinical trial for nicotine addiction. While abstinent rates were higher than placebo (36% vs. 21%), the dosage required (20 mg per day) resulted in enough side effects to make it unattractive as a medication for nicotine addiction.[620]

8.3.2 D_3 Antagonists

Researchers at GlaxoSmithKline have investigated the development of D_3 antagonists for the treatment of addiction. Out of this work, the compound GSK598809 arose and was selected for clinical trials. A detailed report on the SAR leading to the compound is available. In brief, it was decided to substitute an azabicyclo[3.1.0]hexane ring for a benzodiazepine ring that existed in their existing series of D_3 antagonists. During the SAR study, all new compounds were assayed for their agonistic and antagonistic properties using a functional [^{35}S]GTPγS assay expressing the human D_3 receptor. Key criteria for a compound to proceed were at least 100-fold selectivity versus D_2 and

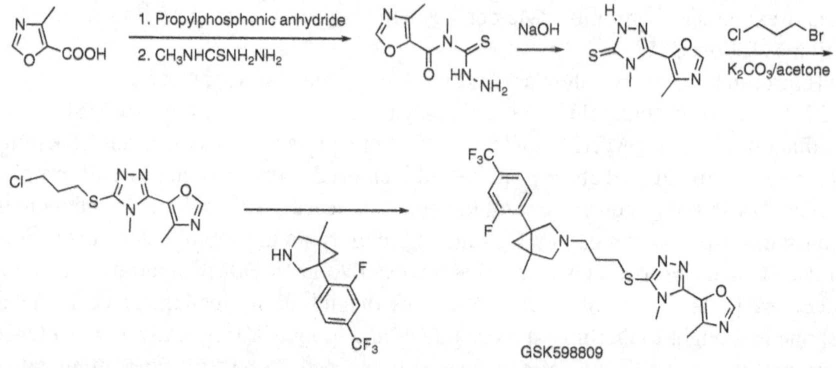

FIGURE 8.7 Synthesis of GSK598809

histamine H_1 receptors in functional assays and 100-fold selectivity versus the hERG ion channel. The synthesis and testing of over 100 compounds led to GSK598809 (compound 74 in the original article). It is a selective and potent competitive D_3R antagonist with good physicochemical, pharmacokinetic, and safety properties. It is over 500 times more selective for the D_3 than D_2 receptor with K_i's of 1.5–6.2 nM and 740 nM, respectively. Results from several receptor and enzyme screening panels showed the compound is at least 100-fold selective for the D_3 receptor versus a wide range of targets. The compound reduced the time spent in both nicotine-paired and cocaine-paired compartments in a dose-dependent fashion in conditioned place preference behavioral assays.[621]

A PET study from GlaxoSmithKline in humans correlated D_3 receptor occupancy with feelings of craving. An oral dose of 75 mg GSK598809 resulted in a maximum concentration in the blood of 520 ng/mL. At a T_{max} of 1.5 hours, it was determined the drug occupied 89% of the D_3 receptors in the brain. Occupancy slowly decreased with 80% occupancy after 17 hours. This level of occupancy transiently alleviated craving in smokers after overnight abstinence. The authors of the study conclude that the data suggest that either higher D_3 occupancy is required for a full effect in humans or that nicotine-seeking behavior in CPP rats only partially translates into craving for cigarettes in short-term abstinent smokers.[622]

The synthesis is outlined in Figure 8.7 and starts with a coupling reaction mediated by polyphosphoric anhydride (T3P) to form the amide that is cyclized to the triazolethione. Addition of the linker allows the introduction of the bicyclic amine resulting in GSK598809. The compound existed as enantiomers that were separated by chiral chromatography and pharmacological testing was performed on each enantiomer. Note that the azabicyclo[3.1.0]hexane compound used is the same as the DOV series of SERT/DAT/NET triple reuptake inhibitors.[623]

8.3.3   VMAT2: Lobeline

A Cochrane review has examined the use of lobeline in smoking cessation. The conclusion was, there is currently no evidence supporting the use of lobeline in aiding chronic smoking cessation.[527]

8.3.4   GABA$_A$ Agonist and AMPA Antagonist: Topiramate

The AMPA/kainate antagonist and GABA agonist topiramate have shown some promise in increasing smoking cessation rates in small clinical trials.[603] While acute topiramate use may enhance the subjective feelings of nicotine, long-term use appears to do the opposite. There are currently two potential clinical trials (NCT01182766, NCT00802412) in the recruiting phase examining the effect of topiramate on nicotine dependence on patients also with alcohol dependence.

8.3.5   PPAR-γ Agonists: Pioglitazone

Pioglitazone at doses up to 45 mg per day will be investigated in a clinical trial to study its effects on smoking (NCT 01395797). The trial is sponsored by the New York Psychiatric Institute in collaboration with NIDA and Omeros Corporation. The primary outcomes of the trial will be the number of drug choices, and if the participants choose to self-administer cigarettes, they will also look at the effect of the drug on mood, as well as on physiological effects.

9

MEDICATION DEVELOPMENT FOR MARIJUANA ADDICTION

Marijuana use is a complicated issue. While there is no doubt tetrahydrocannabinol (THC) is addictive and marijuana users can suffer from drug addiction problems, there are also bona fide medical uses such as an analgesic. In fact, as of February 2014, use of marijuana was legal in the states of Colorado and Washington and the use of medical marijuana was allowed in 16 states. Globally at least four countries allow the use of marijuana for either medical or personal purposes. The adverse health effects of smoking marijuana can arise from the effects of THC on the nervous system and also from the act of inhaling a foreign substance, and the resultant effects on the cardiovascular system. Apparently, occasional use (one joint a week) does not appear to have noticeable harmful effects on lung function.[624]

Marijuana, or more specifically Δ^9-THC, is addictive. Commonly said to be "psychologically" addictive, abstinence can cause withdrawal symptoms in humans similar in scope and severity to nicotine withdrawal; craving, irritability, anxiety, depressed mode, decreased appetite, and sleep difficulties.[112] The DSM-V includes diagnoses for cannabis use disorder, cannabis intoxication, cannabis withdrawal, and cannabis-induced disorders. Cannabis-induced disorders represent maladies such as cannabis-induced psychosis, cannabis-induced anxiety and cannabis-induced sleep disorder.

It is becoming increasingly clear that the human brain continues developing until the early 20's. More studies on the long-term use of cannabis during the adolescent years of life are being done. From these, it is becoming apparent that the use of cannabis in adolescence can have profound effects on the development of IQ. In a remarkable 25-year study involving 1037 participants who were born in 1972

Drug Discovery for the Treatment of Addiction: Medicinal Chemistry Strategies, First Edition. Brian S. Fulton.
© 2014 John Wiley & Sons, Inc. Published 2014 by John Wiley & Sons, Inc.

and 1973 in Dunedin, New Zealand, the effect of cannabis use, age, and IQ was investigated.[625] The individuals involved were enrolled as infants in the longitudinal Dunedin Multidisciplinary Health and Development Study. Their families represented the range of socioeconomic statuses in that region.

The IQ of each participant was determined four times up to age 13. They were asked about past-year cannabis use at ages 18, 21, 26, 32, and 38; and their IQ was determined again at age 38. The researchers used the Wechsler Intelligence Scale for Children-Revised and the Wechsler Adult Intelligence Scale-IV to assess IQ in childhood and adulthood, respectively. The team averaged each participant's four childhood IQ scores and compared that number with his or her score at age 38.

Changes in the participants' IQ scores from childhood to age 38 correlated with the number of assessments at which they reported having used cannabis regularly (at least four times weekly). Those who reported regular use at one of the five drug assessments (i.e., ages 18, 21, 26, 32, and 38) scored three IQ points lower on average at age 38 than they had in childhood. Those who reported regular use at three or more assessments scored five IQ points lower. In contrast, the scores of participants who reported no cannabis use throughout the study increased slightly. The researchers found similar correlations between participants' IQ scores and the number of assessments in which they met diagnostic criteria for cannabis dependence.

They saw greater IQ declines among the adolescent-onset users. Individuals who were dependent on cannabis before age 18 and in a total of three or more assessments lost eight IQ points, on average, whereas individuals who developed dependence as adults did not exhibit IQ declines in relation to their cannabis dependence. The researchers state that the eight-point decline observed among the most persistent adolescent-onset users would move an individual who started at the 50th percentile with an IQ of 100 to the 29th percentile. This could have a dramatic effect on the individual's life/work experiences. (Summarized from http://www.drugabuse.gov/news-events/nida-notes/2013/08/early-onset-regular-cannabis-use-linked-to-iq-decline)

9.1   PHARMACOLOGY OF MARIJUANA ADDICTION

The psychoactive substance in marijuana (*Cannabis sativa*) is the natural product Δ^9-THC. It binds to two cannabinoid GPCR receptors, CB_1 and CB_2. Inhibition binding constants (K_i) vary from 5-80 nM for the CB_1 receptor to 3-75 nM for the CB_2 receptor. Cannabinoid receptors are activated by endogenous ligands that include N-arachidonoylethanolamine (anandamide), N-homo-g-linolenoylethanolamide, N-docosatetra-7,10,13,16-enoylethanolamine, and 2-arachidonoylglycerol. The endogenous ligands are derived from fatty acids and as a group are called endocannabinoids. The term "cannabinoid" is now used for any compound that can mimic the action of plant-derived cannabinoids or are similar in structure to plant cannabinoids.

It is thought there may be a functional relationship between lipid metabolism and psychiatric disorders and metabolic co-morbidities.[626] METSY, a 4-year, €5.94 million multisite project will investigate this using PET and MRI techniques to study endocannabinoid metabolism in early psychosis. The broad goal is to develop blood and PET-MRI biomarker methodology for predicting and monitoring psychiatric disorders and co-morbid metabolic disorders.[627]

Of the two cannabinoid receptors, the CB_1 receptor is predominately localized in the brain and will be the focus of our attention. The CB_1 receptor is widely scattered throughout the brain and in fact is the most abundant GPCR in the mammalian brain. Highest levels of CB_1 receptor have been found in the cerebellum, hippocampus, and the striatum. Some of its functions are the control of motor function, motivation, memory, and pain processing. It is located on presynaptic GABAergic and glutaminergic neurons where activation of the receptor has an *inhibitory* effect on the release of GABA and glutamic acid. There is an overall disinhibitory effect on the reward system. A primary response to CB_1 agonists is inhibition of voltage-gated calcium ion channels and activation of several potassium ion channels. This can lead to a multitude of effects depending on where the receptor (and ion channel) is located. The exact role this may play in drug addiction is not yet clear.

The medicinal chemistry and pharmacology of cannabinoids and cannabinoid receptors have been extensively reviewed in several books and are summarized as follows.[628–630]

Chronic use of cannabis leads to cannabinoid tolerance. This is manifested by CB_1 down-regulation, perturbation of the second messenger signaling pathways, and by changes in endogenous cannabinoid concentrations. As the CB_1 receptor is $G_{i/o}$ coupled Δ^9-THC binding inhibits cAMP production. The down-regulation occurs throughout the brain but the magnitude and rate of decreased CB_1 receptors varies in different brain regions. The hippocampus and cerebellum seem particularly affected. *Ex vivo* autoradiographic studies have shown that chronic use of Δ^9-THC in mice (21 days) leads to time-dependent and region-specific down-regulation and desensitization of brain CB_1 receptors.[631]

Down-regulation is a common phenomenon where the receptor density on the surface of a cell decreases. The exact mechanisms of this are an active field of research. It involves a complex sequence of receptor activation, receptor internalization, receptor metabolic degradation, second messenger signaling resulting in decreased protein synthesis and reduced translocation of receptors to the cell surface. Much of the second messenger process relies on the phosphorylation of proteins by kinases. The central role of these kinases, and more importantly, if they can serve as potential targets of the treatment of addiction, is unknown. A complicating factor is that cannabinoids can act through diverse signaling pathways. Of importance is that on chronic exposure to agonists there is not an apparent direct correlation between receptor down-regulation and changes in signal transduction. The chronic changes are selective for specific regions of the brain.

Desensitization follows a similar pattern to down-regulation. The most rapid changes have been reported in the hippocampus. Desensitization has been linked to down-regulation but it needs to be emphasized that many factors are involved such as receptor-G-protein uncoupling, altered G-protein expression, or a combination of these processes. In all the role of protein kinase A appears particularly important,[632] especially in tolerance.[633]

Precipitated cannabinoid withdrawal studies have found an increase in enzymatic adenylyl cyclase activity in the cerebellum of mice. The increased adenylyl cyclase activity will lead to higher concentrations of cAMP implicating cAMP second messenger systems in the withdrawal process. This also points toward the cerebellum as an important region of the brain in cannabinoid withdrawal.

Animal models of cannabinoid addiction have been difficult to develop. Interestingly, low doses of Δ^9-THC are required for the successful execution of self-administration and conditioned place preference studies. A common *in vivo* assay in cannabinoid research is the mouse tetrad assay. Four effects produced by CB₁ agonists are measured: hypokinesia, catalepsy, hypothermia, and antinociception in the tail-flick or hot plate test. The assays are common to many drugs that are not CB₁ agonists but the power of the assay is that only CB₁ agonists can successfully produce positive results in all four. This is an excellent method of validating CB₁ agonism, as while as brain exposure and bioavailability.

Medicinal chemistry strategies for the treatment of marijuana addiction have focused on the development of CB₁ antagonists or agonists and modulation of endocannabinoid catabolism.

9.2 CB₁ ANTAGONIST: RIMONABANT

Rimonabant (SR141716) was the first selective CB₁ antagonist developed and commercialized. It was first reported in 1994 by researchers at Sanofi (now Aventis).[634] It gained recognition by being developed by Aventis for the treatment of obesity but its development was discontinued due to adverse psychiatric effects. Clinical data indicated a significant increase in suicide from patients taking the drug. Similar to bupropion, it is synthesized from a chloropropiophenone, in this case the isomeric 4-chloro-propiophenone as shown in Figure 9.1. Pyrazole formation occurs to afford only the 1,5-diarylpyrazole by attack of the hydrazine primary amine on the ketone

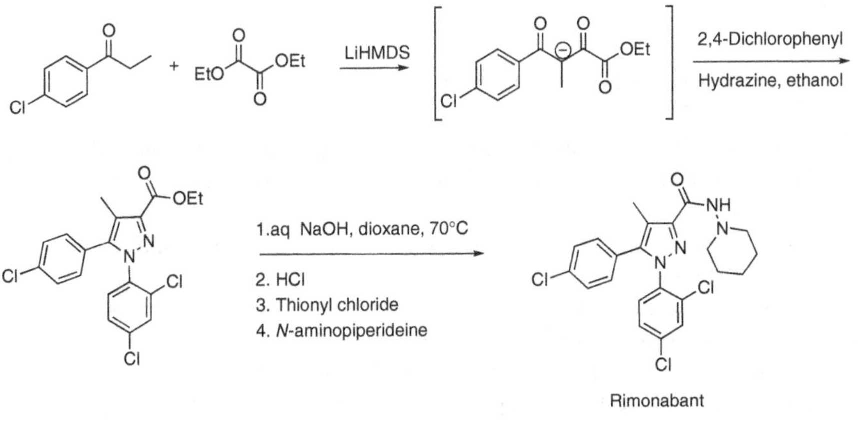

FIGURE 9.1 Synthesis of rimonabant

of the α-ketoester, none of the 1,3-diarylpyrazole is found.[635] Murray and Wachter discovered that reaction of arylhydrazines with the *lithium* salt of aroylpyruvates favors the formation of the 1,5-diarylpyrazoles versus the 1,3-diarylpyrazoles in a greater than 10:1 ratios.[636]

The development of CB_1 receptor antagonists or inverse agonists is still of interest for medicinal chemists in endocrinology though the effort now is to develop peripheral-restricted compounds, that is, unable to pass the blood-brain barrier. Rimonabant is now one of the pharmacological tools of choice in the study of cannabinoid pharmacology. It is extremely selective for the CB_1 receptor with low nanomolar binding of $K_D = 0.19$–1.20 nM and around 1 µM binding to CB_2 receptor. As an antagonist, it is used in precipitating cannabinoid withdrawal in animal studies.

Rimonabant has been determined to act as an antagonist and as an inverse agonist. Inverse agonism occurs when the CB_1 receptors can be constitutively active where the receptor is constantly bound to the $G_{i/o}$ protein. The concept of inverse agonism assumes that two states of the receptor are present. It is presumed that there exists a constitutively active "on" state where the receptor is coupled to its G-protein even in the absence of a ligand and a constitutively inactive "off" state where the receptor is not coupled to a G-protein. In this model, agonists increase the amount of receptor in the "on" state and inverse agonists increase the amount of receptor in the "off" state. Evidence supporting inverse agonism with regard to the CB_1 receptor has been reviewed.[637] Binding of rimonabant thus reduces G-protein binding. One study has shown though that *in vitro* in cell cultures the antagonist effect is much more potent than the inverse agonist effect. Constitutive basal [^{35}S]GTPγS binding was reduced by 50% ($EC_{50} = 4.4$ µM), whereas the inhibition of agonist-induced [^{35}S]GTPγS binding by rimonabant was in the low nanomolar range ($K_e = 0.6$ nM).[638] This suggests that the *in vivo* pharmacological activity of rimonabant is due to its action as an antagonist.

Rimonabant has not been thoroughly studied in humans for the treatment of cannabis addiction. Although it is routinely used to precipitate withdrawal symptoms in rodents, at doses of 20-40 mg it does *not* appear to precipitate withdrawal symptoms in humans.[639]

9.3   MEDICATION DEVELOPMENT

The field of discovering cannabinoid agonists, antagonists, and functional variations within is well developed. However, it has not led to a compound that has shown promise in the treatment of cannabinoid addiction.

Several drugs used in the treatment of other addictions have been investigated for treating cannabinoid addiction. The antidepressants bupropion, nefazodone, fluoxetine and the mood stabilizer divalproex all showed limited promise.[632] To date only oral Δ^9-THC has shown positive clinical effects.[640]

In Chapter 5, we discussed the effects of Δ^9-THC on the activity of opioid agonists. What is the effect of an opioid agonist or antagonist in marijuana addiction? Opioid antagonists appear to potentiate the effects of cannabis. In one study,

29 daily marijuana smokers participated in a within-subject, randomized, double-blind, placebo-controlled study. Naltrexone (0, 12, 25, 50, or 100 mg) was administered before active or inactive marijuana (3.27 or 0% THC) was smoked. This wide range of naltrexone doses generally increased marijuana-induced subjective and cardiovascular effects.[641] Does this mean that an opioid agonist would attenuate the marijuana-induced subjective effects?

9.3.1 CB$_1$ Agonists

Dronabinol   Dronabinol (Marinol®) is an oral formulation of Δ^9-THC indicated for the treatment of anorexia associated with weight loss in patients with AIDS and for the treatment of nausea and vomiting associated with cancer chemotherapy. The drug is administered in soft gelatin capsules containing 2.5, 5, or 10 mg of Δ^9-THC. The drug is well tolerated and does not produce the adverse effects such as anxiety and panic seen from cannabis smoking. Presumably, this is due to the lack of sudden and high Δ^9-THC concentrations that would occur on smoking. The FDA-approved product Marinol® is a Schedule III substance under the Controlled Substances Act. Schedule III drugs are classified as having less potential for abuse than the drugs or substances in Schedules I and II and have a currently accepted medical use in treatment in the United States, and abuse of the drug may lead to moderate or low physical dependence or psychological dependence.

The effect of dronabinol on cannabis use was investigated in a recent clinical trial where 156 cannabis-dependent adults were enrolled in a randomized, double-blind, placebo-controlled, 12-week trial. After a 1-week placebo lead-in phase, the participants were randomized to receive dronabinol 20 mg twice a day or placebo. Doses were maintained constant until the end of week 8 and then tapered off over 2 weeks. Although patients receiving dronabinol showed a higher rate of treatment retention and reduced withdrawal symptoms, there was unfortunately little difference in abstinent rates after the 2-week tapering period compared with placebo.[642]

Sativex®   Sativex® is an oromucosal mouth spray that consists of a cannabinoid extract whose main components are Δ^9-THC and cannabidiol. The drug was developed by the British firm GW Pharmaceuticals for the treatment of spasticity due to multiple sclerosis and is being investigated for use as an analgesic for cancer pain and neuropathic pain. Sativex® is approved for use in many European countries, Australia, New Zealand, and Canada, but is not yet approved for use in the United States. Sativex® is undergoing Phase III trials in the United States for cancer pain but no regulatory application by GW Pharmaceuticals has been made for multiple sclerosis spasticity.

According to the Summary of Products Characterization sheet for Sativex® each milliliter of Sativex contains 38-44 mg and 35-42 mg of two extracts (as soft extracts) from *Cannabis sativa* L., folium cum flore (Cannabis leaf and flower) corresponding to 27 mg Δ^9-THC and 25 mg cannabidiol. The materials are extracted using liquid

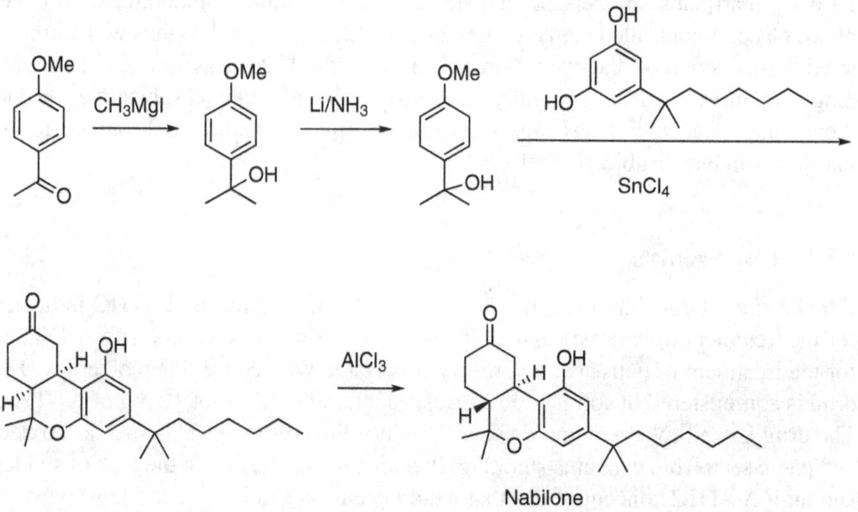

FIGURE 9.2   Synthesis of nabilone

carbon dioxide. Each 100 μL spray contains 2.7 mg Δ^9-THC, 2.5 mg cannabidiol, and up to 40 mg of alcohol.

Nabilone   Nabilone (Cesamet^{TM}) is approved in the United States for the treatment of nausea and vomiting caused by cancer chemotherapy and is a Schedule II Controlled Substance. Nabilone was prepared at Eli Lilly in the early 1970s as part of Lilly's cannabinoid research program.

It is a synthetic compound readily prepared from *p*-methoxyacetophenone in about five steps as shown in Figure 9.2

The key cyclization step is quite interesting and during the development of the synthesis was found to be very sensitive to the nature of the acidic catalyst, solvent, temperature, and reaction time. Optimal yields were obtained when one mole equivalent of water was added to the tin tetrachloride-catalyzed cyclization. This key piece of information along with the identification of intermediates led the authors to suggest the following mechanism depicted in Figure 9.3.[643]

Nabilone is a nonselective CB agonist with binding constants to cloned hCB receptors of K_i (CB$_1$) = 2.2 nM and K_i (CB$_2$) = 1.8 nM.[644] The chemistry and pharmacology of nabilone have been well reviewed.[645]

In the spirit of potential partial agonist therapy, nabilone is being investigated for use in cannabis addiction. In small pilot trials, nabilone has been shown to be well tolerated and to produce effects similar to dronabinol. Nabilone has a longer time to peak effects than dronabinol, and effects were more dose-related, suggesting improved bioavailability. Nabilone, as a stronger agonist than Δ^9-THC, required less dosage than dronabinol, for example, 2-8 mg per day versus 10-20 mg per day, respectively.[646]

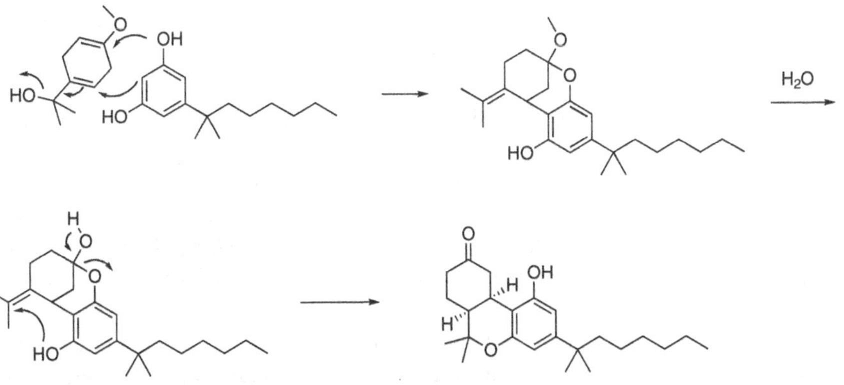

FIGURE 9.3   Mechanism of cyclization

A clinical trial by the University of British Columbia where nabilone at 1 mg per day for 21 days was tested for outpatient management of acute marijuana withdrawal (clinicalTrials.gov Identifier: NCT01025700) was completed in 2011, but no results have yet been published.

A new study (ClinicalTrials.gov Identifier: NCT01347762) from McLean Hospital and NIDA is currently (January 2014) ongoing. This pilot study consists of a 10-week treatment trial with monitoring of the participants by functional MRI brain scans. Nabilone will be dosed at 1 mg twice a day. The investigators propose to assess the relationship of nabilone, when added to behavioral treatment, on cannabis use patterns in cannabis-dependent patients. The investigators also aim to determine the effects of nabilone on performance on neuropsychological tests and to assess the correlation of neuropsychological performance to brain changes using functional MRI brain scans. The investigators hypothesize that patients receiving nabilone will reduce their use of cannabis more than patients receiving placebo during this 10-week treatment trial.

Cannabidiol   Cannabidiol (Figure 9.4) is one of the natural products of *Cannabis sativa* and is recognized as a CB_1 antagonist and a CB_2 inverse agonist/antagonist so it lacks the psychotropic effects of THC.[647]

Beyond its interactions with cannabinoids receptors, the compound shows a complex pharmacological profile with binding affinities in the nanomolar to micromolar range with at least 17 other receptor systems. Cannabidiol inhibits the reuptake and hydrolysis of anandamide and does not cause psychotic effects. In numerous human

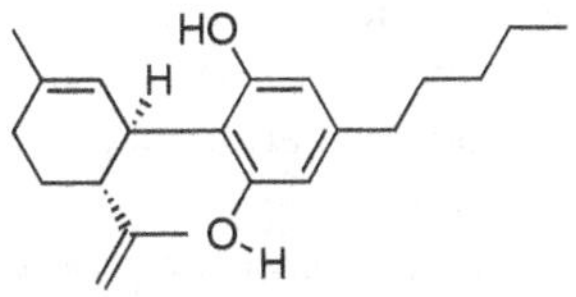

Cannabidiol (CBD)

FIGURE 9.4   Structure of cannabidiol

trials, cannabidiol has found to have beneficial effects on several disorders as while as being effective in reducing some of the psychotic effects of Δ^9-THC and appears to be well tolerated.

A Phase I clinical trial sponsored by Yale University is examining the interactive effects of cannabidiol (5 mg per day) and Δ^9-THC (0.035 mg/kg i.v.) in healthy adults (ClinicalTrials.gov identifier: NCT01180374). A pilot study to investigate the use of Sativex (cannabidiol and Δ^9-THC) for cannabis withdrawal and prevention of relapse is to be conducted by the Centre for Addiction and Mental Health and NIDA (ClinicalTrials.gov identifier: NCT01747850). They hope to study 45 patients over a 6-month period.

Researchers interested in the use of cannabis for therapeutic indications can obtain cannabis from NIDA. Being a controlled substance, one can imagine that the process of doing so is not trivial. Further information on obtaining cannabis and funding possibilities can be found at: www.drugabuse.gov/marijuana-research-nida.

9.3.2   GABA$_B$ Agonist: Baclofen

Baclofen is a GABA$_B$ agonist that inhibits the release of dopamine, noradrenalin, glutamate, and serotonin. In a limited clinical trial ($n = 6$), baclofen at 40 mg per day was shown to reduce the severity of withdrawal and perhaps delay relapse.[648] The male subjects were of age 20-33 and had smoked 6-10 joints of marijuana per day for 1-6 years. All were abstinent for 30-390 days with negative toxicity screens for cannabis. One subject suffered relapse upon withdrawal of baclofen.

However, a second study looked at the effect of baclofen (60, 90 mg per day) on 10 daily marijuana smokers (average of 9.4 marijuana cigarettes per day) for 16 days. Baclofen produced a dose-dependent reduction in craving but there was little effect on mood during abstinence and did not decrease relapse.[649]

9.3.3   Glutamate Modulation: Gabapentin

Gabapentin (Pfizer-Neurontin®) is approved for the control of seizures in persons with epilepsy and as an analgesic for the treatment of postherpetic neuralgia (e.g., shingles).

It has a complex mechanism of action but one clear mode of action is its binding to the $\alpha_2\delta$ subunit ($IC_{50} = 140$ nM) present in presynaptic voltage-gated calcium channels. Binding reduces calcium influx that in turns reduces glutamate release into synapses located in the dorsal horn. Glutamate mediates the fast transmission of the pain sensation from primary sensory neurons to the brain. A reduction in dorsal horn synaptic glutamate thus reduces the sensation of pain. Gabapentin also attenuates ($IC_{50} = 8.9$ μM) the release of norepinephrine induced by high levels of potassium ions or *in vitro* electrical stimulation.[650]

With regard to cannabis addiction, it has been proposed that the neuronal peptide hormone CRF can activate the release of GABA in the amygdala. It is known that during cannabis (and alcohol) withdrawal, there is an increased level of CRF in the amygdala. This suggests that chronic marijuana use upsets the normal regulation of the stress circuitry in the brain. Gabapentin appears to moderate this and restore homeostasis in brain stress systems.

A 12-week Phase II(a) study investigated the use of gabapentin (1200 mg per day) in patients ($n = 50$) seeking treatment for the use of cannabis. Cannabis use was tested by weekly urine analyses and self-reporting. Relative to the placebo group, gabapentin reduced cannabis use, decreased withdrawal symptoms, and increased performance in an executive function test.[651] The researchers hypothesized that gabapentin would help relieve the stress incurred during marijuana withdrawal thus reducing the occurrence of relapse. This limited study suggests this hypothesis was validated.

Gabapentin arose out of search in the 1970s to discover GABA mimetics that could act at GABA receptors and cross the blood-brain barrier. It was known at the time that GABA could reduce convulsions in rats so its use in the treatment of epilepsy was being explored. An extensive amount of SAR studies was performed by the original researchers at Goedecke (Germany) and then later by researchers at Parke-Davis (MI, USA). Excellent case histories of the discovery and summary of the SAR results[652] and of the different syntheses of gabapentin[653] are available. In brief, the presence of the gem methylene primary amino and methylene carboxylic acid groups is required for full activity. Of the 5-8 membered cycloalkanes a cyclohexane ring is preferred and ring substitution is not well tolerated except for a methyl group at position 3 of the cyclohexane ring. Gabapentin is quite polar with a log $D_{pH = 7.4} = -1.25$ and enters the brain via the LAT-1 amino acid transporter and has an oral bioavailability of 76%.

The original synthesis of gabapentin as shown in Figure 9.5 started by condensing cyclohexanone with ethyl cyanoacetate in the presence of ammonia to form a cyclic imide. Subsequent hydrolysis of the imide gave a 1,5-dicarboxylic acid that was transformed in five steps to the zwitterionic gabapentin.

9.3.4 α_2-Adrenergic Receptor Agonists: Clonidine and Lofexidine

As was discussed with the treatment of opioid withdrawal, the α_2-adrenergic agonist clonidine has shown some promise in alleviating precipitated cannabis withdrawal symptoms in mice and rhesus monkeys.[654,655]

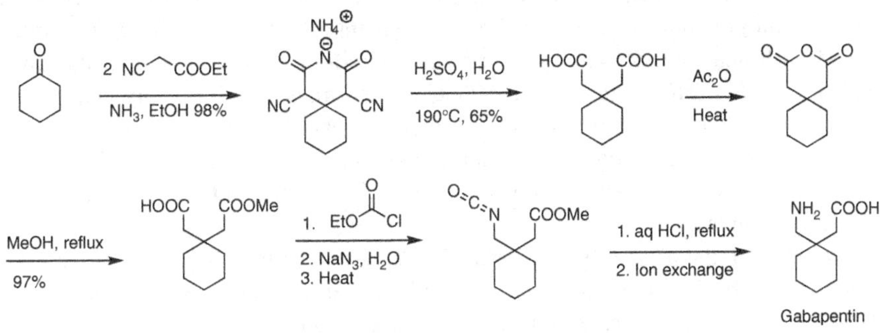

FIGURE 9.5   Synthesis of gabapentin

In humans, the effect of Δ^9-THC, lofexidine, and Δ^9-THC+lofexidine on marijuana withdrawal and relapse was studied on eight nontreatment-seeking male subjects who averaged 12 marijuana cigarettes per day.[656] The combination of Δ^9-THC (60 mg per day) and lofexidine (2.4 mg per day) decreased the symptoms of marijuana withdrawal as while as decreasing craving and relapse. Neither drug alone produced these results.

Marijuana withdrawal, as is opioid withdrawal, is known to potentiate noradrenergic activity, that is, increase norepinephrine release. Agonists, such as lofexidine and clonidine, of the α_2-adgenergic receptor reduce noradrenergic cell firing and release of norephinephrine.[657]

9.3.5   FAAH Inhibition: PF-04457845

An interesting component of cannabinoid neurochemistry is that endocannabinoids are synthesized in postsynaptic neurons and are transported to the presynaptic neuron where they bind to CB_1 receptors. The end result is the inhibition of synaptic transmission. It has been proposed that two of the endocannabinoids, arachidonoylethanolamide (anandamide) and 2-archidonoylglycerol (2-AG) could be classified as neurotransmitters.[112] After anandamide binds to the presynaptic CB_1 receptor, it is deactivated by uptake into the neuron where it is hydrolyzed by fatty-acid amide hydrolase (FAAH) to arachidonic acid and ethanolamine. In the original 1996 paper describing the isolation and purification of FAAH, the rate of hydrolysis of anandamide was measured as 333 nmol/min/mg.[658] A recent deuterium-labeled isotope liquid chromatography-tandem mass spectrometry assay measured the enzymatic activity for the hydrolysis of anandamide by recombinant human FAAH as $K_M = 12.3$ µM and a V_{max} of 27.6 nmol/min/mg.[659]

As low doses of Δ^9-THC can attenuate withdrawal symptoms (via a partial agonist effect), a strategy was proposed that prevention of anandamide hydrolysis would increase CB_1 receptor occupancy by anandamide, resulting in an effect similar to a low dosage of Δ^9-THC. This could be accomplished by inhibition of FAAH and there is pharmacological evidence that an FAAH inhibitor could be useful for the treatment of addiction. Knockout $FAAH^{-/-}$ and wild-type mice were treated for 5.5 days with subchronic doses of anandamide (50 mg/kg twice daily) or Δ^9-THC (50 mg/kg twice daily). Upon precipitated withdrawal by rimonabant (10 mg/kg), the $FAAH^{-/-}$ mice treated with anandamide showed less severe withdrawal symptoms than the mice treated with Δ^9-THC. Unlike the Δ^9-THC-treated mice, no significant changes in CB_1 receptor levels or G-protein activity were observed in the anandamide-treated mice. Taken together, it suggests that increased levels of anandamide resulting from FAAH inhibition would not result in symptoms of tolerance and dependence as seen with CB_1 agonists.

FAAH is a membrane-bound serine hydrolase enzyme. In 2002, the crystal structure of FAAH complexed with methoxy arachidonyl fluorophosphonate, an active-site-directed irreversible inhibitor was reported.[660] Now amendable to structure-guided inquiry, FAAH has since become an extremely attractive target

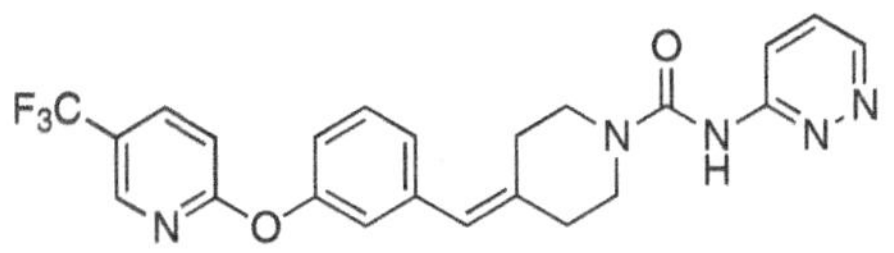

PF-04457845

FIGURE 9.6   Structure of PF-04457845

for medicinal chemists. In general, FAAH inhibitors contain a carbonyl group that interacts with the serine in the active site. The carbonyl group often exists as a urea, carbamate, or an α-keto heterocycle. Several reviews have been written on the synthesis and pharmacological activity of FAAH inhibitors.[661–667] The neurochemistry and medicinal chemistry of fatty-acid amide signaling molecules and their inhibition have been well reviewed.[668]

From clinicaltrials.gov, there is one FAAH inhibitor that has progressed into clinical trials for addiction. The FAAH inhibitor PF-04457845 (Figure 9.6) will be studied for its effect on cannabis withdrawal and relapse in cannabis-dependent subjects in a Phase II double-blind, randomized, controlled, proof-of-concept study. The study (NCT01618656) is sponsored by Yale University and NIDA, at the time of this writing the study was recruiting subjects.

PF-04457845 was discovered by Pfizer and is a covalent irreversible inhibitor of FAAH. The SAR of PF-04457845 has been fully explored and reported.[373] Its mechanism of action is by carbamylation of the active-site serine nucleophile of FAAH with high *in vitro* potency of $k_{inact}/K_i = 40,300 \text{ M}^{-1} \text{ s}^{-1}$ and $IC_{50} = 7.2 \text{ nM}$ for human FAAH. The researchers used the second-order rate constant k_{inact}/K_i as a measure of enzyme inhibition as they state that unlike IC_{50} values, k_{inact}/K_i values are independent of preincubation times and substrate concentrations and are considered the best measure of potency for irreversible inhibitors. The drug is highly selective for FAAH and was shown to not inhibit over 200 other serine hydrolase enzymes at a concentration of 100 μM.

It appears that its initial development was for use in inflammatory pain, but development appears to have stopped according to the November 8, 2012 summary of the Pfizer pipeline.

10

DESIGNER DRUGS

This last chapter briefly looks at what are referred to as designer drugs. These are completely synthetic drugs with the goal to mimic effects of natural products. Some have been discovered by accident, others designed to achieve a certain effect. With regard to medication development for drug addiction they do not prevent new challenges, as their mode of action is the same as we have discussed. However, as they have the potential of becoming an extreme problem for law enforcement due to the seemingly endless number of potential psychoactive agents, it is worth examining them.

Carroll expertly summarizes the current situation of designer drugs as follows: "there are numerous medicinal chemistry reports in the literature describing the pharmacological properties of thousands of narcotics, stimulants, hallucinogens, sedative-hypnotic drugs, cannabinoids, and other psychoactive substances as well as synthetic methods for their preparations. This information, while essential for the advancement of science, has been used by clandestine chemists to manufacture and market an endless variety of analogs of the so-called designer drugs. Clandestine chemists use the principles of medicinal chemistry to design molecules, referred to as designer drugs, that elicit the effects of opioids, amphetamine and analogs, cannabinoids, and phencyclidine analogs while circumventing the law."[669]

Drug Discovery for the Treatment of Addiction: Medicinal Chemistry Strategies, First Edition. Brian S. Fulton.
© 2014 John Wiley & Sons, Inc. Published 2014 by John Wiley & Sons, Inc.

10.1   CATHINONE DRUGS

10.1.1   Khat

Cathinone is a natural product contained in the leaves of the khat plant (*Catha edulis*) that is native to the Horn of Africa and the Arabian Peninsula. Leaves of the plant have been chewed for centuries due to their stimulant effects and use of khat appears to be an important cultural component of Yemeni life.[670] The effects produced by the drug include excitation, hypersensitivity, anorexia, insomnia, euphoria, increased respiration, and hyperthermia. The stimulant and euphoric effects of the leaves rapidly decrease with ageing and drying. The most potent effects are from chewing fresh leaves from the tips of the branches. The juice is swallowed while the solid plant material is expectorated. A typical dosage is 100–200 g of plant material.[671]

The khat tree is a perennial green plant that is grown by grafting, usually reaching 6–7 meters in height. However, under optimal conditions it may grow to 15–20 meters in height. It grows at altitudes of 1500–2500 meters. In places experiencing severe frost in winter, the aerial parts of the plants are either stunted or killed, and the tree never develops to a height of more than 1.5 meters. It is allowed to grow for 3–4 years before harvesting the leaves. The trunk is straight and slender and the bark is thin, smooth, and grayish brown in appearance. The plant has a taproot that grows to a depth of 3 meters or more. The khat plant is polymorphic and the branches have either opposite or alternate leaves. The leaves are 2–5 cm wide and 5–10 cm long. The shapes of the leaves range from ovate-lanceolate to elliptical and have serrated edges. Old leaves are leathery in texture, highly polished on their upper surface and deep green in color. The leaf peduncle is around 3–7 mm long. The leaf odor is faintly aromatic and the taste is astringent and slightly sweet.[672]

The structure of cathinone is very similar to amphetamine as can be seen in Figure 10.1. The simplicity of the structure coupled with a rich literature on the synthesis of phenethylamines and their pharmacological properties makes it an attractive target for clandestine chemists.

Cathinone can exist as two enantiomers with the S enantiomer being more active than the R enantiomer. Studies at the United Nations Narcotic Laboratory in Geneva, Switzerland, in the 1970s showed that the active constituent was the S enantiomer. During the studies on the isolation and identification of the active constituent, 1.5 kg of fresh leaves produced 0.5 g of pure cathinone, isolated as the oxalate salt. The original work has been described in a report titled *Studies on the Chemical Composition of Khat III. Investigations on the phenylalkylamine fraction*. The report is commonly referenced as MNAR/11/75, but is very difficult to obtain a copy though the normal scientific literature. Interested readers can obtain the report though by contacting the United Nations Office on Drug and Crime. A description of the chemistry of khat by Szendrei is available from the United Nations Office on Drugs and Crime Bulletin on Narcotics 1980 Issue 3.[673] In 1982, Carroll reported the synthesis of both enantiomers.[674]

Kalix has reviewed aspects of the pharmacology of khat and cathinone that are summarized as follows.[671,675] Chewing khat produces a stimulatory effect along with

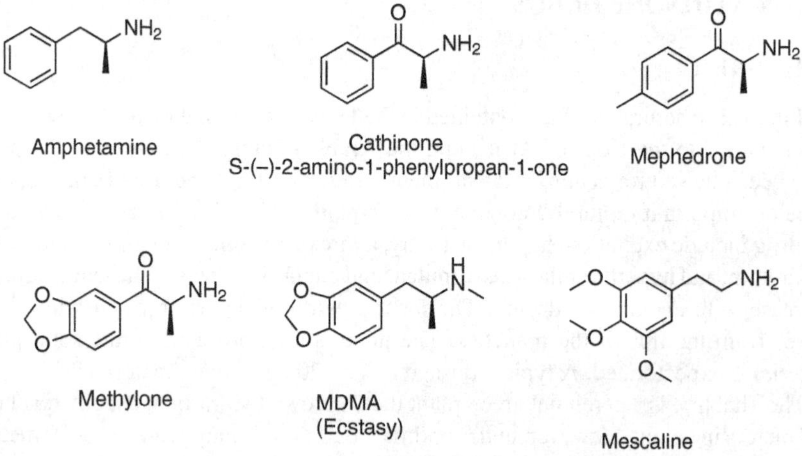

Amphetamine

Cathinone
S-(–)-2-amino-1-phenylpropan-1-one

Mephedrone

Methylone

MDMA
(Ecstasy)

Mescaline

FIGURE 10.1   Structures of abused phenethylamines

a mild euphoric effect. The normal dosage of 100-200 g of leaves has been reported to give effects similar to 5 mg of amphetamine. In animals, cathinone substitutes for amphetamine and can have reinforcing effects similar to cocaine. In monkeys trained to self-administer cocaine, cathinone was found to have a reinforcing effect and to elicit higher rates of cocaine responding than amphetamine. Cathinone is about 75% as potent as (+)-amphetamine in the efflux of [^{3}H]-dopamine from prelabeled rabbit caudate nucleus tissue slices. It was proposed that chronic use of khat or an overdose should be treated the same as amphetamine intoxication.

10.1.2   Mephedrone

Around 2009, the use of "bath salts" or "plant food" became popular as a "legal" drug. They contain a mixture of the cathinone analogs mephedrone and methylone (Figure 10.1) and could be purchased online and at head shops with a 50-mg package selling for $25 to $50. The effect of the drugs is said to be similar to 3,4-methylenedioxy-methamphetamine (MDMA) (Ecstasy). Of concern is the long-term effects of the drugs in humans, which are unknown. A recent extensive study compared mephedrone, methylone, methamphetamine, and MDMA's pharmacological and behavioral real time effects in moving rats. The ability to cause radiolabeled MPP$^+$ (1-methyl-4-phenylpyridinium salt) to be released from NET- and DAT-containing synaptosomes and for [^{3}H]-serotonin to be released from SERT-containing synaptosomes was measured. All the drugs were found to be good substrates for the monoamine transporters. Mephedrone was a substrate for NET (EC_{50} = 63 nM) and DAT (EC_{50} = 49 nM), with slightly lower potency at SERT (EC_{50} = 118.3 nM). Methylone displayed a selectivity profile similar to mephedrone but was about half as potent. MDMA was about equipotent at all three transporters with EC_{50}'s of

50–54 nM. Methamphetamine though was much more potent at NET (EC_{50} = 14 nM) and DAT (EC_{50} = 9 nM) with poor potency at SERT (EC_{50} = 1.3 μM).

In vivo microdialysis measurement of serotonin and dopamine levels in rat extracellular brain fluid showed that doses of 0.3 mg/kg of mephedrone and 1.0 mg/kg of methylone increased dopamine and serotonin levels in the brain with potency similar to MDMA. The greatest effect was seen for serotonin with up to a 1200% increase in serotonin levels relative to saline. Analysis of rats' brain tissue showed that all three drugs caused transporters in cellular membranes to release dopamine and serotonin into the extracellular space. This effect reverses the normal activity of the transporters, which is to draw the neurotransmitters out of the extracellular space.

Important pharmacological and behavior differences between mephedrone/ methylone and MDMA were observed. Repeated high doses of MDMA reduced serotonin concentrations in the cortex and striatum while this was not observed with mephedrone or methylone. Acute doses of all three drugs resulted in locomotor stimulation and hyperthermia, but repeated high doses of the drugs produced different behavioral responses. Specifically, instead of the flattened body posture and forepaw treading seen with MDMA, mephedrone and methylone induced rearing behavior.[676]

Williams, Dybdal-Hargreaves, Holder, Ottoson, and Sweeney have written a review detailing the public health risk and pharmacology of mephedrone.[677] Their introduction is reproduced in part below as it gives an excellent overview of the rise and use of mephedrone. What they wrote about mephedrone can be readily extended to all designer drugs.

"During the mid-2000s, a law enforcement crackdown on MDMA led to a drug shortage in many major countries. As law enforcement agencies placed stronger restrictions on MDMA precursors, drug makers turned to novel products to satisfy the demand for MDMA. The change in MDMA precursors led to a marked decrease in drug purity, with the number of pills sold as MDMA actually containing the compound dropping from 90% to 50%. The most common MDMA replacement was the synthetic cathinone mephedrone. At the time of their initial use, mephedrone and related compounds were legal designer drugs; however, legislation in April 2010 reclassified mephedrone as a Class B restricted substance in the United Kingdom and President Barak Obama signed a law banning mephedrone in the USA in July 2012. Many other countries are adopting similar legislation, often citing a small but growing mephedrone literature base and comparing the designer drug with well researched drugs of abuse; however, these bans have been unable to stop the spread of designer drugs. From 2008 to 2009, online searches for mephedrone rose exponentially, particularly in the UK and Scandinavia (Psychonaut Web Mapping Research Group, 2009). Along with street dealing, the market for mephedrone has extended to the Internet. Multiple websites sell drugs labeled "mephedrone" as well as other synthetic cathinones under the premise that they are a "legal-high." Street names and pseudonyms for mephedrone range from "bath salts" to "plant food" to "bubbles" (Psychonaut Web Mapping Research Group, 2009). The absence of research on the synthetic drugs made initial legislation difficult, and gave short-term truth to the "legal-high" status of mephedrone.

While the drug was legal, mephedrone users reported taking between 0.5 and 1 g of drug in a typical session, most commonly by snorting or ingestion. Typically, the drug takes effect 15–45 minutes after administration and shows clinical effects for approximately 2–5 hours. Compared to other recreational drugs with similar effects, users felt that mephedrone had a longer lasting, better high with a reduced risk of addiction. These subjective measures are validated by empirical evidence that suggests that the effects of cocaine last for 5–30 minutes while the effects of MDMA last for hours."

10.2   MDMA—ECSTASY

We will have a very general discussion of MDMA taken from NIDA Drug Facts of 2013, detailed information on MDMA can be found in a recent review by Parrott.[678] 3,4-Methylenedioxy-methamphetamine (Figure 10.1), popularly known as ecstasy or, more recently, as "Molly," is a synthetic, psychoactive drug that has similarities to both the stimulant amphetamine and the hallucinogen mescaline. It produces feelings of increased energy, euphoria, emotional warmth and empathy toward others, and distortions in sensory and time perception.

MDMA was initially popular among White adolescents and young adults in the nightclub scene or at "raves" (long dance parties), but the drug now affects a broader range of users and ethnicities. MDMA is taken orally, usually as a capsule or tablet. The popular term "Molly" (slang for "molecular") refers to the pure crystalline powder form of MDMA, usually sold in capsules. Ecstasy is readily bioavailable from the small intestine. Although basic, the more alkaline environment in the small intestine coupled with the small intestine's large surface area allows diffusion through the membrane into the systemic circulation. Ecstasy is predominantly in its non-polar form in blood and therefore it crosses the barrier into the brain very easily. It will take about 15 minutes for ecstasy to reach the brain if taken on an empty stomach.

The drug's effects last approximately 3-6 hours, although it is not uncommon for users to take a second dose of the drug as the effects of the first dose begin to fade. It is commonly taken in combination with other drugs. For example, some urban gay and bisexual men report using MDMA as part of a multiple-drug experience that includes cocaine, γ-hydroxybutyrate (GHB), methamphetamine, ketamine, and the erectile-dysfunction drug sildenafil (Viagra®).

As mentioned above, MDMA is a DAT/NET/SERT reuptake inhibitor. The emotional and pro-social effects of MDMA are likely caused directly or indirectly by the release of large amounts of serotonin into the neocortex and parts of the limbic system, which influences mood as well as other functions such as appetite and sleep. Serotonin also triggers the release of the hormones oxytocin and vasopressin, which play important roles in love, trust, sexual arousal, and other social experiences. This may account for the characteristic feelings of emotional closeness and empathy produced by the drug; studies in both rats and humans have shown that MDMA raises the levels of these hormones.

However, the surge of serotonin caused by taking MDMA depletes the brain of this important chemical, causing negative after effects—including confusion, depression, sleep problems, drug craving, and anxiety—that may occur soon after taking the drug or during the days or even weeks thereafter. Some heavy MDMA users experience long-lasting confusion, depression, sleep abnormalities, and problems with attention and memory, although it is possible that some of these effects may be due to the use of other drugs in combination with MDMA (especially marijuana).

MDMA can have many of the same physical effects as other stimulants like cocaine and amphetamines. These include increases in heart rate and blood pressure, which are particularly risky for people with circulatory problems or heart disease. MDMA users may experience other symptoms such as muscle tension, involuntary teeth clenching, nausea, blurred vision, faintness, and chills or sweating.

In high doses, MDMA can interfere with the body's ability to regulate temperature. On rare but unpredictable occasions, this can lead to a sharp increase in body temperature (hyperthermia), which can result in liver, kidney, or cardiovascular system failure or even death. MDMA can interfere with its own metabolism, causing potentially harmful levels to build up in the body if it is taken repeatedly within short periods of time.

Because ecstasy is synthesized in clandestine laboratories, its purity can vary substantially from laboratory to laboratory. Compounding the risks is the fact that ecstasy tablets and even capsules of supposedly pure "Molly" sometimes actually contain other drugs instead or in addition. These may include ephedrine (a stimulant), dextromethorphan (a cough suppressant), ketamine, caffeine, cocaine, methamphetamine, or even, most recently, synthetic cathinones. These substances are harmful alone and may be particularly dangerous if mixed with MDMA. Users who intentionally or unknowingly combine such a mixture with additional substances such as marijuana and alcohol may be putting themselves at even higher risk for adverse health effects. Additionally, the closeness-promoting effects of MDMA and its use in sexually charged contexts (and especially in combination with sildenafil) might encourage unsafe sex, which is a risk factor for contracting or spreading HIV and hepatitis.

For the interested reader, a review of animal models used in the study of serotonergic psychedelics is available.[679]

10.3 CANNABINOID DESIGNER DRUGS

10.3.1 Spice

Smokable herbal blends marketed as being "legal" and providing a marijuana-like high have also become increasingly popular, particularly among teens and young adults. They are easily available and, in many cases, they are more potent and dangerous than marijuana. These products consist of plant material that has been coated with synthetic psychoactive cannabinoids that mimic THC. Just as with the synthetic cathinones, synthetic cannabinoids are sold at a variety of retail outlets, in head shops and over the Internet. Brands such as "Spice," "K2," "Blaze," and "Red X Dawn" are labeled as incense to mask their intended purpose.

Because the chemicals used in Spice have a high potential for abuse and no medical benefit, the Drug Enforcement Administration (DEA) has designated the five active chemicals most frequently found in Spice as Schedule I controlled substances, making it illegal to sell, buy, or possess them. The five compounds are 1-pentyl-3-(1-naphthoyl)indole (JWH-018), 1-butyl-3-(1-naphthoyl)indole (JWH-073), 1-[2-(4-morpholinyl)ethyl]-3-(1-naphthoyl)indole (JWH-200), 5-(1,1-dimethylheptyl)-2-(3-hydroxycyclohexyl)-phenol (CP-47,497), and 5-(1,1- dimethyloctyl)-2-(3-hydroxycyclohexyl)-phenol (cannabicyclohexanol, CP-47,497 C8 homologue) including their salts, isomers, and salts of isomers whenever the existence of such salts, isomers, and salts of isomers is possible. Note these are all compounds that were originally synthesized in drug discovery programs. Manufacturers of Spice products attempt to evade these legal restrictions by substituting different chemicals in their mixtures, while the DEA continues to monitor the situation and evaluate the need for updating the list of banned cannabinoids.

To give an example of the severity of the problem, the DEA conducted a nationwide law-enforcement action in collaboration with other law-enforcement agencies in July 2012 called "Operation Log Jam." This resulted in more than five million packets of finished designer synthetic drugs being seized, 90 individuals being arrested, and more than $36 million in cash being seized. More than 4.8 million packets of synthetic cannabinoids (i.e., K2, Spice) and the products to produce nearly 13.6 million more, as well as 167,000 packets of synthetic cathinones (e.g., bath salts), and the products to produce an additional 392,000 were seized. While many of the designer drugs being marketed today that were seized as part of Operation Log Jam are not specifically prohibited in the Controlled Substances Act, the Controlled Substance Analogue Enforcement Act of 1986 allows these drugs to be treated as controlled substances if they are proven to be chemically and/or pharmacologically similar to a Schedule I or Schedule II controlled substance. A number of cases that are part of Operation Log Jam will be prosecuted federally under this analogue provision, which specifically exists to combat these new and emerging designer drugs.[680]

CONCLUSION

Drug addiction is a multifaceted disease, yet common underlying themes exist across all the abused drugs. A single-drug, single-target approach has not worked and most likely will not. It seems clear that a drug with binding affinities and functional activities across several protein targets and neurocircuits will be required. Exactly what those criteria are remains to be determined. Much emphasis has been placed on the opioid and dopamine neurocircuits. Mixed modulation of the cholinergic, GABAergic, and glutaminergic systems may be more fruitful. Of course these systems control a wide range of homeostatic conditions so the opportunities for side effects abound. The research being conducted on memory reconsolidation appears very intriguing. Could pharmacological intervention of memory reconsolidation prevent relapse? Of the different stages of addiction, relapse is the most difficult for the patient to control, hence the most important to focus future efforts on. Though existing drugs that can attenuate the effects of withdrawal improvements are always possible. I would think that as with major depression and schizophrenia, individuals trying not to succumb to relapse may be required to take a drug, may be for years and possibly their entire life. Hopefully, it would not come to that, but it is a matter that needs to be addressed.

Beyond therapeutic development, addiction research offers many opportunities for fundamental research. For chemists, an obvious subject worth addressing is the development of protein-protein interaction modulators that could be used to determine the importance of receptor dimerization. If receptor dimerization proved to be of importance *in vivo* and, as some suggest, lead to differing functional activities than

Drug Discovery for the Treatment of Addiction: Medicinal Chemistry Strategies, First Edition. Brian S. Fulton.
© 2014 John Wiley & Sons, Inc. Published 2014 by John Wiley & Sons, Inc.

the separate receptors, a new set of targets could arise. However, any drug developed to act on such a set of targets would still need to be a small molecule that can be taken orally. As with all mental disorders, there is a striking absence of diagnostic tools to determine whether an illness exists and to what extent it exists. At this point, the development of PET or SPECT ligands could be most useful as new diagnostic tools. Most interesting would be the use of magnetic resonance spectroscopy, as no radiolabeled ligands would be needed.

One will note that the subject of vaccines was not discussed. Kim Janda of Scripps Research Institute has been especially active in developing vaccines against heroin, nicotine, methamphetamine, and cocaine. Vaccines could be useful for maintaining abstinence by blocking the effects of the drug if the individual resumed drug taking. One would imagine though, that a constant titer of the antibodies produced by the vaccine would be required. A multisite clinical trial of a cocaine vaccine is in progress (NCT00969878), so there is an obvious interest. As I decided to focus on small molecules, this subject was not addressed, but it deserves to be closely watched.

APPENDIX A

FURTHER READING FOR CHEMISTS INTERESTED IN A MORE DETAILED UNDERSTANDING OF ADDICTION AND THE CENTRAL NERVOUS SYSTEM

Basic Neurochemistry, 8th ed. (Brady S., Siegel G., Albers R.W., Price D., editors), Elsevier, 2011.

Blood-Brain Barriers: From Ontogeny to Artificial Interfaces (Dermietzel R., Spray D., Nedergaard M., editors), Wiley-VCH, 2006.

Cell Biology of Addiction (Madras B., Colvis C.M., Pollock J.D., Rutter J.L., Shurtleff D., Zastrow M., editors), Cold Spring Harbor Laboratory Press, 2006.

Current Protocols in Neuroscience (Gerfen C., Holmes A., Sibley D., Skolnick P., Wray S., editors), John Wiley & Sons, Inc, 2014.

Current Protocols in Pharmacology (Enna S.J., Williams M., editors), John Wiley & Sons, Inc, 2014.

Drug Discrimination: Applications to Medicinal Chemistry and Drug Studies (Glennon R.A., Young R., editors), John Wiley & Sons, 2011.

Drugs and Behavior, an Introduction to Behavioral Pharmacology, 7th ed. (McKim W.A., Hancock S.), Pearson, 2012.

Neurobiology of Addiction (Koob G.F., Moal M.), Academic Press, 2006.

Neurobiology of Mental Illness, 3rd ed. (Charney D.S., Nestler E.J., editors), Oxford University Press, 2009.

Pathology of Drug Abuse, 4th ed. (Karch S.B.), CRC Press, 2008.

The Human Brain: An Introduction to Its Functional Anatomy, 6th ed. (Nolte J.), Mosby Elsevier, 2009.

The American Disease, Origins of Narcotic Control (Musto D.R.), Oxford University Press, 1999.

Drug Discovery for the Treatment of Addiction: Medicinal Chemistry Strategies, First Edition. Brian S. Fulton.
© 2014 John Wiley & Sons, Inc. Published 2014 by John Wiley & Sons, Inc.

APPENDIX B

PUBLIC DATABASES AND SOURCES OF INFORMATION OF INTEREST TO MEDICINAL CHEMISTRY ADDICTION RESEARCHERS

BRENDA—Comprehensive Enzyme Information. http://www.brenda-enzymes.org

Clinical Trials from NIH. http://clinicaltrials.gov/

Drug Enforcement Agency (DEA). http://www.justice.gov/dea

European Monitoring Centre for Drugs and Drug Addiction (EMCDDA). http://www.emcdda.europa.eu/

International Classification of Diseases (ICD). http://www.who.int/classifications/ icd/en

IUPHAR Committee on Receptor Nomenclature and Drug Classification. http://www.iuphar-db.org

IUPHAR/BPS Guide to Pharmacology. http://www.guidetopharmacology.org

National Institutes of Drug Abuse (NIDA). http://www.drugabuse.gov

NIMH Psychoactive Drug Screening Program. http://pdsp.med.unc.edu

Orange Book. http://www.accessdata.fda.gov/scripts/cder/ob/default.cfm

Pubchem. http://pubchem.ncbi.nlm.nih.gov

Pubmed. http://www.ncbi.nlm.nih.gov/pubmed

United Nations Office on Drugs and Crime (UNODC). http://www.unodc.org

Drug Discovery for the Treatment of Addiction: Medicinal Chemistry Strategies, First Edition. Brian S. Fulton.
© 2014 John Wiley & Sons, Inc. Published 2014 by John Wiley & Sons, Inc.

APPENDIX C

GLOSSARY OF TERMS USED IN ADDICTION RESEARCH

Acetylcholine: A neurotransmitter that functions in the brain to regulate memory and that controls the actions of skeletal and smooth muscle in the peripheral nervous system.

Action potential: Transmission of signal from the cell body to the synaptic terminal at the end of the cell's axon. When the action potential reaches the end of the axon, the neuron releases neurotransmitters or electrical signals.

Acquisition: In addiction research it pertains to the increase in drug responding upon presentation of cues.

Acute: Refers to brief exposure to a drug, often in a high dose. In rodent studies measured in hours or days.

Adaptation: A progressive decrease in sensitivity shown by sensory receptors occurring from a maintained stimulus.

Addiction: A chronic, relapsing disease characterized by compulsive drug seeking and use despite serious adverse consequences, and by long-lasting changes in the brain.

Adrenal glands: Glands, located above each kidney, that secrete hormones, for example, adrenaline.

Affective state: A state pertaining to mood and emotions.

Afferent: Projection going toward a neuron (axon).

Amygdala: A small almond-shaped cluster of cells that is part of the limbic system controlling affective states.

Drug Discovery for the Treatment of Addiction: Medicinal Chemistry Strategies, First Edition. Brian S. Fulton.
© 2014 John Wiley & Sons, Inc. Published 2014 by John Wiley & Sons, Inc.

Anesthetic: An agent that causes insensitivity to pain and is used for surgeries and other medical procedures.

Anxiety disorders: Varied disorders that involve excessive or inappropriate feelings of anxiety or worry. Examples are panic disorder, post-traumatic stress disorder, social phobia, and others.

Attention deficit hyperactivity disorder (ADHD): A disorder that typically presents in early childhood, characterized by inattention, hyperactivity, and impulsivity.

Attenuate: A reduced response to a drug or stimulus.

Axon: The fiber-like extension of a neuron by which the cell carries information to target cells via propagation of an action potential.

Axon terminal: The structure at the end of an axon that produces and releases neurotransmitters to transmit the neuron's message across the synapse.

Basal ganglia: Structures located deep in the brain that play an important role in the initiation of movements. These clusters of neurons include the caudate nucleus, putamen, globus pallidus, and substantia nigra. It also contains the nucleus accumbens, which is the main center of reward in the brain.

Bipolar disorder: A mood disorder characterized by alternating episodes of depression and mania or hypomania.

Brainstem: The major route by which the forebrain sends information to, and receives information from, the spinal cord and peripheral nerves.

Breaking point: The point at which a subject will stop responding to obtain a reinforcer. At this point, it is said that the demand required to obtain the reinforcer is too high.

Cell body (or soma): The central structure of a neuron, which contains the cell nucleus. The cell body contains the molecular machinery that regulates the activity of the neuron.

Central nervous system (CNS): The brain and spinal cord.

Cerebellum: A large structure located in the back of the brain that helps control the coordination of movement by making connections to other parts of the CNS (pons, medulla, spinal cord, and thalamus). It also may be involved in aspects of motor learning.

Cerebral cortex: The outermost layer of the cerebral hemispheres of the brain about 1.5–4.5 mm in thickness. It is largely responsible for conscious experience, including perception, emotion, thought, and planning.

Cerebral hemispheres: The two specialized halves of the brain. The left hemisphere is specialized for speech, writing, language, and calculation; the right hemisphere is specialized for spatial abilities, face recognition in vision, and some aspects of music perception and production.

Cerebrum: The upper part of the brain consisting of the left and right hemispheres.

Chronic: Refers to a disease or condition that persists over a long period of time.

CNS depressants: A class of drugs that slow CNS function (also called sedatives and tranquilizers), some of which are used to treat anxiety and sleep disorders; include barbiturates and benzodiazepines.

Cognitive behavioral therapy (CBT): A form of psychotherapy that teaches people strategies to identify and correct problematic behaviors to enhance self-control, stop drug use, and address a range of other problems that often co-occur with them.

Comorbidity: The occurrence of two disorders or illnesses in the same person also referred to as co-occurring conditions or dual diagnosis. Patients with comorbid illnesses may experience a more severe illness course and require treatment for each or all conditions.

Conditioned place preference: Nonoperant procedure for determining reinforcing efficacy of drugs using a classical conditioning paradigm.

Conditioned stimulus (CS): Generally a motivationally neutral stimulus such as light or auditory tone that gains salience by being predictive of an unconditioned stimulus or reinforcer such as food or electric shock.

Contingency management (CM): A therapeutic management approach based on frequent monitoring of the target behavior and the provision (or removal) of tangible, positive rewards when the target behavior occurs (or does not). CM techniques have shown to be effective for keeping people in treatment and promoting abstinence.

Consolidation: The memory process by which short-term memory representations (or "traces") are made more permanent, leading to memory storage.

Craving: A powerful, often uncontrollable desire for drugs.

Dementia: A condition of deteriorated mental function.

Dendrite: The specialized branches that extend from a neuron's cell body and function to receive messages from other neurons.

Depression: A disorder marked by sadness, inactivity, difficulty with thinking and concentration, significant increase or decrease in appetite and time spent sleeping, feelings of dejection and hopelessness, and sometimes, suicidal thoughts or an attempt to commit suicide.

Disinhibition: An excitation response due to the inhibition of an inhibitory input.

Detoxification: A process in which the body rids itself of a drug (or its metabolites). During this period, withdrawal symptoms can emerge that may require medical treatment. This is often the first step in drug abuse treatment.

Discrimination: Can occur when a response is reinforced only in the presence of a specific stimulus.

Dopamine: A brain chemical, classified as a neurotransmitter, found in regions that regulate movement, emotion, motivation, and pleasure.

Drug discrimination: Procedures to determine the ability of an animal to differentiate (discriminate) the effects of similar drugs.

Efferent: Projection going away from a neuron (dendrite).

Endogenous: Something produced by the brain or body.

Euphoria: A feeling of well-being or elation.

Extinction: When the reinforcer (e.g., food or shock) associated with learning a particular task or response is omitted or withdrawn. The behavior (response) is no longer effective.

Fixed ratio (FR): A fixed number of responses are required to obtain the reinforcer.

Forebrain: The largest division of the brain, which includes the cerebral cortex and basal ganglia. It is credited with the highest intellectual functions.

Frontal cortex: The front part of the brain involved with reasoning, planning, problem solving, and other higher cognitive functions.

Frontal lobe: One of the four divisions of each cerebral hemisphere. The frontal lobe is important for controlling movement and associating the functions of other cortical areas.

γ-aminobutyric acid (GABA): The main inhibitory neurotransmitter in the central nervous system. GABA provides the needed counterbalance to the actions of other systems, particularly the excitatory neurotransmitter glutamate.

Glia cells: Non-neuronal cells responsible for metabolic, electrical, and mechanical support in the brain.

Glutamate: An excitatory neurotransmitter found throughout the brain that influences the reward system and is involved in learning and memory among other functions.

Hallucinations: Perceptions of something (such as a visual image or a sound) that does not really exist. Hallucinations usually arise from a disorder of the nervous system or in response to drugs.

Hallucinogens: A diverse group of drugs that alter perceptions, thoughts, and feelings.

Hedonic: An affective state marked by pleasure.

Hippocampus: A seahorse-shaped structure located within the brain that is considered an important part of the limbic system. One of the most studied areas of the brain, the hippocampus plays key roles in learning, memory, and emotion.

Homeostasis: The tendency of living systems to maintain or restore equilibrium.

Hormone: A chemical substance formed in glands in the body and carried in the blood to organs and tissues, where it influences function, structure, and behavior.

Hypothalmic-pituitary-adrenal (HPA) axis: A brain-body circuit that plays a critical role in the body's response to stress.

Hypothalamus: The part of the brain that controls many bodily functions, including feeding, drinking, and the release of many hormones.

Impulse: An electrical communication signal sent between neurons by which neurons communicate with each other.

Interneuron: Neurons that form connections with other neurons within a local area.

Intracerebroventricular (i.c.v.): Administration directly into the brain via a syringe.

Intraperitoneal (ip): Injection into the body cavity.

Limbic system: A set of brain structures that generates our feelings, emotions, and motivations. It is also important in learning and memory.

Long-term depression: Reduction in signal transmission, efficacy of neuronal synapses, lasts hours or longer following a long patterned stimulus.

Long-term potentiation: Long-lasting enhancement in signal transmission, efficacy of neuronal synapses, between two neurons that result from stimulating them synchronously.

Maintenance: Used to describe the continuation of operant conditioning.

Major depressive disorder (MDD): A mood disorder having a clinical course of one or more serious depression episodes that last two or more weeks. Episodes are characterized by a loss of interest or pleasure in almost all activities; disturbances in appetite, sleep, or psychomotor functioning; a decrease in energy; difficulties in thinking or making decisions; loss of self-esteem or feelings of guilt; and suicidal thoughts or attempts.

Mania: A mood disorder characterized by abnormally and persistently elevated, expansive, or irritable mood; mental and physical hyperactivity; and/or disorganization of behavior.

Mesocorticolimbic: A set of neuronal fibers that originate in the ventral tegmental area and travel to the cerebral cortex and the limbic area.

Mental disorder: A mental condition marked primarily by sufficient disorganization of personality, mind, and emotions to seriously impair the normal psychological or behavioral functioning of the individual. Addiction is a mental disorder.

Motivational enhancement therapy (MET): A systematic form of intervention designed to produce rapid, internally motivated change. MET does not attempt to treat the person, but rather mobilize their own internal resources for change and engagement in treatment.

Myelin: Fatty material that surrounds and insulates axons of most neurons.

Negative reinforcement: Occurs when a behavior (response) is followed by the removal of an aversive stimulus, thereby increasing that behavior's frequency.

Neural circuit: A network of neurons and their interconnections.

Neuron (nerve cell): A specialized type of cell found in the brain that can process and transmit information.

Neurotransmission: The process that occurs when a neuron releases neurotransmitters to communicate with another neuron across the synapse.

Neurotransmitter: A chemical produced by neurons that carry messages from one nerve cell to another.

Norepinephrine (NE): A neurotransmitter present in the brain and the peripheral (sympathetic) nervous system; and a hormone released by the adrenal glands. Norepinephrine is involved in attention, responses to stress, and it regulates smooth muscle contraction, heart rate, and blood pressure.

Nucleus: A cluster or group of nerve cells that is dedicated to performing its own special function(s). Nuclei are found in all parts of the brain but are called cortical fields in the cerebral cortex.

Nucleus accumbens (NAc): A part of the brain reward system, located in the limbic system, which processes information related to motivation and reward. Virtually all drugs of abuse act on the nucleus accumbens to reinforce drug taking.

Occipital lobe: The lobe of the cerebral cortex at the back of the head that includes the visual cortex.

Operant: A response or set of response that can be defined and measured.

Operant conditioning: Learned behavior as a result of a respondent or operant (e.g., food). Differs from classical (respondent) conditioning as it uses reinforcement or punishment to change behavior.

Parietal lobe: One of the four subdivisions of the cerebral cortex; it is involved in sensory processes, attention, and language.

Physical dependence: A physiological state that occurs with regular drug use and results in a withdrawal syndrome when drug use is stopped; usually occurs with tolerance.

Plasticity: The capacity of the brain to change its structure and function within certain limits. Plasticity underlies brain functions such as learning and allows the brain to generate normal, healthy responses to long-lasting environmental changes.

po (per os): Oral administration.

Polyneuropathy: Permanent change or malfunction of nerves.

Post-traumatic stress disorder (PTSD): A disorder that develops after exposure to a highly stressful event (e.g., wartime combat, physical violence, or natural disaster). Symptoms include sleeping difficulties, hypervigilance, avoiding reminders of the event, and re-experiencing the trauma through flashbacks or recurrent nightmares.

Potentiate: In pharmacology, the interaction of two or more drugs producing an effect greater than the sum of each individual response.

Prefrontal cortex (PFC): A highly developed area at the front of the brain that, in humans, plays a role in executive functions such as judgment, decision making, and problem solving, as well as emotional control and memory.

Priming: New exposure to a formerly abused substance. This exposure can precipitate rapid resumption of abuse at previous levels or at higher levels.

Progressive ratio (PR): A schedule of reinforcement in which the number of responses required for a reinforcer is increased to the "break point." The greater the number of responses required reaching the break point, the greater the reinforcing property of the drug.

Psychoactive: Having a specific effect on the mind.

Psychosis: A mental disorder characterized by delusional or disordered thinking detached from reality; symptoms often include hallucinations.

Reinforcer: Any event that increases the probability of a response.

Reinstatement: Induced (cues, drug, stress) resumption of drug-taking behavior.

Relapse: In drug abuse, relapse is the resumption of drug use after trying to stop taking drugs. Relapse is a common occurrence in many chronic disorders, including addiction, that require behavioral adjustments to treat effectively.

Resting membrane potential: The difference in electrical charge between the inside and the outside of a nerve cell when the cell is not firing. The inside of a resting neuron has a greater negative charge than the outside of the neuron.

Reward: The process that reinforces behavior. It is mediated at least in part by the release of dopamine into the nucleus accumbens. Human subjects report that reward is associated with feelings of pleasure.

Reward system (or brain reward system): A brain circuit that, when activated, reinforces behaviors. The circuit includes the dopamine-containing neurons of the ventral tegmental area, the nucleus accumbens, and part of the prefrontal cortex. The activation of this circuit causes feelings of pleasure.

Rush: A surge of pleasure (euphoria) that rapidly follows the administration of some drugs.

Salient: It is used to distinguish a factor, emotion, stimulus, environment, etc., that is, considered most important in controlling what is being observed.

Schedules of reinforcement: Schedules of reinforcement are rules that control the delivery of reinforcement. The rules specify either the time that reinforcement is to be made available, or the number of responses to be made, or both.

Schizophrenia: A psychotic disorder characterized by symptoms that fall into two categories: (1) positive symptoms, such as distortions in thoughts (delusions), perception (hallucinations), and language and thinking and (2) negative symptoms, such as flattened emotional responses and decreased goal-directed behavior.

Schizophreniform disorders: Similar to schizophrenia, but of shorter duration and possibly lesser severity.

Sedatives: Drugs that suppress anxiety and promote sleep; the NSDUH classification includes benzodiazepines, barbiturates, and other types of CNS depressants.

Self-medication: The use of a substance to lessen the negative effects of stress, anxiety, or other mental disorders (or side effects of their pharmacotherapy). Self-medication may lead to addiction and other drug- or alcohol-related problems.

Sensitization: The increase in the expected effect of a drug after repeated administration. May be one of the neurobiological mechanisms involved in craving and relapse.

Serotonin (5-HT): A neurotransmitter used in widespread parts of the brain, which is involved in sleep, movement, and emotions.

Stimulants: A class of drugs that enhance the activity of monoamines in the brain, increasing arousal, heart rate, blood pressure, and respiration, and decreasing appetite; includes some medications used to treat attention deficit hyperactivity disorder (e.g., methylphenidate and amphetamines), as well as cocaine and methamphetamine.

Synapse: The site where presynaptic and postsynaptic neurons communicate with each other.

Synaptic space (or synaptic cleft): The intercellular space between the presynaptic and postsynaptic neurons.

Temporal lobe: The lobe of the cerebral cortex at the side of the head that hears and interprets music and language.

Thalamus: Located deep within the brain, the thalamus is the key relay station for sensory information flowing into the brain, filtering out important messages from the mass of signals entering the brain.

Tolerance: A condition in which higher doses of a drug are required to produce the same effect achieved during initial use; often associated with physical dependence.

Toxic: Causing temporary or permanent effects detrimental to the functioning of a body organ or group of organs.

Unconditioned stimulus (US): generally a reinforcer such as food or shock that elicits unconditioned responses (such as salivation to food) and is subject to Pavlovian conditioning following several pairings with a predictive conditioned stimulus.

Ventral striatum: An area of the brain that is part of the basal ganglia and becomes activated and flooded with dopamine in the presence of salient stimuli. The release of this chemical also occurs during physically rewarding activities such as eating, sex, and taking drugs, and is a key factor behind our desire to repeat these activities.

Ventral tegmental area (VTA): The group of dopamine-containing neurons that make up a key part of the brain reward system. These neurons extend axons to the nucleus accumbens and the prefrontal cortex.

Vesicle: A membranous sac within an axon terminal that stores and releases neurotransmitter.

Withdrawal: Adverse symptoms that occur after chronic use of a drug are reduced or stopped.

APPENDIX D

GLOSSARY OF TERMS USED IN MEDICINAL CHEMISTRY

IUPAC Recommendations © IUPAC 1998[681]
IUPAC Recommendations © IUPAC 2013[682]

Active transport: Active transport is the carriage of a solute across a biological membrane from low to high concentration that requires the expenditure of (metabolic) energy.

ADMET: Acronym referring to the absorption, distribution, metabolism, excretion, and toxicity profile or processes for a xenobiotic upon its administration *in vivo*. *Note*: ADME is also used to delineate these selected parameters within the context of a xenobiotics pharmacokinetic profile. Because any of the five characteristics may become hurdles during drug development, ADMET behaviour is typically studied and optimized among efficacious analogues during the early drug discovery stage by using *in vitro* models that attempt to predict such behaviors in clinical studies.

Adverse effect, adverse event, adverse drug event: (1) (In medicinal chemistry) Undesirable reaction in response to the administration of a drug or test compound. *Note*: In most instances such effects result from off-*target* interactions. (2) (In toxicokinetics) Change in biochemistry, morphology, physiology, growth, development, or lifespan of an organism, which results in impairment of functional capacity or impairment of capacity to compensate for additional stress or increase in susceptibility to other environmental influences.

Drug Discovery for the Treatment of Addiction: Medicinal Chemistry Strategies, First Edition. Brian S. Fulton.
© 2014 John Wiley & Sons, Inc. Published 2014 by John Wiley & Sons, Inc.

Affinity: Affinity is the tendency of a molecule to associate with another. The affinity of a drug is its ability to bind to its biological target (receptor, enzyme, transport system, etc.). For pharmacological receptors, it can be thought of as the frequency with which the drug, when brought into the proximity of a receptor by diffusion, will reside at a position of minimum free energy within the force field of that receptor.

Agonist: An agonist is an endogenous substance or a drug that can interact with a receptor and initiate a physiological or a pharmacological response characteristic of that receptor (contraction, relaxation, secretion, enzyme activation, etc.).

Allosteric antagonist: Compound that binds to a receptor at a site separate from, but actively coupled to, that of the endogenous agonist to reduce actively receptor signals. *Note*: The terms "allosteric antagonist" and "noncompetitive antagonist" are often synonymous but not necessarily so. See also noncompetitive antagonist.

Allosteric enzyme: An allosteric enzyme is an enzyme that contains a region to which small, regulatory molecules ("effectors") may bind in addition to and separate from the substrate-binding site and thereby affect the catalytic activity. On binding the effector, the catalytic activity of the enzyme toward the substrate may be enhanced, in which case the effector is an activator, or reduced, in which case it is a deactivator or inhibitor.

Analogue, Analog: Chemical compound having structural similarity to a reference compound. *Note*: Despite the structural similarity, an analogue may display different chemical and/or biological properties, as is often intentionally the case during design and synthesis to optimize either efficacy or ADMET properties within a given series. See also analogue-based drug discovery, congener, follow-on drug.

Analog-based drug discovery: Strategy for drug discovery and/or optimization in which structural modification of an existing drug provides a new drug with improved chemical and/or biological properties. *Note*: Within the context of analogue-based drug discovery, three categories of drug analogues are recognized: Compounds possessing structural, chemical, and pharmacological similarities, termed "direct analogues" and sometimes referred to as "me-too" drugs; compounds possessing structural and often chemical but not pharmacological similarities, termed "structural analogues;" and structurally different compounds displaying similar pharmacological properties, termed "pharmacological analogues." See also ligand-based drug design.

Antagonist: An antagonist is a drug or a compound that opposes the physiological effects of another. At the receptor level, it is a chemical entity that opposes the receptor-associated responses normally induced by another bioactive agent.

ATP binding cassette protein, ABC protein: Large gene family of transporter proteins that bind ATP and use the energy to transport substrates (e.g., sugars, amino acids, metal ions, peptides, proteins, and a large number of hydrophobic compounds and metabolites) across lipid membranes. *Note*: These proteins have an important role in limiting oral absorption and brain penetration of xenobiotics. See also efflux pump, P-glycoprotein.

Atropisomer: Stereoisomer resulting from hindered rotation about a single bond in which steric hindrance to rotation is sufficient to allow isolation of individual isomers.

Attrition rate: Rate of loss of drug candidates during progression through the optimization and developmental stages while on route to the marketplace. *Note*: It has been estimated that for every 10,000 compounds examined during the early stages of biological testing just one reaches the market.

Autoinduction: Capacity of a drug to induce enzymes that mediate its own metabolism. *Note*: This often results in lower, often sub-therapeutic, drug exposure on prolonged or multiple dosing.

Autoreceptor: An autoreceptor, present at a nerve ending, is a receptor that regulates, via positive or negative feedback processes, the synthesis and/or release of its own physiological ligand.

Back-up compound: Molecule selected as a replacement for the lead drug candidate, should it subsequently fail during further preclinical evaluation or in clinical studies *Note*: Ideally, a back-up should be pharmacologically equivalent to the lead drug but have significant structural differences. Possession of a distinct core scaffold is optimal.

Best-in-class: Drug acting on a specific molecular target that provides the best balance between efficacy and adverse effects.

Bioassay: A bioassay is a procedure for determining the concentration, purity, and/or biological activity of a substance (e.g., vitamin, hormone, plant growth factor, antibiotic, enzyme) by measuring its effect on an organism, tissue, cell, and enzyme or receptor preparation compared with a standard preparation.

Bioinformatics: Discipline encompassing the development and utilization of computational tools to store, analyze, and interpret biological data. *Note*: Typically protein or DNA sequence or 3D information.

Bioisostere: A bioisostere is a compound resulting from the exchange of an atom or of a group of atoms with another, broadly similar, atom or group of atoms. The objective of a bioisosteric replacement is to create a new compound with similar biological properties to the parent compound. The bioisosteric replacement may be physicochemical or topologically based.

Biological agent: A biopolymer-based pharmaceutical, such as a protein, that is applicable to the prevention, treatment, or cure of diseases or injuries to man. *Note*: Biological agents may be any virus, therapeutic serum, toxin, antitoxin, vaccine, blood component or derivative, allergenic product, or analogous products.

Biomarker: Indicator signaling an event or condition in a biological system or sample and giving a measure of exposure, effect, or susceptibility.

Blockbuster drug: Drug that generates annual sales of USD 1 billion or more.

Blood–brain barrier (BBB): Layer of endothelial cells that line the small blood vessels of the brain. *Note 1*: These cells form "tight junctions" that restrict the free exchange of substances between the blood and the brain. Such cells are rich in P-glycoprotein, which serves to pump substrates back to the peripheral side of the vasculature. *Note 2*: Passive diffusion across the BBB is highly dependent on drug lipophilicity, and very few orally active agents acting in the central nervous system have a polar surface area greater than 0.9 nm^2.

Carcinogen: Agent (chemical, physical, or biological) that is capable of increasing the incidence of malignant neoplasms, thus causing cancer.

Chemical biology: Application of chemistry to the study of molecular events in biological systems, often using tool compounds. *Note*: Distinguished from medicinal chemistry, which is focused on the design and optimization of compounds for specific molecular targets.

Chemical database: Specific electronic repository for storage and retrieval of chemical information. *Note 1*: Chemical structural information is sometimes stored in string notation such as the InChI or SMILES notations. *Note 2*: Such databases can be searched to retrieve structural information and data on specific or related molecules. *Note 3*: A free database of chemical structures of small organic molecules and information on their biological activities is available from PubChem.

Chemical library, compound library, compound collection: (1) Collection of samples (e.g., chemical compounds, natural products, over-expression library of a microbe) available for biological screening. (2) Set of compounds produced through combinatorial chemistry or other means, which expands around a single core structure or scaffold.

Chemical space: Set of all possible stable molecules based on a specific chemical entity that interacts at one or more specific molecular targets.

Cheminformatics, chemoinformatics: Use of computational, mathematical, statistical, and information techniques to address chemistry-related problems.

Chemogenomics, chemical genomics: Systematic screening of chemical libraries of congeneric compounds against members of a target family of proteins.

Chemokine: Member of a superfamily of proteins with the primary function to control leukocyte activity and trafficking through tissues.

CLOGP values: Calculated 1-octanol/water partition coefficients. *Note*: Frequently used in structure–property correlation or quantitative structure–activity relationship studies. See also log P, log D.

Cluster: Group of compounds that are related by structural, physicochemical, or biological properties. *Note*: Organizing a set of compounds into clusters is often used to assess the diversity of those compounds, or to develop structure–activity relationship models.

Co-drug, mutual prodrug: Two chemically linked synergistic drugs designed to improve the drug delivery properties of one or both drugs. *Note*: The constituent drugs are indicated for the same disease, but may exert different therapeutic effects via disparate mechanisms of action.

Comparative molecular field analysis (CoMFA): Comparative molecular field analysis (CoMFA) is a 3D-QSAR method that uses statistical correlation techniques for the analysis of the quantitative relationship between the biological activities of a set of compounds with a specified alignment, and their three-dimensional electronic and steric properties. Other properties such as hydrophobicity and hydrogen bonding can also be incorporated into the analysis.

Congener: Substance structurally related to another and linked by origin or function. *Note*: Congeners may be analogues or vice versa but not necessarily. The term "congener," while most often a synonym for "homologue," has become somewhat more diffuse in meaning so that the terms "congener" and "analogue" are frequently used interchangeably in the literature. See also analogue, follow-on drug.

Constitutive activity: Receptor or enzymatic function displayed in the absence of an agonist or activator.

Contract research organization (CRO): Commercial organization that can be engaged to undertake specifically defined chemical, biological, safety, or clinical studies. *Note*: Typically, such studies are subject to confidentially agreements.

Covalent drug: Ligand that binds irreversibly to its molecular target through the formation of a new chemical bond.

Cytochrome P450 (CYP450): Member of a superfamily of heme-containing monooxygenases involved in xenobiotic metabolism, cholesterol biosynthesis, and steroidogenesis, in eukaryotic organisms found mainly in the endoplasmic reticulum and inner mitochondrial membrane of cells.

Designed multiple ligands: Compounds conceived and synthesized to act on two or more molecular targets.

Diversity: Unrelatedness of a set of molecules (e.g., building blocks or members of a compound library), as measured by properties such as atom connectivity, physical properties, or computationally generated descriptors. *Note*: Inverse of molecular similarity.

Diversity-oriented synthesis (DOS): Efficient production of a range of structures and templates with skeletal and stereochemical diversity as opposed to the synthesis of a specific target molecule.

Double-blind study: A double-blind study is a clinical study of potential and marketed drugs, where neither the investigators nor the subjects know which subjects will be treated with the active principle and which ones will receive a placebo.

Drug: A drug is any substance presented for treating, curing, or preventing disease in human beings or in animals. A drug may also be used for making a medical diagnosis or for restoring, correcting, or modifying physiological functions (e.g., the contraceptive pill).

Drug cocktail: (1) (In drug therapy) Administration of two or more distinct pharmacological agents to achieve a combination of their individual effects. *Note 1*: The combined effect may be additive, synergistic, or designed to reduce side effects. *Note 2*: This term is often used synonymously with that of "drug combination" but is preferred to avoid confusion with medications in which different drugs are included in a single formulation. (2) (In drug testing) Administration of two or more distinct compounds to test simultaneously their individual behaviors (e.g., pharmacological effects in high-throughput screens or drug metabolism.

Drug delivery: Process by which a drug is administered to its intended recipient. *Note*: Examples include administration orally, intravenously, or by inhalation. See also drug distribution, targeted drug delivery.

Drug distribution: Measured amounts of an administered compound in various parts of the organism to which it is given. See also drug delivery.

Drug disposition: Drug disposition refers to all processes involved in the absorption, distribution metabolism, and excretion of drugs in a living organism.

Druggability: Capacity of a molecular target to be modulated in a favorable manner by a small-molecule drug. *Note*: It is estimated that only around 10% of the human genome affords druggable targets.

Drug-like(ness): Physical and chemical properties in a small molecule that makes it likely to perform efficiently as a drug.

Drug repurposing, drug repositioning, drug reprofiling: Strategy that seeks to discover new applications for an existing drug that was not previously referenced and not currently prescribed or investigated. *Note*: Various additional synonymous terms have been used to describe the process of drug repurposing. All appear to be used interchangeably.

Drug safety: Assessment of the nontolerable biological effects of a drug. *Note*: Because the nontolerable effects of a drug are directly related to its concentration or dose, safety is generally ranked relative to the dose required to obtain the desirable effect. See also therapeutic index.

Dual binding site: The presence of two distinct ligand-binding sites on the same molecular target.

Effective concentration (EC): Concentration of a substance that produces a defined magnitude of response in a given system. *Note 1*: EC_{50} is the median dose that causes 50% of the maximal response. *Note 2*: The term usually refers to an agonist in a receptor system effect and could represent either an increase or a decrease in a biological function. See also IC_{50}, effective dose.

Effective dose (ED): Dose of a substance that causes a defined magnitude of response in a given system. *Note*: ED_{50} is the median dose that causes 50% of the maximal response. See also IC_{50}, effective concentration.

Efficacy: Efficacy describes the relative intensity with which agonists vary in the response they produce even when they occupy the same number of receptors and with the same affinity. Efficacy *is not* synonymous to intrinsic activity. Efficacy is the property that enables drugs to produce responses. It is convenient to differentiate the properties of drugs into two groups, those that cause them to associate with the receptors (affinity) and those that produce stimulus (efficacy). This term is often used to characterize the level of maximal responses induced by agonists. In fact, not all agonists of a receptor are capable of inducing identical levels of maximal responses. Maximal response depends on the efficiency of receptor coupling, that is, from the cascade of events, which, from the binding of the drug to the receptor, leads to the observed biological effect.

Efflux pump: Transporter protein located in the membrane of cells that utilizes active transport to move a compound from the internal to the external environment.

Elimination: Elimination is the process achieving the reduction in the concentration of a xenobiotic including its metabolism.

Epigenetic(s): Phenotypic change(s) in an organism brought about by alteration in the expression of genetic information without any change in the genomic sequence itself. *Note*: Common examples include changes in nucleotide base methylation and changes in histone acetylation. Changes of this type may become heritable.

Equilibrium solubility: Analytical composition of a mixture or solution that is saturated with one of the components in the designated mixture or solution. *Note 1*: Solubility may be expressed in any units corresponding to quantities that denote relative composition, such as mass, amount concentration, molality, etc. *Note 2*: The mixture or solution may involve any physical state: solid, liquid, gas, vapor, supercritical fluid. *Note 3*: The term "solubility" is also often used in a more general sense to refer to processes and phenomena related to dissolution. See also intrinsic solubility, kinetic solubility, solubility, supersaturated solution.

Fast follower: Compound selected as a rapid successor to a lead drug candidate. *Note*: Fast followers usually possess a marked increase in one or more pharmacological/pharmaceutical characteristics such as potency, efficacy, therapeutic index, or physicochemical parameters (e.g., solubility).

Fingerprint: Representation of a compound or chemical library by attributes (descriptors) such as atom connectivity, 3D structure, or physical properties.

First-in-class: First drug acting on a hitherto unaddressed molecular target to reach the market.

Follow-on drug: Drug having a similar mechanism of action to an existing drug. *Note*: Compounds may be of the same or different chemical class. A therapeutic advantage over first-in-class drugs must be demonstrated for regulatory approval.

Fragment: Low-molar-mass ligand (typically smaller than 200 Da) that binds to a target with low affinity but high ligand efficiency. *Note*: Typically fragments have affinities in a concentration interval from 0.1 to 1.0 mM.

Fragment-based lead discovery: Screening libraries of low-molar-mass compounds (typically 120–250 Da) using sensitive biophysical techniques capable of detecting weakly binding lead compounds. *Note*: X-ray structures are frequently used to drive the optimization of fragment hits to leads. See also fragment, ligand efficiency.

Frequent hitter: Structural feature that regularly results in a positive response in a variety of high-throughput or primary screens. *Note*: Such compounds often exert their actions through nonspecific mechanisms and are therefore unreliable leads.

Genomics: Science of using DNA- and RNA-based technologies to demonstrate alterations in gene expression.

Good laboratory practice (GLP): Set of principles that provide a framework within which laboratory studies are planned, performed, monitored, recorded, reported, and archived. *Note*: These studies are undertaken to generate data by which the hazards and risks to users, consumers, and third parties, including the environment, can be assessed for pharmaceuticals (only preclinical studies), agrochemicals, cosmetics, food additives, feed additives and contaminants, novel foods, biocides, detergents, etc. GLP helps assure regulatory authorities that the data submitted are a true reflection of the results obtained during the study and can therefore be relied upon when making risk/safety assessments.

Good manufacturing practice (GMP): Quality assurance process that ensures that medicinal products are consistently produced and controlled to the standards appropriate to their intended use. *Note*: Quality standards are those required under marketing authorization or product specification. GMP is concerned with both production and quality control.

G-protein, guanine binding protein: Member of a family of membrane-associated proteins that on activation by cellular receptors lead to signal transduction. See also G-protein-coupled receptor.

G-protein-coupled receptor (GPCR): Large family of cell surface receptors in which seven portions of the protein cross the cellular membrane and are linked to internal G-proteins. *Note 1*: Interaction of these receptors with extracellular ligands activates signal transduction pathways and, ultimately, cellular responses. *Note 2*: G-protein-coupled receptors are found only in eukaryotes. See also G-protein.

Green chemistry: Invention, design, and application of chemical products and processes to reduce or eliminate the use and generation of hazardous substances.

Hydrogen-bond acceptor (HBA): Typically N or O with a free lone pair of electrons.

Hydrogen-bond donor (HBD): A N-H or O-H functional group.

High-throughput screening (HTS): Method for the rapid assessment of the activity of samples from large compound collections. *Note 1*: Typically, these assays are carried out in microplates of at least 96 wells using automated or robotic techniques *Note 2*: The rate of at least 10^5 assays per day has been termed "ultra-high-throughput screening" (UHTS).

Hit: Molecule that produces reproducible activity above a defined threshold in a biological assay and whose structural identity has been established. *Note*: Hits typically derive from high-throughput screening initiatives or other relatively extensive primary assays and do not become true hits until fully validated.

Hit expansion: Generation of additional compound sets that contain chemical motifs and scaffolds that have activity in the primary screen. *Note*: This methodology permits the identification of additional hits and new scaffolds and develops structure–activity relationships around existing hits.

Hit-to-lead chemistry: Process by which a proven molecule or series derived from high-throughput screening or primary screens is chemically optimized to a viable lead or series.

Homology model: Computational representation of a protein built from the 3D structure of a similar protein or proteins using alignment techniques and homology arguments.

Hydrophilicity: Hydrophilicity is the tendency of a molecule to be solvated by water.

Hydrophobic fragmental constant: Representation of the lipophilicity contribution of a constituent part of a structure to the total lipophilicity.

Hydrophobic interaction: Entropically driven favorable interaction between non-polar substructures or surfaces in aqueous solution.

Hydrophobicity: Hydrophobicity is the association of nonpolar groups or molecules in an aqueous environment that arises from the tendency of water to exclude nonpolar molecules.

IC_{50} (inhibitory concentration 50): The concentration of an enzyme inhibitor or receptor antagonist that reduces the enzyme activity or agonist response by 50%. *Note*: IC_{50} values are influenced by experimental conditions (e.g., substrate or agonist concentration, which should be specified). Related terms: inhibition constant, K_i.

Inhibition constant, K_i: (1) Equilibrium dissociation constant of an enzyme-inhibitor complex: $K_i = [E][I]/[EI]$. (2) The equilibrium dissociation constant of a receptor-ligand complex. *Note*: This value is usually obtained through competition binding experiments, where the K_i is determined after the IC_{50} obtained in a competition assay performed in the presence of a known concentration of labeled reference ligand. Related term: IC_{50}.

International chemical identifier (InChI): Nonproprietary identifier for chemical substances that can be used in printed and electronic data sources to enable easier linking of diverse data compilations. *Note*: This IUPAC notation frequently replaces the earlier SMILES notation.

Intercalation: Thermodynamically favorable, reversible inclusion of a molecule (or group) between two other molecules (or groups). Examples include DNA intercalation.

Intrinsic activity: Intrinsic activity is the maximal stimulatory response induced by a compound in relation to that of a given reference compound (see also partial agonist). This term has evolved with common usage. It was introduced by Ariëns as a proportionality factor between tissue response and receptor occupancy. The numerical value of intrinsic activity (alpha) could range from unity (for full agonists, i.e., agonist inducing the tissue maximal response) to zero (for antagonists), the fractional values within this range denoting partial agonists. Ariëns' original definition equates the molecular nature of alpha to maximal response only when response is a linear function of receptor occupancy. This function has been verified. Thus, intrinsic activity, which is a drug and tissue parameter, cannot be used as a characteristic drug parameter for classification of drugs or drug receptors. For this purpose, a proportionality factor derived by null methods, namely, relative efficacy, should be used. Finally, "intrinsic activity" should not be used instead of "intrinsic efficacy." A "partial agonist" should be termed "agonist with intermediate intrinsic efficacy" in a given tissue.

Intrinsic solubility: Equilibrium solubility of the uncharged form of an ionizable compound at a pH where it is fully unionized.

$$HA(s) = HA(aq)$$
$$HA(aq) = H^+ + A^-$$

The intrinsic solubility can be determined from the analytical composition at a pH, where [HA] is very much greater than [A$^-$]. See also equilibrium solubility, kinetic solubility, solubility, supersaturated solution.

Inverse agonist: An inverse agonist is a drug which acts at the same receptor as that of an agonist, yet produces an opposite effect. Also called negative antagonists.

Investigational new drug (IND): Drug not yet approved for general use by the national authority, such as the Food and Drug Administration of the United States of America, but undergoing clinical investigation to assess its safety and efficacy.

Ionotropic receptor: Transmembrane ion channel that opens or closes in response to the binding of a ligand.

Kinetic solubility, turbidimetric solubility: Composition of a solution with respect to a compound when it's induced precipitate first appears. See also equilibrium solubility, intrinsic solubility, solubility, supersaturated solution.

Lead: Compound (or compound series) that satisfies predefined minimum criteria for further structure and activity optimization. *Note*: Typically, a lead will demonstrate appropriate activity, selectivity, and tractable structure–activity relationship and have confirmed activity in a relevant cell-based assay. See lead validation.

Lead discovery: Lead discovery is the process of identifying active new chemical entities, which by subsequent modification may be transformed into a clinically useful drug.

Lead generation: Lead generation is the term applied to strategies developed to identify compounds that possess a desired but non-optimized biological activity.

Lead optimization: Lead optimization is the synthetic modification of a biologically active compound, to fulfill all stereoelectronic, physicochemical, pharmacokinetic, and toxicologic requirements for clinical usefulness.

Lead validation: Process by which a lead compound is authenticated by the confirmation of its expected pharmacological properties. *Note*: Usually, a cluster of structurally similar compounds showing discernable structure–activity relationships will support the validation process.

Ligand: Ion or molecule that binds to a molecular target to elicit, block, or attenuate a biological response.

Ligand-based drug design: Method of drug discovery and/or optimization in which the pursuit of new structures and/or structural modifications is based upon one or more ligands known to interact with the molecular target of interest. *Note*: This approach is applicable even when no structural detail of the target is known. In such cases, a series of analogues is usually prepared and tested to produce structure–activity relationship data that can be extrapolated to indirectly derive a topographical map of the biological surface. See also analogue-based drug discovery, structure-based drug design.

Ligand efficiency (LE): Measure of the free energy of binding per heavy atom count (i.e., non-hydrogen) of a molecule. *Note 1*: It is used to rank the quality of molecules in drug discovery, particularly in fragment-based lead discovery. *Note 2*: An LE value of 1.25 kJ mol^{-1} (non-hydrogen atom)$^{-1}$ is the minimum requirement of a good lead or fragment.

Ligand lipophilic efficiency (LLE), lipophilic efficiency: Parameter used to identify ligands with a high degree of specific interaction toward the desired molecular target. *Note 1*: The potency of a ligand toward a molecular target may be dominated by nonspecific partitioning from the aqueous phase. It can be advantageous to separate out the nonspecific component of the potency to identify more specific interactions; typically using an equation such as: LLE, symbol E_{LL}, is defined by the logarithm of the potency minus a lipophilicity measure, where a typical example would be

$$E_{LL} = -\log(IC_{50}) - \log P.$$

Note 2: E_{LL} can be regarded as part of a thermodynamic cycle used as a complementary measure to potency in the search for specific target interactions. In this case, the dissociation constant, K_d, is a more appropriate measure than IC50 since it refers to the Gibbs energy of the binding process

$$E_{LL} = -\log K_d - \log P.$$

Lipophilicity: Lipophilicity represents the affinity of a molecule or a moiety for a lipophilic environment. It is commonly measured by its distribution behavior in a biphasic system, either liquid-liquid (e.g., partition coefficient in octanol/water) or solid/liquid (retention on reversed-phase high performance liquid chromatography (RP-HPLC) or thin-layer chromatography (TLC) system).

log D, lg D: Logarithm of the apparent partition coefficient at a specified pH. See CLOGP, log P.

log P, lg P: Measure of the lipophilicity of a compound by its partition coefficient between an apolar solvent (e.g., 1-octanol) and an aqueous buffer. Thus, P is the quotient of the concentration of non-ionized drug in the solvent divided by the respective concentration in buffer. See CLOGP, log D.

Markush structure: Generalized formula or description for a related set of chemical compounds used in patent applications and chemical papers.

Metabotropic receptor: Receptor that modulates electric potential-gated channels via G-proteins. *Note*: The interaction can occur entirely within the membrane or by the generation of diffusible second messengers. The involvement of G-proteins causes the activation of these receptors to last tens of seconds to minutes, in contrast with the brief effect of ionotropic receptors.

Microarray: Planar surface where assay reagents and samples are distributed as sub-microliter drops. *Note*: This screening format is a direct offshoot of genomic microarray technologies and makes use of ultra-low-volume miniaturization provided by nanodispensing technologies.

microRNA (miRNA): Small single-stranded RNA molecules that play a significant role in the post-transcriptional regulation of gene expression. *Note*: MicroRNA usually comprises approximately 22 nucleotides.

Molecular descriptor: Parameter that characterizes a specific structural or physicochemical aspect of a molecule.

Molecular dynamics: Computational simulation of the motion of atoms in a molecule or of individual atoms or molecules in solids, liquids, and gases, according to Newton's laws of motion. *Note*: The forces acting on the atoms, required to simulate their motions, are generally calculated using molecular mechanics force fields.

Molecular similarity: Measure of the coincidence or overlap between the structural and physicochemical profiles of compounds.

Molecular target: Protein (e.g., receptor, enzyme, or ion channel), RNA, or DNA that is implicated in a clinical disorder or the propagation of any untoward event. *Note*: Usually, biochemical, pharmacological, or genomic information supporting the role of such a target in disease will be available.

Multidrug resistance (MDR): Characteristic of cells that confers resistance to the effects of several different classes of drugs. *Note*: There are several forms of drug resistance of which each is determined by genes that govern how cells will respond to chemical agents. One type of multidrug resistance involves the ability to eject several drugs out of cells (e.g., efflux pumps such as P-glycoprotein).

Multiparameter optimization (MPO): Drug-likeness penetrability algorithm derived from CLOGP, clogD, molar mass, topological polar surface area, number of hydrogen-bond donors, and pK_a. *Note*: The MPO desirability score is larger or equal to 4 on a scale of 0-6.

Multi-target-directed ligand (MTDL), multi-target drug: Ligand acting on more than one distinct molecular target. Targets may be of the same or different mechanistic classes.

Multi-target drug discovery (MTDD): Deliberate design of compounds that act on more than one molecular target.

Murcko assembly: Core scaffold of a molecule that remains after all chain substituents that do not terminate in a ring are removed. Single atoms connected by a double bond are typically also retained.

Neural network: A statistical analysis procedure based on models of nervous system learning in animals. *Note*: Neural networks have the ability to "learn" from a collection of examples to discover patterns and trends. These data-mining techniques can be used in forecasting or prediction.

Neutral antagonist (in pharmacology): Ligand that blocks the responses of a receptor to both agonists and inverse agonists with the same intensity. It binds to the receptor without evoking any change of conformation or change to the ratio of activated to inactivated conformations. *Note*: Perfect neutral antagonism is difficult to achieve.

New chemical entity: Drug that contains no active moiety previously approved for use by the national authority, such as the US Food and Drug Administration. See also new molecular entity.

New molecular entity: Active ingredient that has never before been approved in any form by the national authority, such as the US Food and Drug Administration.

Noncompetitive antagonist: Functional antagonist that either binds irreversibly to a receptor or to a site distinct from that of the natural agonist. See allosteric antagonist.

Nuclear hormone receptor, nuclear receptor: Ligand-activated transcription factor that regulates gene expression by interacting with specific DNA sequences upstream of its target gene(s).

Obviousness: Term associated with intellectual property wherein the latter's patentability is assessed relative to the combination of more than one item of "prior art." *Note*: To be patentable within the context of medicinal chemistry, a given compound must be: (1) novel, in that its specific arrangement of atoms has never been previously disclosed; (2) non-obvious, in that its specific arrangement of atoms is not readily suggested to be of benefit by a person having ordinary skill in the art upon considering two or more other, previously disclosed structures; and (iii) useful, in that it should have some benefit, the disclosure of the latter encompassing a valid "reduction to practice."

Off-target effect: Pharmacological action induced by any molecule at molecular/biological sites distinct from that for which it was designed. *Note*: Such effects are dose-dependent and may be beneficial, adverse, or neutral.

Orphan disease: Disease for which drug research, development, and marketing is economically unfavorable. *Note 1*: The poor commercial environment could be due to a lack of economic incentives or a lack of understanding of the diseases or a combination of both. *Note 2*: Sometimes the term "rare disease" is used synonymously with orphan disease, although there is a slight difference. For example, a rare disease is so uncommon that there is no drug development effort. *Note 3*: Which diseases are classified as orphan depends strongly on the country that classifies it. In the United States, for example, any disease affecting less than 200,000 people is considered an orphan or rare disease. Europe and countries such as Japan, Australia, and Singapore have a different definition.

Orphan drug: Pharmaceutical agent that has been approved specifically to treat a rare and commercially unfavorable medical condition.

Orphan receptor: Receptor for which an endogenous ligand has yet to be identified.

Parallel synthesis: Simultaneous preparation of sets of discrete compounds in arrays of physically separate reaction vessels or microcompartments without interchange of intermediates during the assembly process.

Partial agonist: A partial agonist is an agonist that is unable to induce maximal activation of a receptor population, regardless of the amount of drug applied.

Patentability: Set of criteria that must be satisfied to achieve commercial exclusivity for an invention. *Note*: These criteria are essentially the same in all major countries and include: suitability, novelty, inventiveness, utility, and the provision of an adequate description.

P-glycoprotein (Pgp) ATP-binding cassette transporter is responsible for the efflux of small molecules from cells. *Note*: P-glycoproteins can play a major role in limiting brain penetration and restricting the intestinal absorption of drugs. Their over-expression in cancer cells becomes a common mechanism of multidrug resistance. See also blood–brain barrier, efflux pump.

Pharmacodynamics: Relates to the specific interaction of a drug with its target.

Pharmacogenetics: Study of inherited differences (variation) in drug metabolism and response.

Pharmacogenomics: General study of all of the many different genes that determine drug behaviour. *Note*: The distinction between the terms pharmacogenetics and pharmacogenomics has blurred with time and they are now frequently used interchangeably.

Pharmacokinetics: The pattern of absorption, distribution, and excretion of a drug over time.

Pharmacophore: A pharmacophore is the ensemble of steric and electronic features that is necessary to ensure the optimal supramolecular interactions with a specific biological target structure and to trigger (or to block) its biological response. A pharmacophore does not represent a real molecule or a real association of functional groups, but a purely abstract concept that accounts for the common molecular interaction capacities of a group of compounds toward their target structure. The pharmacophore can be considered as the largest common denominator shared by a set of active molecules. This definition discards a misuse often found in the medicinal chemistry literature that consists of naming as pharmacophores simple chemical functionalities such as guanidines, sulfonamides or dihydroimidazoles (formerly imidazolines), or typical structural skeletons such as flavones, phenothiazines, prostaglandins, or steroids.

Pharmacophoric descriptors: Pharmacophoric descriptors are used to define a pharmacophore, including H-bonding, hydrophobic and electrostatic interaction sites, defined by atoms, ring centers, and virtual points.

Phase 0 clinical studies, exploratory investigational new drug: Exploratory first-in-human trials that involve microdosing of drug to allow the assessment of pharmacokinetic parameters with limited drug exposure. *Note*: These trials have no therapeutic or diagnostic intent but are designed to assist decision making by providing bioavailability, metabolism, and other limited data from a small number of patients.

Phase I clinical studies: Initial introduction of an investigational new drug into humans. *Note*: These studies are designed to determine the metabolic and pharmacologic actions of the drug in humans, the side effects associated with increasing doses, and, if possible, to gain early evidence on effectiveness.

Phase II clinical studies: Controlled clinical studies conducted in a limited number of individuals to obtain some preliminary data on the effectiveness of an investigational new drug for a particular indication or indications in patients with the disease or condition. *Note*: Phase II clinical trials have two subclasses, II(a) and II(b). Phase II(a) trials are essentially pilot clinical trials designed to evaluate efficacy (and safety) in selected populations of patients with the disease or condition to be treated, diagnosed, or prevented. Phase II(b) trials extend those of Phase II(a) to well-controlled trials that evaluate the same parameters in similar patient populations.

Phase III clinical studies: Expanded controlled and uncontrolled trials in humans. *Note*: These trials are performed after preliminary evidence suggesting effectiveness of the drug has been obtained in Phase II, and are intended to gather the additional information about effectiveness and safety that is needed to evaluate the overall benefit/risk relationship of the drug.

Phase IV clinical studies: Extended post-marketing studies in humans. *Note*: These trials are designed to broaden information concerning treatment risks, benefits, and optimal drug use.

Phenotypic screening: Evaluation of compounds (small molecules, peptides, siRNA, etc.) in cells, tissues, or organisms for their ability to modify the system in a measurable manner. *Note*: Phenotypic screening differs from target-based screens in that it assesses the overall response of the compound under investigation rather than its specific response on a purified molecular target. Advances in molecular biology resulted in a marked shift away from phenotypic screening, although the latter is now regaining popularity.

Pipeline: Discovery and development compound portfolio of a pharmaceutical company or research organization.

Pivotal study: Experiment that provides strong support for or against the drug or molecular target under investigation. See also proof of concept.

Placebo: A placebo is an inert substance or dosage form that is identical in appearance, flavor, and odor to the active substance or dosage form. It is used as a negative control in a bioassay or in a clinical study.

Polar surface area (PSA), topological polar surface area: Surface area over all polar atoms (usually oxygen and nitrogen), including any attached hydrogen atoms, of a molecule. *Note*: Polar surface area is a commonly used metric (c.f. molecular descriptor) for the optimization of cell permeability. Molecules with a PSA of greater than 1.4 nm^2 are usually poor at permeating cell membranes. For molecules to penetrate the blood–brain barrier, the polar surface area should normally be smaller than 0.6 nm^2, although values up to 0.9 nm^2 can be tolerated. See also blood–brain barrier.

Polymorphism: Ability of a compound to exist in more than one crystalline form (polymorph) with each having a different arrangement or conformation of the molecules within the crystal lattice. *Note*: Polymorphs generally differ in their melting points, solubility, and relative intestinal absorption such that optimal polymorphs can markedly enhance the attractiveness of some drugs.

Positron emission tomography (PET):   Imaging technique used to visualize small amounts of a compound in biological tissues by the use of radionuclide labels. These radionuclide's, such as ^{11}C, ^{18}F, ^{13}N, and ^{15}O, are positron emitters.

Potency: Potency is the dose of drug required to produce a specific effect of given intensity as compared to a standard reference. Potency is a comparative rather than an absolute expression of drug activity. Drug potency depends on both affinity and efficacy. Thus, two agonists can be equipotent, but have different intrinsic efficacies with compensating differences in affinity.

Preclinical candidate (PCC), safety assessment candidate: Optimized lead compound successfully passing key screening, selectivity, and physicochemical criteria sufficient to warrant further detailed pharmacological and pharmacokinetic evaluation in animal models. *Note*: Critical studies usually include bioavailability, therapeutic efficacy in an appropriate disease model, and side effect profiling.

Privileged structure: Substructural feature that confers desirable (often drug-like) properties on compounds containing that feature. They often consist of a semi-rigid scaffold that presents multiple hydrophobic residues without undergoing hydrophobic collapse. *Note 1*: Such structures are commonly found to confer activity against different targets belonging to the same receptor family.

Prodrug: A prodrug is any compound that undergoes biotransformation before exhibiting its pharmacological effects. Prodrugs can thus be viewed as drugs containing specialized non-toxic protective groups used in a transient manner to alter or to eliminate undesirable properties in the parent molecule.

Proof of concept (in pharmacology):   Procedure by which a specific therapeutic mechanism or treatment/diagnostic paradigm is shown to be beneficial. *Note*: Similar to, and often simultaneously associated with, that for a new drug candidate. This process usually involves an early supporting step prior to clinical testing and final validation within human studies. See also lead validation, pivotal study.

Protein data bank (PDB):   Repository for the 3D structural data of large biological molecules including proteins and nucleic acids. *Note*: These high-resolution structures, generated predominantly by X-ray or NMR spectroscopic techniques, provide a major resource for structural biology.

Protein–protein interaction (PPI):   Association of one protein with one or more other proteins to form either homo- or heteromeric proteins. *Note*: Such associations are common in biological systems and are responsible for the regulation of numerous cellular functions in addition to the mediation of disease morphology where aberrant interactions play a significant role.

Prototype drug: Early compound that has biological properties suitable for target validation but may not necessarily be adequate for clinical studies.

QTc interval: Time between the start of the Q wave and the end of the T wave in the heart's electrical cycle. When corrected for individual heart rate it becomes known as the corrected QT, or QTc interval. *Note*: Significant prolongation of the QTc interval by pharmaceutical agents can induce life-threatening ventricular arrhythmia (Torsades de Pointes), typically by interacting with the hERG channel.

Quantitative structure-activity relationships (QSAR): Quantitative structure-activity relationships are mathematical relationships linking chemical structure and pharmacological activity in a quantitative manner for a series of compounds. Methods that can be used in QSAR include various regression and pattern recognition techniques.

Receptor: A receptor is a molecule or a polymeric structure in or on a cell that specifically recognizes and binds a compound acting as a molecular messenger (neurotransmitter, hormone, lymphokine, lectin, drug, etc.).

Retrosynthesis: Process of conceptually deconstructing complex molecules into simpler fragments capable of chemical manipulation to reform the parent compound.

Rule of five: Set of molecular descriptors used to assess the potential oral bioavailability of a compound. *Note 1*: Characterized by a mass of less than 500 Da, less than or equal to 5 hydrogen-bond donors, less than or equal to 10 hydrogen-bond acceptors (usually using the N+O count as a surrogate for the number of hydrogen-bond acceptors), and a CLOGP less than or equal to 5. *Note 2*: Frequently used to profile a chemical library or virtual chemical library with respect to the proportion of drug-like members that it contains. Often used as a surrogate for "drug-likeness." *Note 3*: While these criteria are frequently referred to as Lipinski's rules, the depersonalized term "rule of five" is preferred.

Rule of three: Set of molecular descriptors used to assess the quality of hit or lead molecules. *Note 1*: Most commonly applied to fragments that ideally are characterized by a mass of less than 300 Da, less than or equal to three hydrogen-bond donors, less than or equal to three hydrogen-bond acceptors, and a CLOGP of less than or equal to 3. In addition, the number of rotatable bonds should average or be less than 3 and the polar surface area about 0.6 nm^2. *Note 2*: Often used to distinguish "lead-like" from "drug-like" molecules. See fragment-based lead discovery.

Scaffold, template: Core portion of a molecule common to all members of a chemical library or compound series.

Scaffold hopping: Exchange of one scaffold for another while maintaining molecular features that are important for biological properties.

Second messenger: A second messenger is an intracellular metabolite or ion increasing or decreasing as a response to the stimulation of receptors by agonists, considered as the "first messenger." This generic term usually does not prejudge the rank order of intracellular biochemical events.

Site-directed mutagenesis: Molecular biology technique in which mutations are created at one or more defined sites in a DNA molecule. *Note*: Typically used in molecular target validation and to determine whether specific amino acids are involved at ligand or substrate interaction sites.

SMILES (simplified molecular input line entry system) notation: String notation used to describe the atom type and connectivity of molecular structures. *Note*: Primarily used to input chemical structures into electronic databases and now frequently replaced by the InChI notation.

Solubility: Analytical composition of a mixture or solution that is saturated with one of the components of the mixture or solution, expressed in terms of the proportion of the designated component in the designated mixture or solution. *Note 1*: Solubility may be expressed in any units corresponding to quantities that denote relative composition, such as mass, amount concentration, molality, etc. *Note 2*: The mixture or solution may involve any physical state: solid, liquid, gas, vapor, supercritical fluid. *Note 3*: The term "solubility" is also often used in a more general sense to refer to processes and phenomena related to dissolution. See equilibrium solubility, intrinsic solubility, kinetic solubility, supersaturated solution.

Spare receptor, receptor reserve: Residual binding site still available to an endogenous ligand after sufficient sites have already been filled to elicit the maximal response possible for that particular biological system. *Note*: Xenobiotics such as drugs may similarly interact with such receptors, but by definition, their identification and quantification occur via use of the natural ligand.

Stem cell: Multipotent cell with mitotic potential that may serve as a precursor for many kinds of differentiated cells. *Note*: Unipotent stem cells can differentiate into one mature cell type only.

Structural alert: Chemical features present in a hit or lead molecule indicative of potential toxicity. *Note*: Typically, such features include chemically reactive functionality and components known to metabolize to chemically reactive entities. Examples include anhydrides, aromatic amines, and epoxides.

Structure-activity relationship (SAR): Structure-activity relationship is the relationship between chemical structure and pharmacological activity for a series of compounds.

Structure-based design: Structure-based design is a drug design strategy based on the 3D structure of the target obtained by X-ray or NMR.

Structure-property correlations (SPC): Structure-property correlations refer to all statistical mathematical methods used to correlate any structural property to any other property (intrinsic, chemical, or biological), using statistical regression and pattern recognition techniques.

Supersaturated solution: Solution that has a greater composition of a solute than one that is in equilibrium with undissolved solute at specified values of temperature and pressure. See also equilibrium solubility, intrinsic solubility, kinetic solubility.

Systems biology: Integration of high-throughput biology measurements with computational models that study the projection of the mechanistic characteristics of metabolic and signaling pathways onto physiological and pathological phenotypes.

Targeted drug delivery: Approach to target a drug to a specific tissue or molecular target using a prodrug or antibody recognition systems.

Target validation: Process by which a protein, RNA, or DNA is implicated in a biological pathway thought to be of relevance to a disease or adverse pathology. *Note*: Typically, validation will involve location of the molecular target in relevant cells, organs, or tissues, evidence for its up-regulation/activation in the disorder, and the ability to attenuate adverse responses by agents known to interfere with the target.

Tautomer: Structural isomer that can readily convert to another form that differs only by the attachment position of a hydrogen atom and the location of double bond(s). *Note*: In most cases, these isomers are formed by a proton shift to or from heteroatoms such as O, N, or S as typified by the enol and keto forms of carbonyl compounds. Tautomers rapidly interconvert by proton transfer and are usually in equilibrium with one another.

Tautomerism: Reversible interconversion of two different tautomers.

Therapeutic index, therapeutic ratio: Ratio of the exposure/concentration of a therapeutic agent that causes beneficial effects to that which causes the first observed adverse effect. *Note*: A commonly used measure of therapeutic index is the toxic dose of a drug for 50% of a population divided by the minimum effective dose for 50% of a population.

Three-dimensional quantitative structure-activity relationship (3D-QSAR): A three-dimensional quantitative structure-activity relationship is the analysis of the quantitative relationship between the biological activity of a set of compounds and their spatial properties using statistical methods.

Training set: Specific group of compounds selected for characterization of both the molecular descriptors and the measured values of the targeted property. *Note*: Statistical methods applied to the set are used to derive a function between the molecular descriptors and the targeted property.

Unmet medical need: Term used for diseases or other disorders for which no optimal therapeutic options exist.

Virtual chemical library: Collection of chemical structures constructed solely in electronic form or on paper. *Note*: The building blocks required for such a library may not exist, and the chemical steps for such a library may not have been tested. These libraries are used in the design and evaluation of possible libraries.

Virtual screening, in silico screening: Evaluation of compounds using computational methods. *Note*: The source of the model could be a macromolecular structure or based on physicochemical parameters or ligand structure–activity relationships.

Volume of distribution (V_d)**:** Apparent (hypothetical) volume of fluid required to contain the total amount of a substance in the body at the same concentration as that present in the plasma assuming equilibrium has been attained.

Wild-type receptor: Receptor that occurs naturally in human and other species.

Xenobiotic: A xenobiotic is a compound foreign to an organism (xenos [Greek] = foreign).

REFERENCES

1. NIDA: Comorbidity: Addiction and Other Mental Disorders. NIH Publication Number 10-5771. Available from http://www.drugabuse.gov/publications/research-reports/comorbidity-addiction-other-mental-illnesses. Last Accessed September, 2010.

2. American Psychiatric Association. http://www.psych.org/ Last accessed February, 2014.

3. O'Brien, C. P.; Volkow, N.; Li, T. K. What's in a word? Addiction versus dependence in DSM-V. *Am J Psychiat* **2006**, 163, 764–5.

4. NIDA: Understanding Drug Abuse and Addiction. Available from http://www.drugabuse.gov/publications/drugfacts/understanding-drug-abuse-addiction. Last accessed November, 2012.

5. NIDA: Understanding Drug Abuse and Addiction. DrugFacts. Available at http://www.drugabuse.gov/publications/drugfacts/understanding-drug-abuse-addiction. Last accessed November, 2012.

6. Skinner, B. F. *Science and Human Behavior*. Free Press, 1953.

7. Nestler, E. J. Is there a common molecular pathway for addiction? *Nat Neurosci* **2005**, 8, 1445–9.

8. George F.; Koob, M. L. M. *Neurobiology of Addiction*. Academic Press, 2006.

9. Warner, L. A.; Kessler, R. C.; Hughes, M.; Anthony, J. C.; Nelson, C. B. Prevalence and correlates of drug use and dependence in the United States. Results from the National Comorbidity Survey. *Arch Gen Psychiat* **1995**, 52, 219–29.

10. McKim, W. A. *Drugs and Behavior*, 5th edition. Prentice Hall, 2003.

11. Coe, J. W.; Brooks, P. R.; Vetelino, M. G.; Wirtz, M. C.; Arnold, E. P.; Huang, J.; Sands, S. B.; Davis, T. I.; Lebel, L. A.; Fox, C. B.; et al. Varenicline: an alpha4beta2 nicotinic receptor partial agonist for smoking cessation. *J Med Chem* **2005**, 48, 3474–7.

Drug Discovery for the Treatment of Addiction: Medicinal Chemistry Strategies, First Edition. Brian S. Fulton.
© 2014 John Wiley & Sons, Inc. Published 2014 by John Wiley & Sons, Inc.

12. Badiani, A.; Belin, D.; Epstein, D.; Calu, D.; Shaham, Y. Opiate versus psychostimulant addiction: the differences do matter. *Nat Rev Neurosci* **2011**, 12, 685–700.

13. Agrawal, A.; Nurnberger, J. I., Jr.; Lynskey, M. T. Cannabis involvement in individuals with bipolar disorder. *Psychiat Res* **2011**, 185, 459–61.

14. Semple, D. M.; McIntosh, A. M.; Lawrie, S. M. Cannabis as a risk factor for psychosis: systematic review. *J Psychopharmacol* **2005**, 19, 187–94.

15. Wobrock, T.; Hasan, A.; Malchow, B.; Wolff-Menzler, C.; Guse, B.; Lang, N.; Schneider-Axmann, T.; Ecker, U. K.; Falkai, P. Increased cortical inhibition deficits in first-episode schizophrenia with comorbid cannabis abuse. *Psychopharmacology (Berl)* **2010**, 208, 353–63.

16. Sturgess, J. E.; George, T. P.; Kennedy, J. L.; Heinz, A.; Muller, D. J. Pharmacogenetics of alcohol, nicotine and drug addiction treatments. *Addict Biol* **2011**, 16, 357–76.

17. Khokhar, J. Y.; Ferguson, C. S.; Zhu, A. Z.; Tyndale, R. F. Pharmacogenetics of drug dependence: role of gene variations in susceptibility and treatment. *Annu Rev Pharmacol Toxicol* **2010**, 50, 39–61.

18. Mroziewicz, M.; Tyndale, R. F. Pharmacogenetics: a tool for identifying genetic factors in drug dependence and response to treatment. *Addict Sci Clin Pract* **2010**, 5, 17–29.

19. Kendler, K. S.; Myers, J.; Prescott, C. A. Specificity of genetic and environmental risk factors for symptoms of cannabis, cocaine, alcohol, caffeine, and nicotine dependence. *Arch Gen Psychiat* **2007**, 64, 1313–20.

20. Palmer, R. H.; Button, T. M.; Rhee, S. H.; Corley, R. P.; Young, S. E.; Stallings, M. C.; Hopfer, C. J.; Hewitt, J. K. Genetic etiology of the common liability to drug dependence: evidence of common and specific mechanisms for DSM-IV dependence symptoms. *Drug Alcohol Depend* **2012**, 123 Suppl 1, S24–32.

21. Agrawal, A.; Lynskey, M. T. Candidate genes for cannabis use disorders: findings, challenges and directions. *Addiction* **2009**, 104, 518–32.

22. Gelernter, J.; Panhuysen, C.; Wilcox, M.; Hesselbrock, V.; Rounsaville, B.; Poling, J.; Weiss, R.; Sonne, S.; Zhao, H.; Farrer, L.; et al. Genomewide linkage scan for opioid dependence and related traits. *Am J Hum Genet* **2006**, 78, 759–69.

23. Chen, L. S.; Xian, H.; Grucza, R. A.; Saccone, N. L.; Wang, J. C.; Johnson, E. O.; Breslau, N.; Hatsukami, D.; Bierut, L. J. Nicotine dependence and comorbid psychiatric disorders: examination of specific genetic variants in the CHRNA5-A3-B4 nicotinic receptor genes. *Drug Alcohol Depend* **2012**, 123 Suppl 1, S42–51.

24. Robison, A. J.; Nestler, E. J. Transcriptional and epigenetic mechanisms of addiction. *Nat Rev Neurosci* **2011**, 12, 623–37.

25. Russo, S. J.; Dietz, D. M.; Dumitriu, D.; Morrison, J. H.; Malenka, R. C.; Nestler, E. J. The addicted synapse: mechanisms of synaptic and structural plasticity in nucleus accumbens. *Trends Neurosci* **2010**, 33, 267–76.

26. Olds, J.; Milner, P. Positive reinforcement produced by electrical stimulation of septal area and other regions of rat brain. *J Comp Physiol Psychol* **1954**, 47, 419–27.

27. Stein, L. Secondary reinforcement established with subcortical stimulation. *Science* **1958**, 127, 466–7.

28. Carlsson, A.; Lindqvist, M.; Magnusson, T.; Waldeck, B. On the presence of 3-hydroxytyramine in brain. *Science* **1958**, 127, 471.

29. Yokel, R. A.; Wise, R. A. Increased lever pressing for amphetamine after pimozide in rats: implications for a dopamine theory of reward. *Science* **1975**, 187, 547–9.

30. Healy, D. *The Antidepressant Era*. Cambridge, MA: Harvard University Press, 1997.

31. Nolte, J. *The Human Brain: An Introduction to its Functional Anatomy*, 6th edition. Mosby Elsevier, 2009.

32. Haber, S. N.; Knutson, B. The reward circuit: linking primate anatomy and human imaging. *Neuropsychopharmacology* **2010**, 35, 4–26.

33. Koob, G. F.; Volkow, N. D. Neurocircuitry of addiction. *Neuropsychopharmacology* **2010**, 35, 217–38.

34. Koob, G. F. Neural mechanisms of drug reinforcement. *Ann N Y Acad Sci* **1992**, 654, 171–91.

35. Siegel, G. J. *Basic Neurochemistry: Molecular, Cellular, and Medical Aspects*, 7th edition. Amsterdam; Boston: Elsevier, 2006.

36. Hammond, C. *Cellular and Molecular Neurophysiology*, 3rd edition. Academic Press, 2008.

37. Ainscow, E. K.; Mirshamsi, S.; Tang, T.; Ashford, M. L. J.; Rutter, G. A. Dynamic imaging of free cytosolic ATP concentration during fuel sensing by rat hypothalamic neurones: evidence for ATP-independent control of ATP-sensitive K+ channels. *J Physiol* **2002**, 544, 429–45.

38. Zachariou, V.; Duman, R. S.; Nestler, E. J.; In: Brady, S.; Siegel, G.; Albers, R. W.; Price, D. (eds.), *Basic Neurochemistry, 8th edition (Chapter 21)*, New York: Academic Press, 2012; pp. 411–22.

39. Zylbergold, P.; Ramakrishnan, N.; Hebert, T. The role of G proteins in assembly and function of Kir3 inwardly rectifying potassium channels. *Channels (Austin)* **2010**, 4, 411–21.

40. Luscher, C.; Slesinger, P. A. Emerging roles for G protein-gated inwardly rectifying potassium (GIRK) channels in health and disease. *Nat Rev Neurosci* **2010**, 11, 301–15.

41. Lu, Z. Mechanism of rectification in inward-rectifier K+ channels. *Ann Rev Physiol* **2004**, 66, 103–29.

42. Alexander, S. P.; Mathie, A.; Peters, J. A. Guide to Receptors and Channels (GRAC), 5th edition. *Br J Pharmacol* **2011**, 164 Suppl 1, S1–324.

43. Tatsumi, M.; Groshan, K.; Blakely, R. D.; Richelson, E. Pharmacological profile of antidepressants and related compounds at human monoamine transporters. *Eur J Pharmacol* **1997**, 340, 249–58.

44. Seung, S. *Connectome: How the Brain's Wiring Makes Us Who We Are*. Boston: Houghton Mifflin Harcourt, 2012.

45. Kalivas, P. W. The glutamate homeostasis hypothesis of addiction. *Nat Rev Neurosci* **2009**, 10, 561–72.

46. Danbolt, N. C. Glutamate uptake. *Prog Neurobiol* **2001**, 65, 1–105.

47. Cleva, R. F.; Olive, M. F. mGlu receptors and drug addiction. *WIREs Membr Transp Signal* **2012**, 1, 281–95.

48. Muto, T.; Tsuchiya, D.; Morikawa, K.; Jingami, H. Structures of the extracellular regions of the group II/III metabotropic glutamate receptors. *Proc Natl Acad Sci USA* **2007**, 104, 3759–64.

49. Kunishima, N.; Shimada, Y.; Tsuji, Y.; Sato, T.; Yamamoto, M.; Kumasaka, T.; Nakanishi, S.; Jingami, H.; Morikawa, K. Structural basis of glutamate recognition by a dimeric metabotropic glutamate receptor. *Nature* **2000**, 407, 971–7.

50. Pin, J. P.; Duvoisin, R. The metabotropic glutamate receptors: structure and functions. *Neuropharmacology* **1995**, 34, 1–26.

51. Brabet, I.; Parmentier, M. L.; De Colle, C.; Bockaert, J.; Acher, F.; Pin, J. P. Comparative effect of L-CCG-I, DCG-IV and gamma-carboxy-L-glutamate on all cloned metabotropic glutamate receptor subtypes. *Neuropharmacology* **1998**, 37, 1043–51.

52. Kalivas, P. W.; Lalumiere, R. T.; Knackstedt, L.; Shen, H. Glutamate transmission in addiction. *Neuropharmacology* **2009**, 56 Suppl 1, 169–73.

53. Peters, J.; Kalivas, P. W. The group II metabotropic glutamate receptor agonist, LY379268, inhibits both cocaine- and food-seeking behavior in rats. *Psychopharmacology (Berl)* **2006**, 186, 143–9.

54. Marek, G. J. Metabotropic glutamate 2/3 receptors as drug targets. *Curr Opin Pharmacol* **2004**, 4, 18–22.

55. Lee, B.; Platt, D. M.; Rowlett, J. K.; Adewale, A. S.; Spealman, R. D. Attenuation of behavioral effects of cocaine by the metabotropic glutamate receptor 5 antagonist 2-methyl-6-(phenylethynyl)-pyridine in squirrel monkeys: comparison with dizocilpine. *J Pharmacol Exp Ther* **2005**, 312, 1232–40.

56. Chiamulera, C.; Epping-Jordan, M. P.; Zocchi, A.; Marcon, C.; Cottiny, C.; Tacconi, S.; Corsi, M.; Orzi, F.; Conquet, F. Reinforcing and locomotor stimulant effects of cocaine are absent in mGluR5 null mutant mice. *Nat Neurosci* **2001**, 4, 873–4.

57. Traynelis, S. F.; Wollmuth, L. P.; McBain, C. J.; Menniti, F. S.; Vance, K. M.; Ogden, K. K.; Hansen, K. B.; Yuan, H.; Myers, S. J.; Dingledine, R. Glutamate receptor ion channels: structure, regulation, and function. *Pharmacol Rev* **2010**, 62, 405–96.

58. Sobolevsky, A. I.; Rosconi, M. P.; Gouaux, E. X-ray structure, symmetry and mechanism of an AMPA-subtype glutamate receptor. *Nature* **2009**, 462, 745–56.

59. Sanacora, G.; Zarate, C. A.; Krystal, J. H.; Manji, H. K. Targeting the glutamatergic system to develop novel, improved therapeutics for mood disorders. *Nat Rev Drug Discov* **2008**, 7, 426–37.

60. Bunch, L.; Erichsen, M. N.; Jensen, A. A. Excitatory amino acid transporters as potential drug targets. *Expert Opin Ther Target* **2009**, 13, 719–31.

61. Jensen, A. A.; Bräuner-Osborne, H. Pharmacological characterization of human excitatory amino acid transporters EAAT1, EAAT2 and EAAT3 in a fluorescence-based membrane potential assay. *Bioch Pharmacol* **2004**, 67, 2115–27.

62. Baker, D. A.; McFarland, K.; Lake, R. W.; Shen, H.; Tang, X. C.; Toda, S.; Kalivas, P. W. Neuroadaptations in cystine-glutamate exchange underlie cocaine relapse. *Nat Neurosci* **2003**, 6, 743–9.

63. Boudreau, A. C.; Wolf, M. E. Behavioral sensitization to cocaine is associated with increased AMPA receptor surface expression in the nucleus accumbens. *J Neurosci* **2005**, 25, 9144–51.

64. Wise, R. A.; Morales, M. A ventral tegmental CRF-glutamate-dopamine interaction in addiction. *Brain Res* **2009**, 1314, 38–43.

65. Wise, R. A. Ventral tegmental glutamate: a role in stress-, cue-, and cocaine-induced reinstatement of cocaine-seeking. *Neuropharmacology* **2009**, 56 Suppl 1, 174–6.

66. Li, K. Y.; Xiao, C.; Xiong, M.; Delphin, E.; Ye, J. H. Nanomolar propofol stimulates glutamate transmission to dopamine neurons: a possible mechanism of abuse potential? *J Pharmacol Exp Ther* **2008**, 325, 165–74.

67. Squires, R. F. *GABA and Benzodiazepine Receptors*. Boca Raton, FL: CRC Press, 1988.

68. Tan, K. R.; Rudolph, U.; Luscher, C. Hooked on benzodiazepines: GABAA receptor subtypes and addiction. *Trends Neurosci* **2011**, 34, 188–97.

69. Jacob, T. C.; Moss, S. J.; Jurd, R. GABA(A) receptor trafficking and its role in the dynamic modulation of neuronal inhibition. *Nat Rev Neurosci* **2008**, 9, 331–43.

70. Cohen, J. Y.; Haesler, S.; Vong, L.; Lowell, B. B.; Uchida, N. Neuron-type-specific signals for reward and punishment in the ventral tegmental area. *Nature* **2012**, 482, 85–8.

71. Cousins, M. S.; Roberts, D. C.; de Wit, H. GABA(B) receptor agonists for the treatment of drug addiction: a review of recent findings. *Drug Alcohol Depend* **2002**, 65, 209–20.

72. Brambilla, P.; Perez, J.; Barale, F.; Schettini, G.; Soares, J. C. GABAergic dysfunction in mood disorders. *Mol Psychiat* **2003**, 8, 721–37, 715.

73. Thiele, A. Muscarinic signaling in the brain. *Annu Rev Neurosci* **2013**, 36, 271–94.

74. Sugimoto, H.; Ogura, H.; Arai, Y.; Limura, Y.; Yamanishi, Y. Research and development of donepezil hydrochloride, a new type of acetylcholinesterase inhibitor. *Jpn J Pharmacol* **2002**, 89, 7–20.

75. Haga, T. Molecular properties of muscarinic acetylcholine receptors. *Proc Jpn Acad Ser B Phys Biol Sci* **2013**, 89, 226–56.

76. Haga, K.; Kruse, A. C.; Asada, H.; Yurugi-Kobayashi, T.; Shiroishi, M.; Zhang, C.; Weis, W. I.; Okada, T.; Kobilka, B. K.; Haga, T.; et al. Structure of the human M2 muscarinic acetylcholine receptor bound to an antagonist. *Nature* **2012**, 482, 547–51.

77. Kruse, A. C.; Ring, A. M.; Manglik, A.; Hu, J.; Hu, K.; Eitel, K.; Hubner, H.; Pardon, E.; Valant, C.; Sexton, P. M.; et al. Activation and allosteric modulation of a muscarinic acetylcholine receptor. *Nature* **2013**, 504, 101–6.

78. Kruse, A. C.; Hu, J.; Pan, A. C.; Arlow, D. H.; Rosenbaum, D. M.; Rosemond, E.; Green, H. F.; Liu, T.; Chae, P. S.; Dror, R. O.; et al. Structure and dynamics of the M3 muscarinic acetylcholine receptor. *Nature* **2012**, 482, 552–6.

79. Conn, P. J.; Jones, C. K.; Lindsley, C. W. Subtype-selective allosteric modulators of muscarinic receptors for the treatment of CNS disorders. *Trends Pharmacol Sci* **2009**, 30, 148–55.

80. Brown, D. Muscarinic Acetylcholine Receptors (mAChRs) in the nervous system: some functions and mechanisms. *J Mol Neurosci* **2010**, 41, 340–46.

81. Thomsen, M.; Woldbye, D. P.; Wortwein, G.; Fink-Jensen, A.; Wess, J.; Caine, S. B. Reduced cocaine self-administration in muscarinic M5 acetylcholine receptor-deficient mice. *J Neurosci* **2005**, 25, 8141–9.

82. Basile, A. S.; Fedorova, I.; Zapata, A.; Liu, X.; Shippenberg, T.; Duttaroy, A.; Yamada, M.; Wess, J. Deletion of the M5 muscarinic acetylcholine receptor attenuates morphine reinforcement and withdrawal but not morphine analgesia. *Proc Natl Acad Sci USA* **2002**, 99, 11452–7.

83. Crespo, J. A.; Sturm, K.; Saria, A.; Zernig, G. Activation of muscarinic and nicotinic acetylcholine receptors in the nucleus accumbens core is necessary for the acquisition of drug reinforcement. *J Neurosci* **2006**, 26, 6004–10.

84. Adinoff, B.; Devous, M. D., Sr.; Williams, M. J.; Best, S. E.; Harris, T. S.; Minhajuddin, A.; Zielinski, T.; Cullum, M. Altered neural cholinergic receptor systems in cocaine-addicted subjects. *Neuropsychopharmacology* **2010**, 35, 1485–99.

85. Nair-Roberts, R. G.; Chatelain-Badie, S. D.; Benson, E.; White-Cooper, H.; Bolam, J. P.; Ungless, M. A. Stereological estimates of dopaminergic, GABAergic and glutamatergic neurons in the ventral tegmental area, substantia nigra and retrorubral field in the rat. *Neuroscience* **2008**, 152, 1024–31.

86. Sokoloff, P.; Martres, M. P.; Giros, B.; Bouthenet, M. L.; Schwartz, J. C. The third dopamine receptor (D3) as a novel target for antipsychotics. *Biochem Pharmacol* **1992**, 43, 659–66.

87. Beaulieu, J. M.; Gainetdinov, R. R. The physiology, signaling, and pharmacology of dopamine receptors. *Pharmacol Rev* **2011**, 63, 182–217.

88. Heidbreder, C. A.; Gardner, E. L.; Xi, Z. X.; Thanos, P. K.; Mugnaini, M.; Hagan, J. J.; Ashby, C. R., Jr. The role of central dopamine D3 receptors in drug addiction: a review of pharmacological evidence. *Brain Res Brain Res Rev* **2005**, 49, 77–105.

89. Chien, E. Y.; Liu, W.; Zhao, Q.; Katritch, V.; Han, G. W.; Hanson, M. A.; Shi, L.; Newman, A. H.; Javitch, J. A.; Cherezov, V.; et al. Structure of the human dopamine D3 receptor in complex with a D2/D3 selective antagonist. *Science* **2010**, 330, 1091–5.

90. Gil-Mast, S.; Kortagere, S.; Kota, K.; Kuzhikandathil, E. V. An amino acid residue in the second extracellular loop determines the agonist-dependent tolerance property of the human D3 dopamine receptor. *ACS Chem Neurosci* **2013**, 4, 940–51.

91. Heidbreder, C. A.; Newman, A. H. Current perspectives on selective dopamine D(3) receptor antagonists as pharmacotherapeutics for addictions and related disorders. *Ann N Y Acad Sci* **2010**, 1187, 4–34.

92. Volkow, N. D.; Wang, G. J.; Telang, F.; Fowler, J. S.; Logan, J.; Childress, A. R.; Jayne, M.; Ma, Y.; Wong, C. Cocaine cues and dopamine in dorsal striatum: mechanism of craving in cocaine addiction. *J Neurosci* **2006**, 26, 6583–8.

93. Nader, M. A.; Czoty, P. W. PET imaging of dopamine D2 receptors in monkey models of cocaine abuse: genetic predisposition versus environmental modulation. *Am J Psychiat* **2005**, 162, 1473–82.

94. Weinshenker, D.; Schroeder, J. P. There and back again: a tale of norepinephrine and drug addiction. *Neuropsychopharmacology* **2007**, 32, 1433–51.

95. Guiard, B. P.; El Mansari, M.; Blier, P. Cross-talk between dopaminergic and noradrenergic systems in the rat ventral tegmental area, locus ceruleus, and dorsal hippocampus. *Mol Pharmacol* **2008**, 74, 1463–75.

96. Nichols, D. E.; Nichols, C. D. Serotonin receptors. *Chem Rev* **2008**, 108, 1614–41.

97. Vollenweider, F. X.; Kometer, M. The neurobiology of psychedelic drugs: implications for the treatment of mood disorders. *Nat Rev Neurosci* **2010**, 11, 642–51.

98. Ohno, Y. New insight into the therapeutic role of 5-HT1A receptors in central nervous system disorders. *Cent Nerv Syst Agents Med Chem* **2010**, 10, 148–57.

99. Matsubara, K.; Shimizu, K.; Suno, M.; Ogawa, K.; Awaya, T.; Yamada, T.; Noda, T.; Satomi, M.; Ohtaki, K.; Chiba, K.; et al. Tandospirone, a 5-HT1A agonist, ameliorates movement disorder via non-dopaminergic systems in rats with unilateral 6-hydroxydopamine-generated lesions. *Brain Res* **2006**, 1112, 126–33.

100. Ago, Y.; Koyama, Y.; Baba, A.; Matsuda, T. Regulation by 5-HT1A receptors of the in vivo release of 5-HT and DA in mouse frontal cortex. *Neuropharmacology* **2003**, 45, 1050–6.

101. Pockros, L. A.; Pentkowski, N. S.; Swinford, S. E.; Neisewander, J. L. Blockade of 5-HT2A receptors in the medial prefrontal cortex attenuates reinstatement of cue-elicited cocaine-seeking behavior in rats. *Psychopharmacology (Berl)* **2010**, 213, 307–20.

102. Manvich, D. F.; Kimmel, H. L.; Howell, L. L. Effects of serotonin 2C receptor agonists on the behavioral and neurochemical effects of cocaine in squirrel monkeys. *J Pharmacol Exp Ther* **2012**, 341, 424–34.

103. Pentkowski, N. S.; Duke, F. D.; Weber, S. M.; Pockros, L. A.; Teer, A. P.; Hamilton, E. C.; Thiel, K. J.; Neisewander, J. L. Stimulation of medial prefrontal cortex serotonin 2C (5-HT(2C)) receptors attenuates cocaine-seeking behavior. *Neuropsychopharmacology* **2010**, 35, 2037–48.

104. Miles, T. F.; Bower, K. S.; Lester, H. A.; Dougherty, D. A. A coupled array of noncovalent interactions impacts the function of the 5-HT3A serotonin receptor in an agonist-specific way. *ACS Chem Neurosci* **2012**, 3, 753–60.

105. Hagan, R. M.; Kilpatrick, G. J.; Tyers, M. B. Interactions between 5-HT3 receptors and cerebral dopamine function: implications for the treatment of schizophrenia and psychoactive substance abuse. *Psychopharmacology (Berl)* **1993**, 112, S68–75.

106. Yoshimoto, K.; McBride, W. J.; Lumeng, L.; Li, T. K. Alcohol stimulates the release of dopamine and serotonin in the nucleus accumbens. *Alcohol* **1992**, 9, 17–22.

107. Aronson, S. C.; Black, J. E.; McDougle, C. J.; Scanley, B. E.; Jatlow, P.; Kosten, T. R.; Heninger, G. R.; Price, L. H. Serotonergic mechanisms of cocaine effects in humans. *Psychopharmacology (Berl)* **1995**, 119, 179–85.

108. Eiden, L. E.; Weihe, E. VMAT2: a dynamic regulator of brain monoaminergic neuronal function interacting with drugs of abuse. *Ann N Y Acad Sci* **2011**, 1216, 86–98.

109. Wimalasena, K. Vesicular monoamine transporters: structure-function, pharmacology, and medicinal chemistry. *Med Res Rev* **2011**, 31, 483–519.

110. Partilla, J. S.; Dempsey, A. G.; Nagpal, A. S.; Blough, B. E.; Baumann, M. H.; Rothman, R. B. Interaction of amphetamines and related compounds at the vesicular monoamine transporter. *J Pharmacol Exp Ther* **2006**, 319, 237–46.

111. Wang, W.; Zhou, Y.; Sun, J.; Pan, L.; Kang, L.; Dai, Z.; Yu, R.; Jin, G.; Ma, L. The effect of L-stepholidine, a novel extract of Chinese herb, on the acquisition, expression, maintenance, and re-acquisition of morphine conditioned place preference in rats. *Neuropharmacology* **2007**, 52, 355–61.

112. Clapper, J. R.; Mangieri, R. A.; Piomelli, D. The endocannabinoid system as a target for the treatment of cannabis dependence. *Neuropharmacology* **2009**, 56 Suppl 1, 235–43.

113. Howlett, A. C.; Barth, F.; Bonner, T. I.; Cabral, G.; Casellas, P.; Devane, W. A.; Felder, C. C.; Herkenham, M.; Mackie, K.; Martin, B. R.; et al. International Union of Pharmacology. XXVII. Classification of cannabinoid receptors. *Pharmacol Rev* **2002**, 54, 161–202.

114. Katona, I.; Sperlagh, B.; Sik, A.; Kafalvi, A.; Vizi, E. S.; Mackie, K.; Freund, T. F. Presynaptically located CB1 cannabinoid receptors regulate GABA release from axon terminals of specific hippocampal interneurons. *J Neurosci* **1999**, 19, 4544–58.

115. Lafenetre, P.; Chaouloff, F.; Marsicano, G. Bidirectional regulation of novelty-induced behavioral inhibition by the endocannabinoid system. *Neuropharmacology* **2009**, 57, 715–21.

116. McDonald, J.; Schleifer, L.; Richards, J. B.; de Wit, H. Effects of THC on behavioral measures of impulsivity in humans. *Neuropsychopharmacology* **2003**, 28, 1356–65.

117. Lane, S. D.; Cherek, D. R.; Tcheremissine, O. V.; Lieving, L. M.; Pietras, C. J. Acute marijuana effects on human risk taking. *Neuropsychopharmacology* **2005**, 30, 800–9.

118. Ramaekers, J. G.; Kauert, G.; van Ruitenbeek, P.; Theunissen, E. L.; Schneider, E.; Moeller, M. R. High-potency marijuana impairs executive function and inhibitory motor control. *Neuropsychopharmacology* **2006**, 31, 2296–303.

119. Ehlers, C. L.; Slutske, W. S.; Lind, P. A.; Wilhelmsen, K. C. Association between single nucleotide polymorphisms in the cannabinoid receptor gene (CNR1) and impulsivity in southwest California Indians. *Twin Res Hum Genet* **2007**, 10, 805–11.

120. Miyake, M.; Christie, M. J.; North, R. A. Single potassium channels opened by opioids in rat locus ceruleus neurons. *Proc Natl Acad Sci USA* **1989**, 86, 3419–22.

121. Pradhan, A. A.; Befort, K.; Nozaki, C.; Gaveriaux-Ruff, C.; Kieffer, B. L. The delta opioid receptor: an evolving target for the treatment of brain disorders. *Trends Pharmacol Sci* **2011**, 32, 581–90.

122. Manglik, A.; Kruse, A. C.; Kobilka, T. S.; Thian, F. S.; Mathiesen, J. M.; Sunahara, R. K.; Pardo, L.; Weis, W. I.; Kobilka, B. K.; Granier, S. Crystal structure of the mu-opioid receptor bound to a morphinan antagonist. *Nature* **2012**, 485, 321–6.

123. Wu, H.; Wacker, D.; Mileni, M.; Katritch, V.; Han, G. W.; Vardy, E.; Liu, W.; Thompson, A. A.; Huang, X. P.; Carroll, F. I.; et al. Structure of the human kappa-opioid receptor in complex with JDTic. *Nature* **2012**, 485, 327–32.

124. Granier, S.; Manglik, A.; Kruse, A. C.; Kobilka, T. S.; Thian, F. S.; Weis, W. I.; Kobilka, B. K. Structure of the delta-opioid receptor bound to naltrindole. *Nature* **2012**, 485, 400–4.

125. Ghitza, U. E.; Preston, K. L.; Epstein, D. H.; Kuwabara, H.; Endres, C. J.; Bencherif, B.; Boyd, S. J.; Copersino, M. L.; Frost, J. J.; Gorelick, D. A. Brain mu-opioid receptor binding predicts treatment outcome in cocaine-abusing outpatients. *Biol Psychiat* **2010**, 68, 697–703.

126. Bohn, L. M.; Gainetdinov, R. R.; Lin, F. T.; Lefkowitz, R. J.; Caron, M. G. Mu-opioid receptor desensitization by beta-arrestin-2 determines morphine tolerance but not dependence. *Nature* **2000**, 408, 720–3.

127. Molinari, P.; Vezzi, V.; Sbraccia, M.; Gro, C.; Riitano, D.; Ambrosio, C.; Casella, I.; Costa, T. Morphine-like opiates selectively antagonize receptor-arrestin interactions. *J Biol Chem* **2010**, 285, 12522–35.

128. Rives, M.-L.; Rossillo, M.; Liu-Chen, L.-Y.; Javitch, J. A. 6′-Guanidinonaltrindole (6′-GNTI) Is a G protein-biased κ-opioid receptor agonist that inhibits arrestin recruitment. *J Biol Chem* **2012**, 287, 27050–4.

129. White, K. L.; Scopton, A. P.; Rives, M.-L.; Bikbulatov, R. V.; Polepally, P. R.; Brown, P. J.; Kenakin, T.; Javitch, J. A.; Zjawiony, J. K.; Roth, B. L. Identification of novel functionally selective κ-opioid receptor scaffolds. *Mol Pharmacol* **2014**, 85, 83–90.

130. Witkin, J. M.; Statnick, M. A.; Rorick-Kehn, L. M.; Pintar, J. E.; Ansonoff, M.; Chen, Y.; Tucker, R. C.; Ciccocioppo, R. The biology of Nociceptin/Orphanin FQ

(N/OFQ) related to obesity, stress, anxiety, mood, and drug dependence. *Pharmacol Ther* **2014**, 141, 283–99.

131. Yang, Y.; Miller, K. J.; Zhu, Y.; Hong, Y.; Tian, Y.; Murugesan, N.; Gu, Z.; O'Tanyi, E.; Keim, W. J.; Rohrbach, K. W.; et al. Characterization of a novel and selective CB1 antagonist as a radioligand for receptor occupancy studies. *Bioorg Med Chem Lett* **2011**, 21, 6856–60.

132. Martin-Couce, L.; Martin-Fontecha, M.; Capolicchio, S.; Lopez-Rodriguez, M. L.; Ortega-Gutierrez, S. Development of endocannabinoid-based chemical probes for the study of cannabinoid receptors. *J Med Chem* **2011**, 54, 5265–9.

133. Sovago, J.; Dupuis, D. S.; Gulyas, B.; Hall, H. An overview on functional receptor autoradiography using [35S]GTPgammaS. *Brain Res Brain Res Rev* **2001**, 38, 149–64.

134. Sim, L.; Selley, D.; Dworkin, S.; Childers, S. Effects of chronic morphine administration on mu opioid receptor-stimulated [35S]GTPgammaS autoradiography in rat brain. *J Neurosci* **1996**, 16, 2684–92.

135. Nestler, E. J.; Aghajanian, G. K. Molecular and cellular basis of addiction. *Science* **1997**, 278, 58–63.

136. Nestler, E. J. Transcriptional mechanisms of addiction: role of ΔFosB. *Philos Trans R Soc Lond B Biol Sci* **2008**, 363, 3245–55.

137. Bastle, R. M.; Kufahl, P. R.; Turk, M. N.; Weber, S. M.; Pentkowski, N. S.; Thiel, K. J.; Neisewander, J. L. Novel cues reinstate cocaine-seeking behavior and induce Fos protein expression as effectively as conditioned cues. *Neuropsychopharmacology* **2012**, 37, 2109–20.

138. Nestler, E. J. Historical review: molecular and cellular mechanisms of opiate and cocaine addiction. *Trends Pharmacol Sci* **2004**, 25, 210–8.

139. Freeman, W. M.; Vrana, K. E. Future prospects for biomarkers of alcohol consumption and alcohol-induced disorders. *Alcohol Clin Exp Res* **2010**, 34, 946–54.

140. Lande, R. G.; Marin, B. A comparison of two alcohol biomarkers in clinical practice: ethyl glucuronide versus ethyl sulfate. *J Addict Dis* **2013**, 32, 288–92.

141. Freeman, W. M.; Salzberg, A. C.; Gonzales, S. W.; Grant, K. A.; Vrana, K. E. Classification of alcohol abuse by plasma protein biomarkers. *Biol Psychiat* **2010**, 68, 219–22.

142. Adkins, D. E.; McClay, J. L.; Vunck, S. A.; Batman, A. M.; Vann, R. E.; Clark, S. L.; Souza, R. P.; Crowley, J. J.; Sullivan, P. F.; van den Oord, E. J. C. G.; et al. Behavioral metabolomics analysis identifies novel neurochemical signatures in methamphetamine sensitization. *Genes Brain Behav* **2013**, 12, 780–91.

143. Dudai, Y. The neurobiology of consolidations, or, how stable is the engram? *Annu Rev Psychol* **2004**, 55, 51–86.

144. Tronson, N. C.; Taylor, J. R. Molecular mechanisms of memory reconsolidation. *Nat Rev Neurosci* **2007**, 8, 262–75.

145. Collingridge, G. L.; Peineau, S.; Howland, J. G.; Wang, Y. T. Long-term depression in the CNS. *Nat Rev Neurosci* **2010**, 11, 459–73.

146. Diergaarde, L.; Schoffelmeer, A. N.; De Vries, T. J. Pharmacological manipulation of memory reconsolidation: towards a novel treatment of pathogenic memories. *Eur J Pharmacol* **2008**, 585, 453–7.

147. Schiller, D.; Kanen, J. W.; LeDoux, J. E.; Monfils, M. H.; Phelps, E. A. Extinction during reconsolidation of threat memory diminishes prefrontal cortex involvement. *Proc Natl Acad Sci USA* **2013**, 110, 20040–5.

148. Torregrossa, M. M.; Taylor, J. R. Learning to forget: manipulating extinction and recon-solidation processes to treat addiction. *Psychopharmacology (Berl)* **2012**, 226, 659–72.

149. Ressler, K. J.; Rothbaum, B. O.; Tannenbaum, L.; Anderson, P.; Graap, K.; Zimand, E.; Hodges, L.; Davis, M. Cognitive enhancers as adjuncts to psychotherapy: use of D-cycloserine in phobic individuals to facilitate extinction of fear. *Arch Gen Psychiat* **2004**, 61, 1136–44.

150. Yang, Y.; Zhou, Q. Spine modifications associated with long-term potentiation. *The Neuroscientist* **2009**, 15, 464–76.

151. Drdla-Schutting, R.; Benrath, J.; Wunderbaldinger, G.; Sandkuhler, J. Erasure of a spinal memory trace of pain by a brief, high-dose opioid administration. *Science* **2012**, 335, 235–8.

152. Armario, A. Activation of the hypothalamic-pituitary-adrenal axis by addictive drugs: different pathways, common outcome. *Trends Pharmacol Sci* **2010**, 31, 318–25.

153. Sarnyai, Z.; Shaham, Y.; Heinrichs, S. C. The role of corticotropin-releasing factor in drug addiction. *Pharmacol Rev* **2001**, 53, 209–43.

154. Funder, J. W. Glucocorticoid and mineralocorticoid receptors: biology and clinical rele-vance. *Annu Rev Med* **1997**, 48, 231–40.

155. Barik, J.; Parnaudeau, S.; Saint Amaux, A. L.; Guiard, B. P.; Golib Dzib, J. F.; Bocquet, O.; Bailly, A.; Benecke, A.; Tronche, F. Glucocorticoid receptors in dopaminoceptive neu-rons, key for cocaine, are dispensable for molecular and behavioral morphine responses. *Biol Psychiat* **2010**, 68, 231–9.

156. Walter, M.; Gerber, H.; Kuhl, H. C.; Schmid, O.; Joechle, W.; Lanz, C.; Brenneisen, R.; Schachinger, H.; Riecher-Rossler, A.; Wiesbeck, G. A.; et al. Acute effects of intravenous heroin on the hypothalamic-pituitary-adrenal axis response: a controlled trial. *J Clin Psychopharmacol* **2013**, 33, 193–8.

157. van der Kolk, B. A. Psychobiology of posttraumatic stress disorder. In: *Textbook of Biological Psychiatry*, Panksepp, J., Ed. Wiley-Liss, 2004; pp. 319–34.

158. Zorrilla, E. P.; Koob, G. F. Progress in corticotropin-releasing factor-1 antagonist devel-opment. *Drug Discov Today* **2010**, 15, 371–83.

159. Robbins, T. W.; Murphy, E. R. Behavioural pharmacology: 40+ years of progress, with a focus on glutamate receptors and cognition. *Trends Pharmacol Sci* **2006**, 27, 141–8.

160. Branch, M. N. How research in behavioral pharmacology informs behavioral science. *J Exp Anal Behav* **2006**, 85, 407–23.

161. Belin-Rauscent, A.; Belin, D. Animal Models of Drug Addiction. In: Belin, D. (ed.), *Addictions – From Pathophysiology to Treatment*, 2012. doi: 10.5772/52079.

162. Shippenberg, T.; Koob, G. *Neuropsychopharmacology: The Fifth Generation of Progress* (Chapter 97). Philadelphia: Lippincott Williams & Wilkins, 2002.

163. Gravetter, F. J.; Forzano, L.-A. B. *Research Methods for the Behavioral Sciences*, 4th edition, Australia; Belmont, CA: Wadsworth, 2012.

164. Powell, S. B.; Geyer, M. A. Overview of animal models of schizophrenia. *Curr Protoc Neurosci* **2007**, Chapter 9, Unit 9.24.

165. Ramos, A. Animal models of anxiety: do I need multiple tests? *Trends Pharmacol Sci* **2008**, 29, 493–8.

166. Koob, G. F.; Kenneth Lloyd, G.; Mason, B. J. Development of pharmacotherapies for drug addiction: a Rosetta stone approach. *Nat Rev Drug Discov* **2009**, 8, 500–15.

167. Mello, N. K.; Negus, S. S. Preclinical evaluation of pharmacotherapies for treatment of cocaine and opioid abuse using drug self-administration procedures. *Neuropsychopharmacology* **1996**, 14, 375–424.

168. Thomsen, M.; Caine, S. B. Chronic intravenous drug self-administration in rats and mice. *Curr Protoc Neurosci* **2005**, Chapter 9, Unit 9.20.

169. Platt, D. M.; Carey, G.; Spealman, R. D. Intravenous self-administration techniques in monkeys. *Curr Protoc Neurosci* **2005**, Chapter 9, Unit 9.21.

170. Wang, J.; Wu, X.; Li, C.; Wei, J.; Jiang, H.; Liu, C.; Yu, C.; Carlson, S.; Hu, X.; Ma, H.; et al. Effect of morphine on conditioned place preference in rhesus monkeys. *Addict Biol* **2012**, 17, 539–46.

171. Yahyavi-Firouz-Abadi, N.; See, R. E. Anti-relapse medications: preclinical models for drug addiction treatment. *Pharmacol Ther* **2009**, 124, 235–47.

172. Katz, J. L.; Higgins, S. T. The validity of the reinstatement model of craving and relapse to drug use. *Psychopharmacology (Berl)* **2003**, 168, 21–30.

173. Welberg, L. Addiction: craving: a core issue. *Nat Rev Neurosci* **2013**, 14, 307.

174. Gipson, C.; Kupchik, Y.; Shen, H.; Reissner, K.; Thomas, C.; Kalivas, P. Relapse induced by cues predicting cocaine depends on rapid, transient synaptic potentiation. *Neuron* **2013**, 77, 867–72.

175. Wolf, M. E. The Bermuda Triangle of cocaine-induced neuroadaptations. *Trends Neurosci* **2010**, 33, 391–8.

176. Paterson, N. E. Assessment of substance abuse liability in rodents: self-administration, drug discrimination, and locomotor sensitization. *Curr Protoc Pharmacol* **2012**, Chapter 5, Unit 5.62.

177. Thomsen, M.; Conn, P. J.; Lindsley, C.; Wess, J.; Boon, J. Y.; Fulton, B. S.; Fink-Jensen, A.; Caine, S. B. Attenuation of cocaine's reinforcing and discriminative stimulus effects via muscarinic M1 acetylcholine receptor stimulation. *J Pharmacol Exp Ther* **2010**, 332, 959–69.

178. Glennon, R. A.; Young, R. *Drug discrimination: applications to medicinal chemistry and drug studies.* Hoboken, New Jersey: Wiley, 2011.

179. Olsen, C. M.; Winder, D. G. Operant sensation seeking engages similar neural substrates to operant drug seeking in C57 mice. *Neuropsychopharmacology* **2009**, 34, 1685–94.

180. Lile, J. A.; Nader, M. A. The abuse liability and therapeutic potential of drugs evaluated for cocaine addiction as predicted by animal models. *Curr Neuropharmacol* **2003**, 1, 21–46.

181. Haney, M.; Spealman, R. Controversies in translational research: drug self-administration. *Psychopharmacology (Berl)* **2008**, 199, 403–19.

182. Prisinzano, T. E. Natural products as tools for neuroscience: discovery and development of novel agents to treat drug abuse. *J Nat Prod* **2009**, 72, 581–7.

183. Newman, A. H.; Grundt, P.; Nader, M. A. Dopamine D3 receptor partial agonists and antagonists as potential drug abuse therapeutic agents. *J Med Chem* **2005**, 48, 3663–79.

184. Carroll, F. I. 2002 Medicinal Chemistry Division Award address: monoamine transporters and opioid receptors. Targets for addiction therapy. *J Med Chem* **2003**, 46, 1775–94.

185. Beardsley, P. M.; Thomas, B. F.; McMahon, L. R. Cannabinoid CB1 receptor antagonists as potential pharmacotherapies for drug abuse disorders. *Int Rev Psychiat* **2009**, 21, 134–42.

186. Le Foll, B.; Goldberg, S. R. Cannabinoid CB1 receptor antagonists as promising new medications for drug dependence. *J Pharmacol Exp Ther* **2005**, 312, 875–83.

187. De Vries, T. J.; Schoffelmeer, A. N. Cannabinoid CB1 receptors control conditioned drug seeking. *Trends Pharmacol Sci* **2005**, 26, 420–6.

188. Paneda, C.; Winsky-Sommerer, R.; Boutrel, B.; de Lecea, L. The corticotropin-releasing factor-hypocretin connection: implications in stress response and addiction. *Drug News Perspect* **2005**, 18, 250–5.

189. Harris, G. C.; Wimmer, M.; Aston-Jones, G. A role for lateral hypothalamic orexin neurons in reward seeking. *Nature* **2005**, 437, 556–9.

190. Craft, R. M.; Dykstra, L. A. Agonist and antagonist activity of kappa opioids in the squirrel monkey: II. Effect of chronic morphine treatment. *J Pharmacol Exp Ther* **1992**, 260, 334–42.

191. Craft, R. M.; Dykstra, L. A. Agonist and antagonist activity of kappa opioids in the squirrel monkey: I. Antinociception and urine output. *J Pharmacol Exp Ther* **1992**, 260, 327–33.

192. Rothman, R. B. A review of the role of anti-opioid peptides in morphine tolerance and dependence. *Synapse* **1992**, 12, 129–38.

193. Walsh, S. L.; Strain, E. C.; Abreu, M. E.; Bigelow, G. E. Enadoline, a selective kappa opioid agonist: comparison with butorphanol and hydromorphone in humans. *Psychopharmacology (Berl)* **2001**, 157, 151–62.

194. McLaughlin, J. P.; Marton-Popovici, M.; Chavkin, C. Kappa opioid receptor antagonism and prodynorphin gene disruption block stress-induced behavioral responses. *J Neurosci* **2003**, 23, 5674–83.

195. Beardsley, P. M.; Howard, J. L.; Shelton, K. L.; Carroll, F. I. Differential effects of the novel kappa opioid receptor antagonist, JDTic, on reinstatement of cocaine-seeking induced by footshock stressors vs cocaine primes and its antidepressant-like effects in rats. *Psychopharmacology (Berl)* **2005**, 183, 118–26.

196. Tzschentke, T. M.; Schmidt, W. J. Glutamatergic mechanisms in addiction. *Mol Psychiat* **2003**, 8, 373–82.

197. Kalivas, P. W.; Volkow, N.; Seamans, J. Unmanageable motivation in addiction: a pathology in prefrontal-accumbens glutamate transmission. *Neuron* **2005**, 45, 647–50.

198. Kenny, P. J.; Markou, A. The ups and downs of addiction: role of metabotropic glutamate receptors. *Trends Pharmacol Sci* **2004**, 25, 265–72.

199. Bellone, C.; Luscher, C. Cocaine triggered AMPA receptor redistribution is reversed in vivo by mGluR-dependent long-term depression. *Nat Neurosci* **2006**, 9, 636–41.

200. Dravolina, O. A.; Zakharova, E. S.; Shekunova, E. V.; Zvartau, E. E.; Danysz, W.; Bespalov, A. Y. mGlu1 receptor blockade attenuates cue- and nicotine-induced reinstatement of extinguished nicotine self-administration behavior in rats. *Neuropharmacology* **2007**, 52, 263–9.

201. Kotlinska, J.; Bochenski, M. Comparison of the effects of mGluR1 and mGluR5 antagonists on the expression of behavioral sensitization to the locomotor effect of morphine and the morphine withdrawal jumping in mice. *Eur J Pharmacol* **2007**, 558, 113–8.

202. Olive, M. F.; McGeehan, A. J.; Kinder, J. R.; McMahon, T.; Hodge, C. W.; Janak, P. H.; Messing, R. O. The mGluR5 antagonist 6-methyl-2-(phenylethynyl)pyridine decreases ethanol consumption via a protein kinase C epsilon-dependent mechanism. *Mol Pharmacol* **2005**, 67, 349–55.

203. Paterson, N. E.; Semenova, S.; Gasparini, F.; Markou, A. The mGluR5 antagonist MPEP decreased nicotine self-administration in rats and mice. *Psychopharmacology (Berl)* **2003**, 167, 257–64.

204. Bossert, J. M.; Busch, R. F.; Gray, S. M. The novel mGluR2/3 agonist LY379268 attenuates cue-induced reinstatement of heroin seeking. *Neuroreport* **2005**, 16, 1013–6.

205. Baptista, M. A.; Martin-Fardon, R.; Weiss, F. Preferential effects of the metabotropic glutamate 2/3 receptor agonist LY379268 on conditioned reinstatement versus primary reinforcement: comparison between cocaine and a potent conventional reinforcer. *J Neurosci* **2004**, 24, 4723–7.

206. Slusher, B. S.; Thomas, A.; Paul, M.; Schad, C. A.; Ashby, C. R., Jr. Expression and acquisition of the conditioned place preference response to cocaine in rats is blocked by selective inhibitors of the enzyme N-acetylated-alpha-linked-acidic dipeptidase (NAAL-ADASE). *Synapse* **2001**, 41, 22–8.

207. Ma, Y. Y.; Chu, N. N.; Guo, C. Y.; Han, J. S.; Cui, C. L. NR2B-containing NMDA receptor is required for morphine-but not stress-induced reinstatement. *Exp Neurol* **2007**, 203, 309–19.

208. Tyacke, R. J.; Lingford-Hughes, A.; Reed, L. J.; Nutt, D. J. GABAB receptors in addiction and its treatment. *Adv Pharmacol* **2010**, 58, 373–96.

209. Gerasimov, M. R.; Schiffer, W. K.; Gardner, E. L.; Marsteller, D. A.; Lennon, I. C.; Taylor, S. J.; Brodie, J. D.; Ashby, C. R., Jr.; Dewey, S. L. GABAergic blockade of cocaine-associated cue-induced increases in nucleus accumbens dopamine. *Eur J Pharmacol* **2001**, 414, 205–9.

210. Moore, R. J.; Vinsant, S. L.; Nader, M. A.; Porrino, L. J.; Friedman, D. P. Effect of cocaine self-administration on striatal dopamine D1 receptors in rhesus monkeys. *Synapse* **1998**, 28, 1–9.

211. Caine, S. B.; Negus, S. S.; Mello, N. K.; Bergman, J. Effects of dopamine D(1-like) and D(2-like) agonists in rats that self-administer cocaine. *J Pharmacol Exp Ther* **1999**, 291, 353–60.

212. Khroyan, T. V.; Platt, D. M.; Rowlett, J. K.; Spealman, R. D. Attenuation of relapse to cocaine seeking by dopamine D1 receptor agonists and antagonists in non-human primates. *Psychopharmacology (Berl)* **2003**, 168, 124–31.

213. Haney, M.; Collins, E. D.; Ward, A. S.; Foltin, R. W.; Fischman, M. W. Effect of a selective dopamine D1 agonist (ABT-431) on smoked cocaine self-administration in humans. *Psychopharmacology (Berl)* **1999**, 143, 102–10.

214. Cohen, C.; Perrault, G.; Sanger, D. J. Effects of D1 dopamine receptor agonists on oral ethanol self-administration in rats: comparison with their efficacy to produce grooming and hyperactivity. *Psychopharmacology (Berl)* **1999**, 142, 102–10.

215. Goldman-Rakic, P. S.; Castner, S. A.; Svensson, T. H.; Siever, L. J.; Williams, G. V. Targeting the dopamine D1 receptor in schizophrenia: insights for cognitive dysfunction. *Psychopharmacology (Berl)* **2004**, 174, 3–16.

216. Simon, S. L.; Domier, C. P.; Sim, T.; Richardson, K.; Rawson, R. A.; Ling, W. Cognitive performance of current methamphetamine and cocaine abusers. *J Addict Dis* **2002**, 21, 61–74.

217. Simon, S. L.; Domier, C.; Carnell, J.; Brethen, P.; Rawson, R.; Ling, W. Cognitive impairment in individuals currently using methamphetamine. *Am J Addict* **2000**, 9, 222–31.

218. Mash, D. C. D3 receptor binding in human brain during cocaine overdose. *Mol Psychiat* **1997**, 2, 5–6.

219. Rothman, R. B.; Blough, B. E.; Baumann, M. H. Dual dopamine-5-HT releasers: potential treatment agents for cocaine addiction. *Trends Pharmacol Sci* **2006**, 27, 612–8.

220. Yasuda, R. P.; Ciesla, W.; Flores, L. R.; Wall, S. J.; Li, M.; Satkus, S. A.; Weisstein, J. S.; Spagnola, B. V.; Wolfe, B. B. Development of antisera selective for m4 and m5 muscarinic cholinergic receptors: distribution of m4 and m5 receptors in rat brain. *Mol Pharmacol* **1993**, 43, 149–57.

221. Weiner, D. M.; Levey, A. I.; Brann, M. R. Expression of muscarinic acetylcholine and dopamine receptor mRNAs in rat basal ganglia. *Proc Natl Acad Sci USA* **1990**, 87, 7050–4.

222. Vilaro, M. T.; Palacios, J. M.; Mengod, G. Localization of m5 muscarinic receptor mRNA in rat brain examined by in situ hybridization histochemistry. *Neurosci Lett* **1990**, 114, 154–9.

223. Fink-Jensen, A.; Fedorova, I.; Wortwein, G.; Woldbye, D. P.; Rasmussen, T.; Thomsen, M.; Bolwig, T. G.; Knitowski, K. M.; McKinzie, D. L.; Yamada, M.; et al. Role for M5 muscarinic acetylcholine receptors in cocaine addiction. *J Neurosci Res* **2003**, 74, 91–6.

224. Meunier, J. C.; Mollereau, C.; Toll, L.; Suaudeau, C.; Moisand, C.; Alvinerie, P.; Butour, J. L.; Guillemot, J. C.; Ferrara, P.; Monsarrat, B.; et al. Isolation and structure of the endogenous agonist of opioid receptor-like ORL1 receptor. *Nature* **1995**, 377, 532–5.

225. Reinscheid, R. K.; Nothacker, H. P.; Bourson, A.; Ardati, A.; Henningsen, R. A.; Bunzow, J. R.; Grandy, D. K.; Langen, H.; Monsma, F. J., Jr.; Civelli, O. Orphanin FQ: a neuropeptide that activates an opioid like G protein-coupled receptor. *Science* **1995**, 270, 792–4.

226. Mogil, J. S.; Grisel, J. E.; Reinscheid, R. K.; Civelli, O.; Belknap, J. K.; Grandy, D. K. Orphanin FQ is a functional anti-opioid peptide. *Neuroscience* **1996**, 75, 333–7.

227. NIDA. http://www.drugabuse.gov/about-nida/organization/divisions/division-pharmaco therapies-medical-consequences-drug-abuse-dpmcda/research-programs. Last accessed 25 February, 2014.

228. Neubig, R. R.; Spedding, M.; Kenakin, T.; Christopoulos, A. International Union of Pharmacology Committee on Receptor Nomenclature and Drug Classification. XXXVIII. Update on terms and symbols in quantitative pharmacology. *Pharmacol Rev* **2003**, 55, 597–606.

229. Strange, P. G. Antipsychotic drugs: importance of dopamine receptors for mechanisms of therapeutic actions and side effects. *Pharmacol Rev* **2001**, 53, 119–33.

230. Taylor, D. A. In vitro opioid receptor assays. *Curr Protoc Pharmacol* **2011**, Chapter 4, Unit 4.8.1–4.8.34.

231. DeHaven, R. N.; Cassel, J. A.; Windh, R. T.; DeHaven-Hudkins, D. L. Characterization of opioid and ORL1 receptors. *Curr Protoc Pharmacol* **2005**, 29, 1.4.1–1.4.44.

232. Milligan, G. Constitutive activity and inverse agonists of G protein-coupled receptors: a current perspective. *Mol Pharmacol* **2003**, 64, 1271–6.

233. Kenakin, T. Efficacy as a vector: the relative prevalence and paucity of inverse agonism. *Mol Pharmacol* **2004**, 65, 2–11.

234. Parra, S.; Bond, R. A. Inverse agonism: from curiosity to accepted dogma, but is it clinically relevant? *Curr Opin Pharmacol* **2007**, 7, 146–50.

235. Traynor, J. R.; Clark, M. J.; Remmers, A. E. Relationship between rate and extent of G protein activation: comparison between full and partial opioid agonists. *J Pharmacol Exp Ther* **2002**, 300, 157–61.

236. Grabowski, J.; Shearer, J.; Merrill, J.; Negus, S. S. Agonist-like, replacement pharmacotherapy for stimulant abuse and dependence. *Addict Behav* **2004**, 29, 1439–64.

237. Vocci, F.; Ling, W. Medications development: successes and challenges. *Pharmacol Ther* **2005**, 108, 94–108.

238. Wang, L.; Martin, B.; Brenneman, R.; Luttrell, L. M.; Maudsley, S. Allosteric modulators of g protein-coupled receptors: future therapeutics for complex physiological disorders. *J Pharmacol Exp Ther* **2009**, 331, 340–8.

239. Melancon, B. J.; Hopkins, C. R.; Wood, M. R.; Emmitte, K. A.; Niswender, C. M.; Christopoulos, A.; Conn, P. J.; Lindsley, C. W. Allosteric modulation of seven transmembrane spanning receptors: theory, practice, and opportunities for central nervous system drug discovery. *J Med Chem* **2012**, 55, 1445–64.

240. Leach, K.; Sexton, P. M.; Christopoulos, A. Quantification of allosteric interactions at G protein-coupled receptors using radioligand binding assays. *Curr Protoc Pharmacol* **2011**, Chapter 1, Unit 1.22.

241. Maudsley, S.; Martin, B.; Luttrell, L. M. The origins of diversity and specificity in G protein-coupled receptor signaling. *J Pharmacol Exp Ther* **2005**, 314, 485–94.

242. Kenakin, T.; Christopoulos, A. Signalling bias in new drug discovery: detection, quantification and therapeutic impact. *Nat Rev Drug Discov* **2013**, 12, 205–16.

243. Portoghese, P. S.; Lunzer, M. M. Identity of the putative delta1-opioid receptor as a delta-kappa heteromer in the mouse spinal cord. *Eur J Pharmacol* **2003**, 467, 233–4.

244. Waldhoer, M.; Fong, J.; Jones, R. M.; Lunzer, M. M.; Sharma, S. K.; Kostenis, E.; Portoghese, P. S.; Whistler, J. L. A heterodimer-selective agonist shows in vivo relevance of G protein-coupled receptor dimers. *Proc Natl Acad Sci USA* **2005**, 102, 9050–5.

245. Stockton, S. D., Jr.; Devi, L. A. Functional relevance of mu-delta opioid receptor heteromerization: a role in novel signaling and implications for the treatment of addiction disorders: from a symposium on new concepts in mu-opioid pharmacology. *Drug Alcohol Depend* **2012**, 121, 167–72.

246. Park, P. S.; Lodowski, D. T.; Palczewski, K. Activation of G protein-coupled receptors: beyond two-state models and tertiary conformational changes. *Annu Rev Pharmacol Toxicol* **2008**, 48, 107–41.

247. Gurevich, V. V.; Gurevich, E. V. How and why do GPCRs dimerize? *Trends Pharmacol Sci* **2008**, 29, 234–40.

248. Ferre, S.; Ciruela, F.; Woods, A. S.; Lluis, C.; Franco, R. Functional relevance of neurotransmitter receptor heteromers in the central nervous system. *Trends Neurosci* **2007**, 30, 440–6.

249. Casado, V.; Cortes, A.; Ciruela, F.; Mallol, J.; Ferre, S.; Lluis, C.; Canela, E. I.; Franco, R. Old and new ways to calculate the affinity of agonists and antagonists interacting with G-protein-coupled monomeric and dimeric receptors: the receptor-dimer cooperativity index. *Pharmacol Ther* **2007**, 116, 343–54.

250. Prinster, S. C.; Hague, C.; Hall, R. A. Heterodimerization of G protein-coupled receptors: specificity and functional significance. *Pharmacol Rev* **2005**, 57, 289–98.

251. George, S. R.; O'Dowd, B. F.; Lee, S. P. G-protein-coupled receptor oligomerization and its potential for drug discovery. *Nat Rev Drug Discov* **2002**, 1, 808–20.

252. George, S. R.; Fan, T.; Xie, Z.; Tse, R.; Tam, V.; Varghese, G.; O'Dowd, B. F. Oligomer-ization of mu- and delta-opioid receptors. Generation of novel functional properties. *J Biol Chem* **2000**, 275, 26128–35.

253. Gomes, I.; Gupta, A.; Filipovska, J.; Szeto, H. H.; Pintar, J. E.; Devi, L. A. A role for heterodimerization of mu and delta opiate receptors in enhancing morphine analgesia. *Proc Natl Acad Sci USA* **2004**, 101, 5135–9.

254. He, L.; Lee, N. M. Delta opioid receptor enhancement of mu opioid receptor-induced antinociception in spinal cord. *J Pharmacol Exp Ther* **1998**, 285, 1181–6.

255. Malmberg, A. B.; Yaksh, T. L. Isobolographic and dose-response analyses of the interac-tion between intrathecal mu and delta agonists: effects of naltrindole and its benzofuran analog (NTB). *J Pharmacol Exp Ther* **1992**, 263, 264–75.

256. Heyman, J. S.; Vaught, J. L.; Mosberg, H. I.; Haaseth, R. C.; Porreca, F. Modulation of mu-mediated antinociception by delta agonists in the mouse: selective potentiation of morphine and normorphine by [D-Pen2,D-Pen5]enkephalin. *Eur J Pharmacol* **1989**, 165, 1–10.

257. Bowen, C. A.; Negus, S. S.; Zong, R.; Neumeyer, J. L.; Bidlack, J. M.; Mello, N. K. Effects of mixed-action kappa/mu opioids on cocaine self-administration and cocaine discrimination by rhesus monkeys. *Neuropsychopharmacology* **2003**, 28, 1125–39.

258. Calderon, S. N.; Rice, K. C.; Rothman, R. B.; Porreca, F.; Flippen-Anderson, J. L.; Kayakiri, H.; Xu, H.; Becketts, K.; Smith, L. E.; Bilsky, E. J.; et al. Probes for narcotic receptor mediated phenomena. 23. Synthesis, opioid receptor binding, and bioassay of the highly selective delta agonist (+)-4-[(alpha R)-alpha-((2S,5R)-4-Allyl-2,5-dimethyl-1-piperazinyl)-3-methoxybenzyl]-N,N-diethylbenzamide (SNC 80) and related novel non-peptide delta opioid receptor ligands. *J Med Chem* **1997**, 40, 695–704.

259. Stevenson, G. W.; Folk, J. E.; Rice, K. C.; Negus, S. S. Interactions between delta and mu opioid agonists in assays of schedule-controlled responding, thermal nocicep-tion, drug self-administration, and drug versus food choice in rhesus monkeys: stud-ies with SNC80 [(+)-4-[(alphaR)-alpha-((2S,5R)-4-allyl-2,5-dimethyl-1-piperazinyl)-3-meth oxybenzyl]-N,N-diethylbenzamide] and heroin. *J Pharmacol Exp Ther* **2005**, 314, 221–31.

260. Stevenson, G. W.; Folk, J. E.; Linsenmayer, D. C.; Rice, K. C.; Negus, S. S. Opioid interactions in rhesus monkeys: effects of delta + mu and delta + kappa agonists on schedule-controlled responding and thermal nociception. *J Pharmacol Exp Ther* **2003**, 307, 1054–64.

261. Banks, M.; Roma, P.; Folk, J.; Rice, K.; Negus, S. S. Effects of the delta-opioid agonist SNC80 on the abuse liability of methadone in rhesus monkeys: a behavioral economic analysis. *Psychopharmacology* **2011**, 216, 431–9.

262. Lopez-Moreno, J. A.; Lopez-Jimenez, A.; Gorriti, M. A.; de Fonseca, F. R. Functional interactions between endogenous cannabinoid and opioid systems: focus on alcohol, genetics and drug-addicted behaviors. *Curr Drug Targets* **2010**, 11, 406–28.

263. Williams, I. J.; Edwards, S.; Rubo, A.; Haller, V. L.; Stevens, D. L.; Welch, S. P. Time course of the enhancement and restoration of the analgesic efficacy of codeine and morphine by Δ9-tetrahydrocannabinol. *Eur J Pharmacol* **2006**, 539, 57–63.

264. da Fonseca Pacheco, D.; Klein, A.; de Castro Perez, A.; da Fonseca Pacheco, C. M.; de Francischi, J. N.; Duarte, I. D. The mu-opioid receptor agonist morphine, but not agonists at delta- or kappa-opioid receptors, induces peripheral antinociception mediated by cannabinoid receptors. *Br J Pharmacol* **2008**, 154, 1143–9.

265. Braida, D.; Pozzi, M.; Parolaro, D.; Sala, M. Intracerebral self-administration of the cannabinoid receptor agonist CP 55,940 in the rat: interaction with the opioid system. *Eur J Pharmacol* **2001**, 413, 227–34.

266. Yamaguchi, T.; Hagiwara, Y.; Tanaka, H.; Sugiura, T.; Waku, K.; Shoyama, Y.; Watanabe, S.; Yamamoto, T. Endogenous cannabinoid, 2-arachidonoylglycerol, attenuates naloxone-precipitated withdrawal signs in morphine-dependent mice. *Brain Res* **2001**, 909, 121–6.

267. Durroux, T. Principles: a model for the allosteric interactions between ligand binding sites within a dimeric GPCR. *Trends Pharmacol Sci* **2005**, 26, 376–84.

268. Vauquelin, G. Simplified models for heterobivalent ligand binding: when are they applicable and which are the factors that affect their target residence time. *Naunyn Schmiedebergs Arch Pharmacol* **2013**, 1–14.

269. Daniels, D. J.; Lenard, N. R.; Etienne, C. L.; Law, P. Y.; Roerig, S. C.; Portoghese, P. S. Opioid-induced tolerance and dependence in mice is modulated by the distance between pharmacophores in a bivalent ligand series. *Proc Natl Acad Sci USA* **2005**, 102, 19208–13.

270. Lenard, N. R.; Daniels, D. J.; Portoghese, P. S.; Roerig, S. C. Absence of conditioned place preference or reinstatement with bivalent ligands containing mu-opioid receptor agonist and delta-opioid receptor antagonist pharmacophores. *Eur J Pharmacol* **2007**, 566, 75–82.

271. Borioni, A.; Bastanzio, G.; Delfini, M.; Mustazza, C.; Sciubba, F.; Tatti, M.; Giudice, M. R. D. High resolution NMR conformational studies of new bivalent NOP receptor antagonists in model membrane systems. *Bioorganic Chemistry* **2011**, 39, 59–66.

272. Kroeze, W. K.; Roth, B. L. Polypharmacological drugs: "Magic Shotguns" for psychiatric diseases. In: *Polypharmacology in Drug Discovery*, Peters, J.; Ed. John Wiley & Sons, 2012; pp. 135–48.

273. Gray, J. A.; Roth, B. L. The pipeline and future of drug development in schizophrenia. *Mol Psychiat* **2007**, 12, 904–22.

274. Peters, J. U. Polypharmacology – foe or friend? *J Med Chem* **2013**, 56, 8955–71.

275. Xie, L.; Kinnings, S. L.; Bourne, P. E. Novel computational approaches to polypharmacology as a means to define responses to individual drugs. *Annu Rev Pharmacol Toxicol* **2012**, 52, 361–79.

276. Wong, E. H.; Tarazi, F. I.; Shahid, M. The effectiveness of multi-target agents in schizophrenia and mood disorders: relevance of receptor signature to clinical action. *Pharmacol Ther* **2010**, 126, 173–85.

277. Cavalli, A.; Bolognesi, M. L.; Minarini, A.; Rosini, M.; Tumiatti, V.; Recanatini, M.; Melchiorre, C. Multi-target-directed ligands to combat neurodegenerative diseases. *J Med Chem* **2008**, 51, 347–72.

278. Morphy, R.; Rankovic, Z. The physicochemical challenges of designing multiple ligands. *J Med Chem* **2006**, 49, 4961–70.

279. Abbott, N. J.; Patabendige, A. A. K.; Dolman, D. E. M.; Yusof, S. R.; Begley, D. J. Structure and function of the blood-brain barrier. *Neurobiol Dis* **2010**, 37, 13–25.

280. Mehdipour, A. R.; Hamidi, M. Brain drug targeting: a computational approach for overcoming blood-brain barrier. *Drug Discov Today* **2009**, 14, 1030–6.

281. Naik, P.; Cucullo, L. In vitro blood-brain barrier models: current and perspective technologies. *J Pharm Sci* **2012**, 101, 1337–54.

282. Boje, K. M. In vivo measurement of blood-brain barrier permeability. *Curr Protoc Neurosci* **2001**, Chapter 7, Unit 7.19.

283. Banks, W. A. Mouse models of neurological disorders: a view from the blood-brain barrier. *Biochim Biophys Acta* **2010**, 1802, 881–8.

284. Cecchelli, R.; Berezowski, V.; Lundquist, S.; Culot, M.; Renftel, M.; Dehouck, M. P.; Fenart, L. Modelling of the blood-brain barrier in drug discovery and development. *Nat Rev Drug Discov* **2007**, 6, 650–61.

285. Hefti, F. *Drug Discovery for Nervous System Diseases*. Hoboken, N.J.: Wiley-Interscience, 2005; p vii, 319 p.

286. Lipinski, C. A.; Lombardo, F.; Dominy, B. W.; Feeney, P. J. Experimental and computational approaches to estimate solubility and permeability in drug discovery and development settings. *Adv Drug Deliv Rev* **2001**, 46, 3–26.

287. Ghose, A. K.; Herbertz, T.; Hudkins, R. L.; Dorsey, B. D.; Mallamo, J. P. Knowledge-based, Central Nervous System (CNS) lead selection and lead optimization for CNS drug discovery. *ACS Chem Neurosci* **2012**, 3, 50–68.

288. Alelyunas, Y. W.; Empfield, J. R.; McCarthy, D.; Spreen, R. C.; Bui, K.; Pelosi-Kilby, L.; Shen, C. Experimental solubility profiling of marketed CNS drugs, exploring solubility limit of CNS discovery candidate. *Bioorg Med Chem Lett* **2010**, 20, 7312–6.

289. Hitchcock, S. A.; Pennington, L. D. Structure-brain exposure relationships. *J Med Chem* **2006**, 49, 7559–83.

290. Di, L.; Rong, H.; Feng, B. Demystifying brain penetration in central nervous system drug discovery. Miniperspective. *J Med Chem* **2013**, 56, 2–12.

291. Strazielle, N.; Ghersi-Egea, J. F. Physiology of blood-brain interfaces in relation to brain disposition of small compounds and macromolecules. *Mol Pharm* **2013**, 10, 1473–91.

292. Syvanen, S.; Eriksson, J. Advances in PET imaging of P-glycoprotein function at the blood-brain barrier. *ACS Chem Neurosci* **2013**, 4, 225–37.

293. Vlieghe, P.; Khrestchatisky, M. Medicinal chemistry based approaches and nanotechnology-based systems to improve CNS drug targeting and delivery. *Med Res Rev* **2013**, 33, 457–516.

294. Volkow, N. D.; Wang, G. J.; Fowler, J. S.; Tomasi, D. Addiction circuitry in the human brain. *Annu Rev Pharmacol Toxicol* **2012**, 52, 321–36.

295. NIMH. http://www.nimh.nih.gov/research-priorities/therapeutics/cns-radiotracer-table.shtml. Last accessed 25 February, 2014.

296. Ametamey, S. M.; Honer, M.; Schubiger, P. A. Molecular imaging with PET. *Chem Rev* **2008**, 108, 1501–16.

297. Farde, L.; Ehrin, E.; Eriksson, L.; Greitz, T.; Hall, H.; Hedstrom, C. G.; Litton, J. E.; Sedvall, G. Substituted benzamides as ligands for visualization of dopamine receptor binding in the human brain by positron emission tomography. *Proc Natl Acad Sci USA* **1985**, 82, 3863–7.

298. Farde, L.; Hall, H.; Ehrin, E.; Sedvall, G. Quantitative analysis of D2 dopamine receptor binding in the living human brain by PET. *Science* **1986**, 231, 258–61.

299. Miller, P. W. Radiolabelling with short-lived PET (positron emission tomography) isotopes using microfluidic reactors. *J Chem Technol Biotechnol* **2009**, 84, 309–15.

300. Haroun, S.; Sanei, Z.; Jivan, S.; Schaffer, P.; Ruth, T. J.; Li, P. C. H. Continuous-flow synthesis of [11C]raclopride, a positron emission tomography radiotracer, on a microfluidic chip. *Can J Chem* **2013**, 91, 326–32.

301. Henriksen, G.; Willoch, F. Imaging of opioid receptors in the central nervous system. *Brain* **2008**, 131, 1171–96.

302. Greenwald, M.; Johanson, C. E.; Bueller, J.; Chang, Y.; Moody, D. E.; Kilbourn, M.; Koeppe, R.; Zubieta, J. K. Buprenorphine duration of action: mu-opioid receptor availability and pharmacokinetic and behavioral indices. *Biol Psychiat* **2007**, 61, 101–10.

303. Zheng, M. Q.; Nabulsi, N.; Kim, S. J.; Tomasi, G.; Lin, S. F.; Mitch, C.; Quimby, S.; Barth, V.; Rash, K.; Masters, J.; et al. Synthesis and evaluation of 11C-LY2795050 as a kappa-opioid receptor antagonist radiotracer for PET imaging. *J Nucl Med* **2013**, 54, 455–63.

304. Tomasi, G.; Nabulsi, N.; Zheng, M. Q.; Weinzimmer, D.; Ropchan, J.; Blumberg, L.; Brown-Proctor, C.; Ding, Y. S.; Carson, R. E.; Huang, Y. Determination of in vivo Bmax and Kd for 11C-GR103545, an agonist PET tracer for kappa-opioid receptors: a study in nonhuman primates. *J Nucl Med* **2013**, 54, 600–8.

305. Naylor, A.; Judd, D. B.; Lloyd, J. E.; Scopes, D. I.; Hayes, A. G.; Birch, P. J. A potent new class of kappa-receptor agonist: 4-substituted 1-(arylacetyl)-2-[(dialkylamino)methyl]piperazines. *J Med Chem* **1993**, 36, 2075–83.

306. Talbot, P. S.; Narendran, R.; Butelman, E. R.; Huang, Y.; Ngo, K.; Slifstein, M.; Martinez, D.; Laruelle, M.; Hwang, D. R. 11C-GR103545, a radiotracer for imaging kappa-opioid receptors in vivo with PET: synthesis and evaluation in baboons. *J Nucl Med* **2005**, 46, 484–94.

307. Guillaume, P.; Goineau, S.; Froget, G. An overview of QT interval assessment in safety pharmacology. *Curr Protoc Pharmacol* **2013**, Chapter 10, Unit 10.7.

308. Vandenberg, J. I.; Perry, M. D.; Perrin, M. J.; Mann, S. A.; Ke, Y.; Hill, A. P. hERG K(+) channels: structure, function, and clinical significance. *Physiol Rev* **2012**, 92, 1393–78.

309. Beach, S. R.; Celano, C. M.; Noseworthy, P. A.; Januzzi, J. L.; Huffman, J. C. QTc prolongation, torsades de pointes, and psychotropic medications. *Psychosomatics* **2013**, 54, 1–13.

310. Lembke, A. Why doctors prescribe opioids to known opioid abusers. *N Engl J Med* **2012**, 367, 1580–1.

311. *World Drug Report 2010*. United Nations Office on Drugs and Crime, 2010.

312. Karch, S. B. *Pathology of Drug Abuse*, 4th edition. CRC Press, 2009.

313. *Drugs of Abuse*. Drug Enforcement Agency, 2005; p. 22.

314. Todd, A. J. Neuronal circuitry for pain processing in the dorsal horn. *Nat Rev Neurosci* **2010**, 11, 823–36.

315. Sehgal, N.; Smith, H. S.; Manchikanti, L. Peripherally acting opioids and clinical implications for pain control. *Pain Physician* **2011**, 14, 249–58.

316. Moorman-Li, R.; Motycka, C. A.; Inge, L. D.; Congdon, J. M.; Hobson, S.; Pokropski, B. A review of abuse-deterrent opioids for chronic nonmalignant pain. *Pham Therapeut* **2012**, 37, 412–8.

317. Zamloot, M.; Chao, W.; Ling-Ling, K.; Ross, J.; Fu, R. Remoxy: a novel formulation of extended-release oxycodone developed using the ORADUR Technology. *J Appl Res* **2010**, 10, 88–96.

318. Veilleux, J. C.; Colvin, P. J.; Anderson, J.; York, C.; Heinz, A. J. A review of opioid dependence treatment: pharmacological and psychosocial interventions to treat opioid addiction. *Clin Psychol Rev* **2010**, 30, 155–66.

319. TIP 43: Medication-Assisted Treatment for Opioid Addiction in Opioid Treatment Programs. In: Batki, S. L.; Kauffman, J. F.; Marion, I.; Parrino, M. W.; Woody, G. E. (eds.), *Treatment Improvement Protocol Series. Substance Abuse and Mental Health Services Administration*, Rockville, MD, 2005.

320. Bockmühl, M.; Ehrhart, G. Über eine neue Klasse von spasmolytisch und analgetisch wirkenden Verbindungen, I. *Justus Liebigs Annalen der Chemie* **1949**, 561, 52–86.

321. Ferrari, A.; Coccia, C. P.; Bertolini, A.; Sternieri, E. Methadone–metabolism, pharmacokinetics and interactions. *Pharmacol Res* **2004**, 50, 551–9.

322. Gitlow, S. *Substance Use Disorders: A Practical Guide*, 2nd edition. Philadelphia: Lippincott Williams & Wilkins, 2007.

323. Martin, W. R.; Jasinski, D. R.; Haertzen, C. A.; Kay, D. C.; Jones, B. E.; Mansky, P. A.; Carpenter, R. W. Methadone–a reevaluation. *Arch Gen Psychiat* **1973**, 28, 286–95.

324. Kristensen, K.; Christensen, C. B.; Christrup, L. L. The mu1, mu2, delta, kappa opioid receptor binding profiles of methadone stereoisomers and morphine. *Life Sciences* **1994**, 56, 45–50.

325. Raehal, K. M.; Schmid, C. L.; Groer, C. E.; Bohn, L. M. Functional selectivity at the mu-opioid receptor: implications for understanding opioid analgesia and tolerance. *Pharmacol Rev* **2011**, 63, 1001–19.

326. Casy, A. F.; Parfitt, R. T. *Opioid Analgesics: Chemistry and Receptors*. New York: Plenum Press, 1986.

327. Laurel Gorman, A.; Elliott, K. J.; Inturrisi, C. E. The d- and l- isomers of methadone bind to the non-competitive site on the N-methyl-d-aspartate (NMDA) receptor in rat forebrain and spinal cord. *Neurosci Lett* **1997**, 223, 5–8.

328. Davis, A. M.; Inturrisi, C. E. d-Methadone blocks morphine tolerance and N-methyl-D-aspartate-induced hyperalgesia. *J Pharmacol Exp Ther* **1999**, 289, 1048–53.

329. Matsui, A.; Williams, J. T. Activation of mu-opioid receptors and block of Kir3 potassium channels and NMDA receptor conductance by L- and D-methadone in rat locus coeruleus. *Br J Pharmacol* **2010**, 161, 1403–13.

330. Xiao, Y.; Smith, R. D.; Caruso, F. S.; Kellar, K. J. Blockade of rat alpha3beta4 nicotinic receptor function by methadone, its metabolites, and structural analogs. *J Pharmacol Exp Ther* **2001**, 299, 366–71.

331. Wang, J. S.; Ruan, Y.; Taylor, R. M.; Donovan, J. L.; Markowitz, J. S.; DeVane, C. L. Brain penetration of methadone (R)- and (S)-enantiomers is greatly increased by P-glycoprotein deficiency in the blood-brain barrier of Abcb1a gene knockout mice. *Psychopharmacology (Berl)* **2004**, 173, 132–8.

332. Crettol, S.; Digon, P.; Golay, K. P.; Brawand, M.; Eap, C. B. In vitro P-glycoprotein-mediated transport of (R)-, (S)-, (R,S)-methadone, LAAM and their main metabolites. *Pharmacology* **2007**, 80, 304–11.

333. Speeter, M. E.; Byrd, W. M.; Cheney, L. C.; Binkley, S. B. Analgesic carbinols and esters related to amidone (methadon). *J Am Chem Soc* **1949**, 71, 57–60.

334. Wolstein, J.; Gastpar, M.; Finkbeiner, T.; Heinrich, C.; Heitkamp, R.; Poehlke, T.; Scherbaum, N. A randomized, open-label trial comparing methadone and Levo-Alpha-Acetylmethadol (LAAM) in maintenance treatment of opioid addiction. *Pharmacopsychiatry* **2009**, 42, 1–8.

335. Wieneke, H.; Conrads, H.; Wolstein, J.; Breuckmann, F.; Gastpar, M.; Erbel, R.; Scherbaum, N. Levo-alpha-acetylmethadol (LAAM) induced QTc-prolongation – results from a controlled clinical trial. *Eur J Med Res* **2009**, 14, 7–12.

336. Lattin, D. L.; Caviness, B.; Hudson, R. G.; Greene, D. L.; Raible, P. K.; Richardson, J. B. Synthesis of analogues of acetylmethadol and methadol as potential narcotic antagonists. *J Med Chem* **1981**, 24, 903–6.

337. Roberts, S. M.; Price, B. J. *Medicinal Chemistry: the Role of Organic Chemistry in Drug Research*. London, Orlando: Academic Press, 1985.

338. Traynor, J. R.; Nahorski, S. R. Modulation by mu-opioid agonists of guanosine-5'-O-(3-[35S]thio)triphosphate binding to membranes from human neuroblastoma SH-SY5Y cells. *Mol Pharmacol* **1995**, 47, 848–54.

339. Romero, D. V.; Partilla, J. S.; Zheng, Q. X.; Heyliger, S. O.; Ni, Q.; Rice, K. C.; Lai, J.; Rothman, R. B. Opioid peptide receptor studies. 12. Buprenorphine is a potent and selective mu/kappa antagonist in the [35S]-GTP-gamma-S functional binding assay. *Synapse* **1999**, 34, 83–94.

340. Paronis, C. A.; Bergman, J. Buprenorphine and opioid antagonism, tolerance, and naltrexone-precipitated withdrawal. *J Pharmacol Exp Ther* **2011**, 336, 488–95.

341. Mello, N.; Mendelson, J. Buprenorphine suppresses heroin use by heroin addicts. *Science* **1980**, 207, 657–59.

342. Weiss, R. D.; Potter, J. S.; Fiellin, D. A.; Byrne, M.; Connery, H. S.; Dickinson, W.; Gardin, J.; Griffin, M. L.; Gourevitch, M. N.; Haller, D. L.; et al. Adjunctive counseling during brief and extended buprenorphine-naloxone treatment for prescription opioid dependence: a 2-phase randomized controlled trial. *Arch Gen Psychiat* **2011**, 68, 1238–46.

343. Gharagozlou, P.; Demirci, H.; David Clark, J.; Lameh, J. Activity of opioid ligands in cells expressing cloned mu opioid receptors. *BMC Pharmacol* **2003**, 3, 1.

344. Gharagozlou, P.; Hashemi, E.; DeLorey, T. M.; Clark, J. D.; Lameh, J. Pharmacological profiles of opioid ligands at kappa opioid receptors. *BMC Pharmacol* **2006**, 6, 3.

345. Kolinski, M.; Filipek, S. Molecular dynamics of μ opioid receptor complexes with agonists and antagonists. *Open Struct Biol J* **2008**, 2, 8–20.

346. Divin, M. F.; Bradbury, F. A.; Carroll, F. I.; Traynor, J. R. Neutral antagonist activity of naltrexone and 6beta-naltrexol in naive and opioid-dependent C6 cells expressing a mu-opioid receptor. *Br J Pharmacol* **2009**, 156, 1044–53.

347. Comer, S. D.; Sullivan, M. A.; Yu, E.; Rothenberg, J. L.; Kleber, H. D.; Kampman, K.; Dackis, C.; O'Brien, C. P. Injectable, sustained-release naltrexone for the treatment of opioid dependence: a randomized, placebo-controlled trial. *Arch Gen Psychiat* **2006**, 63, 210–8.

348. Diamanti, E.; Del Bello, F.; Carbonara, G.; Carrieri, A.; Fracchiolla, G.; Giannella, M.; Mammoli, V.; Piergentili, A.; Pohjanoksa, K.; Quaglia, W.; et al. Might the observed alpha(2A)-adrenoreceptor agonism or antagonism of allyphenyline analogues be ascribed to different molecular conformations? *Bioorg Med Chem* **2012**, 20, 2082–90.

349. Millan, M. J.; Lejeune, F.; Gobert, A.; Brocco, M.; Auclair, A.; Bosc, C.; Rivet, J. M.; Lacoste, J. M.; Cordi, A.; Dekeyne, A. S18616, a highly potent spiroimidazoline agonist at alpha(2)-adrenoceptors: II. Influence on monoaminergic transmission, motor function, and anxiety in comparison with dexmedetomidine and clonidine. *J Pharmacol Exp Ther* **2000**, 295, 1206–22.

350. Biedermann, J.; Leon-Lomeli, A.; Borbe, H. O.; Prop, G. Two stereoisomeric imidazoline derivatives: synthesis and optical and alpha 2-adrenoceptor activities. *J Med Chem* **1986**, 29, 1183–8.

351. Vartak, A. P.; Crooks, P. A. A scalable, enantioselective synthesis of the α2-adrenergic agonist, lofexidine. *Org Proc Res Dev* **2009**, 13, 415–19.

352. Al-Ghananeem, A. M. Pharmacokinetics of lofexidine hydrochloride in healthy volunteers. *J Pharm Sci* **2009**, 98, 319–26.

353. Yu, E.; Miotto, K.; Akerele, E.; O'Brien, C. P.; Ling, W.; Kleber, H.; Fischman, M. W.; Elkashef, A.; Herman, B. H.; Al-Ghananeem, A. M. Clinical pharmacokinetics of lofexidine, the alpha 2-adrenergic receptor agonist, in opiate addicts plasma using a highly sensitive liquid chromatography tandem mass spectrometric analysis. *Am J Drug Alcohol Abuse* **2008**, 34, 611–6.

354. Gowing, L.; Farrell, M.; Ali, R.; White, J. M. Alpha2-adrenergic agonists for the management of opioid withdrawal. *Cochrane Database Syst Rev* **2009**, CD002024.

355. Del Bello, F.; Mattioli, L.; Ghelfi, F.; Giannella, M.; Piergentili, A.; Quaglia, W.; Cardinaletti, C.; Perfumi, M.; Thomas, R. J.; Zanelli, U.; et al. Fruitful adrenergic alpha(2C)-agonism/alpha(2A)-antagonism combination to prevent and contrast morphine tolerance and dependence. *J Med Chem* **2010**, 53, 7825–35.

356. Grond, S.; Sablotzki, A. Clinical pharmacology of tramadol. *Clin Pharmacokinet* **2004**, 43, 879–923.

357. Raffa, R. B.; Friderichs, E.; Reimann, W.; Shank, R. P.; Codd, E. E.; Vaught, J. L.; Jacoby, H. I.; Selve, N. Complementary and synergistic antinociceptive interaction between the enantiomers of tramadol. *J Pharmacol Exp Ther* **1993**, 267, 331–40.

358. Lofwall, M. R.; Walsh, S. L.; Bigelow, G. E.; Strain, E. C. Modest opioid withdrawal suppression efficacy of oral tramadol in humans. *Psychopharmacology (Berl)* **2007**, 194, 381–93.

359. Shao, L.; Hewitt, M.; Jerussi, T. P.; Wu, F.; Malcolm, S.; Grover, P.; Fang, K.; Koch, P.; Senanayake, C.; Bhongle, N.; et al. In vitro and in vivo evaluation of O-alkyl derivatives of tramadol. *Bioorg Med Chem Lett* **2008**, 18, 1674–80.

360. Riva, E.; Gagliardi, S.; Martinelli, M.; Passarella, D.; Vigo, D.; Rencurosi, A. Reaction of Grignard reagents with carbonyl compounds under continuous flow conditions. *Tetrahedron* **2010**, 66, 3242–7.

361. Evans, G. R.; Fernandez, P. D.; Henshilwood, J. A.; Lloyd, S.; Nicklin, C. Development of highly efficient resolutions of racemic tramadol using mandelic acid. *Org Proc Res Dev* **2002**, 6, 729–37.

362. Itov, Z.; Meckler, H. A Practical procedure for the resolution of (+)- and (-)-tramadol. *Org Proc Res Dev* **2000**, 4, 291–4.

363. Limapichat, W.; Yu, W. Y.; Branigan, E.; Lester, H. A.; Dougherty, D. A. Key binding interactions for memantine in the NMDA receptor. *ACS Chem Neurosci* **2013**, 4, 255–60.

364. Chen, H. S.; Lipton, S. A. Pharmacological implications of two distinct mechanisms of interaction of memantine with N-methyl-D-aspartate-gated channels. *J Pharmacol Exp Ther* **2005**, 314, 961–71.

365. Gerzon, K.; Krumkalns, E. V.; Brindle, R. L.; Marshall, F. J.; Root, M. A. The adamantyl group in medicinal agents. I. Hypoglycemic N-arylsulfonyl-N'-adamantylureas. *J Med Chem* **1963**, 6, 760–3.

366. Stetter, H.; Mayer, J.; Schwarz, M.; Wulff, K. Über Verbindungen mit Urotropin-Struktur, XVI. Beiträge zur Chemie der Adamantyl-(1)-Derivate. *Chemische Berichte* **1960**, 93, 226–30.

367. Sani, G.; Serra, G.; Kotzalidis, G. D.; Romano, S.; Tamorri, S. M.; Manfredi, G.; Caloro, M.; Telesforo, C. L.; Caltagirone, S. S.; Panaccione, I.; et al. The role of memantine in the treatment of psychiatric disorders other than the dementias: a review of current preclinical and clinical evidence. *CNS Drugs* **2012**, 26, 663–90.

368. Ramesh, D.; Ross, G. R.; Schlosburg, J. E.; Owens, R. A.; Abdullah, R. A.; Kinsey, S. G.; Long, J. Z.; Nomura, D. K.; Sim-Selley, L. J.; Cravatt, B. F.; et al. Blockade of endo-cannabinoid hydrolytic enzymes attenuates precipitated opioid withdrawal symptoms in mice. *J Pharmacol Exp Ther* **2011**, 339, 173–85.

369. Mor, M.; Rivara, S.; Lodola, A.; Plazzi, P. V.; Tarzia, G.; Duranti, A.; Tontini, A.; Piersanti, G.; Kathuria, S.; Piomelli, D. Cyclohexylcarbamic acid 3'- or 4'-substituted biphenyl-3-yl esters as fatty acid amide hydrolase inhibitors: synthesis, quantitative structure-activity relationships, and molecular modeling studies. *J Med Chem* **2004**, 47, 4998–5008.

370. Haller, V. L.; Stevens, D. L.; Welch, S. P. Modulation of opioids via protection of anandamide degradation by fatty acid amide hydrolase. *Eur J Pharmacol* **2008**, 600, 50–8.

371. Long, J. Z.; Jin, X.; Adibekian, A.; Li, W.; Cravatt, B. F. Characterization of tunable piperidine and piperazine carbamates as inhibitors of endocannabinoid hydrolases. *J Med Chem* **2010**, 53, 1830–42.

372. Ahn, K.; Johnson, D. S.; Mileni, M.; Beidler, D.; Long, J. Z.; McKinney, M. K.; Weera-pana, E.; Sadagopan, N.; Liimatta, M.; Smith, S. E.; et al. Discovery and characterization of a highly selective FAAH inhibitor that reduces inflammatory pain. *Chem Biol* **2009**, 16, 411–20.

373. Johnson, D. S.; Stiff, C.; Lazerwith, S. E.; Kesten, S. R.; Fay, L. K.; Morris, M.; Beidler, D.; Liimatta, M. B.; Smith, S. E.; Dudley, D. T.; et al. Discovery of PF-04457845: a highly potent, orally bioavailable, and selective urea FAAH inhibitor. *ACS Med Chem Lett* **2011**, 2, 91–6.

374. Chu, L. F.; Liang, D. Y.; Li, X.; Sahbaie, P.; D'Arcy, N.; Liao, G.; Peltz, G.; David Clark, J. From mouse to man: the 5-HT3 receptor modulates physical dependence on opioid narcotics. *Pharmacogenet Genomics* **2009**, 19, 193–205.

375. Porras, G.; De Deurwaerdere, P.; Moison, D.; Spampinato, U. Conditional involvement of striatal serotonin3 receptors in the control of in vivo dopamine outflow in the rat striatum. *Eur J Neurosci* **2003**, 17, 771–81.

376. Imperato, A.; Angelucci, L. 5-HT3 receptors control dopamine release in the nucleus accumbens of freely moving rats. *Neurosci Lett* **1989**, 101, 214–7.

377. van Wijngaarden, I.; Hamminga, D.; van Hes, R.; Standaar, P. J.; Tipker, J.; Tulp, M. T.; Mol, F.; Olivier, B.; de Jonge, A. Development of high-affinity 5-HT3 receptor antagonists. Structure-affinity relationships of novel 1,7-annelated indole derivatives. *J Med Chem* **1993**, 36, 3693–9.

378. Butler, A.; Hill, J. M.; Ireland, S. J.; Jordan, C. C.; Tyers, M. B. Pharmacological properties of GR38032F, a novel antagonist at 5-HT3 receptors. *Br J Pharmacol* **1988**, 94, 397–412.

379. Popik, P.; Layer, R. T.; Skolnick, P. 100 years of ibogaine: neurochemical and pharma-cological actions of a putative anti-addictive drug. *Pharmacol Rev* **1995**, 47, 235–53.

380. Glick, S. D.; Kuehne, M. E.; Maisonneuve, I. M.; Bandarage, U. K.; Molinari, H. H. 18-Methoxycoronaridine, a non-toxic iboga alkaloid congener: effects on morphine and cocaine self-administration and on mesolimbic dopamine release in rats. *Brain Res* **1996**, 719, 29–35.

381. Kuehne, M. E.; He, L.; Jokiel, P. A.; Pace, C. J.; Fleck, M. W.; Maisonneuve, I. M.; Glick, S. D.; Bidlack, J. M. Synthesis and biological evaluation of 18-methoxycoronaridine congeners. Potential antiaddiction agents. *J Med Chem* **2003**, 46, 2716–30.

382. Glick, S. D.; Maisonneuve, I. M.; Kitchen, B. A.; Fleck, M. W. Antagonism of alpha 3 beta 4 nicotinic receptors as a strategy to reduce opioid and stimulant self-administration. *Eur J Pharmacol* **2002**, 438, 99–105.

383. Antonio, T.; Childers, S. R.; Rothman, R. B.; Dersch, C. M.; King, C.; Kuehne, M.; Bornmann, W. G.; Eshleman, A. J.; Janowsky, A.; Simon, E. R.; et al. Effect of iboga alkaloids on μ-opioid receptor-coupled G protein activation. *PLoS One* **2013**, 8, e77262.

384. Sweetnam, P. M.; Lancaster, J.; Snowman, A.; Collins, J. L.; Perschke, S.; Bauer, C.; Ferkany, J. Receptor binding profile suggests multiple mechanisms of action are responsible for ibogaine's putative anti-addictive activity. *Psychopharmacology (Berl)* **1995**, 118, 369–76.

385. Paling, F. P.; Andrews, L. M.; Valk, G. D.; Blom, H. J. Life-threatening complications of ibogaine: three case reports. *Neth J Med* **2012**, 70, 422–4.

386. Brown, T. K. Ibogaine in the treatment of substance dependence. *Curr Drug Abuse Rev* **2013**, 6, 3–16.

387. *Drugs of Abuse.* Drug Enforcement Agency, 2011.

388. Swerdlow, J. L. *Nature's Medicine: Plants That Heal.* Washington, D.C.: National Geographic Society, 2000; p. 400.

389. Llosa, T. *Oral Cocaine in Addictions*, 2nd edition. Publisher Llosa, T; COCA MEDICA: Lima, Peru, 2007.

390. Kimmel, H. L.; Negus, S. S.; Wilcox, K. M.; Ewing, S. B.; Stehouwer, J.; Goodman, M. M.; Votaw, J. R.; Mello, N. K.; Carroll, F. I.; Howell, L. L. Relationship between rate of drug uptake in brain and behavioral pharmacology of monoamine transporter inhibitors in rhesus monkeys. *Pharmacol Biochem Behav* **2008**, 90, 453–62.

391. Bortolotti, F.; Gottardo, R.; Pascali, J.; Tagliaro, F. Toxicokinetics of cocaine and metabolites: the forensic toxicological approach. *Curr Med Chem* **2012**, 19, 5658–63.

392. Methamphetamine: NIDA Research Report, 2013, NIH Publication Number 13-4210.

393. Rothman, R. B.; Baumann, M. H.; Dersch, C. M.; Romero, D. V.; Rice, K. C.; Carroll, F. I.; Partilla, J. S. Amphetamine-type central nervous system stimulants release norepinephrine more potently than they release dopamine and serotonin. *Synapse* **2001**, 39, 32–41.

394. Lewin, A. H.; Miller, G. M.; Gilmour, B. Trace amine-associated receptor 1 is a stereoselective binding site for compounds in the amphetamine class. *Bioorg Med Chem* **2011**, 19, 7044–8.

395. Xi, Z.-X. Medication development for the treatment of cocaine addiction – progress at preclinical and clinical levels. In: *Addictions – From Pathophysiology to Treatment*, Belin, D., Ed. InTech, 2012; pp. 311–54.

396. Gibson, L. C.; Hastings, S. F.; McPhee, I.; Clayton, R. A.; Darroch, C. E.; Mackenzie, A.; Mackenzie, F. L.; Nagasawa, M.; Stevens, P. A.; Mackenzie, S. J. The inhibitory profile

of Ibudilast against the human phosphodiesterase enzyme family. *Eur J Pharmacol* **2006**, 538, 39–42.

397. Snider, S. E.; Hendrick, E. S.; Beardsley, P. M. Glial cell modulators attenuate metham-phetamine self-administration in the rat. *Eur J Pharmacol* **2013**, 701, 124–30.

398. Beardsley, P. M.; Shelton, K. L.; Hendrick, E.; Johnson, K. W. The glial cell modulator and phosphodiesterase inhibitor, AV411 (ibudilast), attenuates prime- and stress-induced methamphetamine relapse. *Eur J Pharmacol* **2010**, 637, 102–8.

399. Lolis, E.; Bucala, R. Macrophage migration inhibitory factor. *Expert Opin Ther Targets* **2003**, 7, 153–64.

400. Aloisi, A. M.; Pari, G.; Ceccarelli, I.; Vecchi, I.; Ietta, F.; Lodi, L.; Paulesu, L. Gender-related effects of chronic non-malignant pain and opioid therapy on plasma levels of macrophage migration inhibitory factor (MIF). *Pain* **2005**, 115, 142–51.

401. Allcock, R. W.; Blakli, H.; Jiang, Z.; Johnston, K. A.; Morgan, K. M.; Rosair, G. M.; Iwase, K.; Kohno, Y.; Adams, D. R. Phosphodiesterase inhibitors. Part 1: Synthesis and structure-activity relationships of pyrazolopyridine-pyridazinone PDE inhibitors devel-oped from ibudilast. *Bioorg Med Chem Lett* **2011**, 21, 3307–12.

402. CPP. http://ir.catalystpharma.com/releasedetail.cfm?ReleaseID=560401. Last accessed 25 November, 2013.

403. Partners, C. P. http://ir.catalystpharma.com/releasedetail.cfm?ReleaseID=531584. Last accessed 25 November, 2013.

404. Pan, Y.; Gerasimov, M. R.; Kvist, T.; Wellendorph, P.; Madsen, K. K.; Pera, E.; Lee, H.; Schousboe, A.; Chebib, M.; Brauner-Osborne, H.; et al. (1S, 3S)-3-amino-4-difluoromethylenyl-1-cyclopentanoic acid (CPP-115), a potent gamma-aminobutyric acid aminotransferase inactivator for the treatment of cocaine addiction. *J Med Chem* **2011**, 55, 357–66.

405. Silverman, R. B. The 2011 E. B. Hershberg award for important discoveries in medic-inally active substances: (1S,3S)-3-amino-4-difluoromethylenyl-1-cyclopentanoic acid (CPP-115), a GABA aminotransferase inactivator and new treatment for drug addiction and infantile spasms. *J Med Chem* **2012**, 55, 567–75.

406. Pan, Y.; Qiu, J.; Silverman, R. B. Design, synthesis, and biological activity of a difluoro-substituted, conformationally rigid vigabatrin analogue as a potent gamma-aminobutyric acid aminotransferase inhibitor. *J Med Chem* **2003**, 46, 5292–3.

407. Malpass, J. R.; Tweddle, N. J. Reaction of chlorosulphonyl isocyanate with 1,3-dienes. Control of 1,2- and 1,4-addition pathways and the synthesis of aza- and oxa-bicyclic systems. *J Chem Soc Perkin Trans 1* **1977**, 874–84.

408. Palmer, C. F.; Parry, K. P.; Roberts, S. M.; Sik, V. Rearrangement of 2-azabicyclo[2.2.1]hept-5-en-3-ones: synthesis of cis-3-aminocyclopentane carboxylic acid derivatives. *J Chem Soc Perkin Trans 1* **1992**, 1021–8.

409. Carroll, K. M.; Nich, C.; Shi, J. M.; Eagan, D.; Ball, S. A. Efficacy of disulfiram and Twelve Step Facilitation in cocaine-dependent individuals maintained on methadone: a randomized placebo-controlled trial. *Drug Alcohol Depend* **2012**, 126, 224–31.

410. Haile, C. N.; De La Garza, R., II; Mahoney, J. J., III, Nielsen, D. A.; Kosten, T. R.; Newton, T. F. The impact of disulfiram treatment on the reinforcing effects of cocaine: a randomized clinical trial. *PLoS One* **2012**, 7, e47702.

411. Devoto, P.; Flore, G.; Saba, P.; Cadeddu, R.; Gessa, G. Disulfiram stimulates dopamine release from noradrenergic terminals and potentiates cocaine-induced dopamine release in the prefrontal cortex. *Psychopharmacology* **2012**, 219, 1153–64.

412. Baker, J. R.; Jatlow, P.; McCance-Katz, E. F. Disulfiram effects on responses to intravenous cocaine administration. *Drug Alcohol Depend* **2007**, 87, 202–9.

413. Goldstein, M.; Anagnoste, B.; Lauber, E.; McKereghan, M. R. Inhibition of dopamine-beta-hydroxylase by disulfiram. *Life Sci* **1964**, 3, 763–7.

414. Stanley, W. C.; Li, B.; Bonhaus, D. W.; Johnson, L. G.; Lee, K.; Porter, S.; Walker, K.; Martinez, G.; Eglen, R. M.; Whiting, R. L.; et al. Catecholamine modulatory effects of nepicastat (RS-25560-197), a novel, potent and selective inhibitor of dopamine-beta-hydroxylase. *Br J Pharmacol* **1997**, 121, 1803–9.

415. Kapoor, A.; Shandilya, M.; Kundu, S. Structural insight of dopamine-beta-hydroxylase, a drug target for complex traits, and functional significance of exonic single nucleotide polymorphisms. *PLoS One* **2011**, 6, e26509.

416. Schroeder, J. P.; Alisha Epps, S.; Grice, T. W.; Weinshenker, D. The selective dopamine beta-hydroxylase inhibitor nepicastat attenuates multiple aspects of cocaine-seeking behavior. *Neuropsychopharmacology* **2013**, 38, 1032–8.

417. Vardanyan, R. S.; Hruby, V. J. *Synthesis of Essential Drugs,* 1st edition. Amsterdam; Boston: Elsevier, 2006.

418. Yao, L.; Fan, P.; Arolfo, M.; Jiang, Z.; Olive, M. F.; Zablocki, J.; Sun, H. L.; Chu, N.; Lee, J.; Kim, H. Y.; et al. Inhibition of aldehyde dehydrogenase-2 suppresses cocaine seeking by generating THP, a cocaine use-dependent inhibitor of dopamine synthesis. *Nat Med* **2010**, 16, 1024–8.

419. Dencker, D.; Thomsen, M.; Wortwein, G.; Weikop, P.; Cui, Y.; Jeon, J.; Wess, J.; Fink-Jensen, A. Muscarinic acetylcholine receptor subtypes as potential drug targets for the treatment of schizophrenia, drug abuse and Parkinson's disease. *ACS Chem Neurosci* **2012**, 3, 80–9.

420. Thomsen, M.; Lindsley, C. W.; Conn, P. J.; Wessell, J. E.; Fulton, B. S.; Wess, J.; Caine, S. B. Contribution of both M(1) and M (4) receptors to muscarinic agonist-mediated attenuation of the cocaine discriminative stimulus in mice. *Psychopharmacology (Berl)* **2012**, 220, 673–85.

421. Lebois, E. P.; Bridges, T. M.; Lewis, L. M.; Dawson, E. S.; Kane, A. S.; Xiang, Z.; Jadhav, S. B.; Yin, H.; Kennedy, J. P.; Meiler, J.; et al. Discovery and characterization of novel subtype-selective allosteric agonists for the investigation of M(1) receptor function in the central nervous system. *ACS Chem Neurosci* **2010**, 1, 104–21.

422. Bodick, N. C.; Offen, W. W.; Levey, A. I.; Cutler, N. R.; Gauthier, S. G.; Satlin, A.; Shannon, H. E.; Tollefson, G. D.; Rasmussen, K.; Bymaster, F. P.; et al. Effects of xanomeline, a selective muscarinic receptor agonist, on cognitive function and behavioral symptoms in Alzheimer disease. *Arch Neurol* **1997**, 54, 465–73.

423. Watson, J.; Brough, S.; Coldwell, M. C.; Gager, T.; Ho, M.; Hunter, A. J.; Jerman, J.; Middlemiss, D. N.; Riley, G. J.; Brown, A. M. Functional effects of the muscarinic receptor agonist, xanomeline, at 5-HT1 and 5-HT2 receptors. *Br J Pharmacol* **1998**, 125, 1413–20.

424. Thomsen, M.; Wess, J.; Fulton, B. S.; Fink-Jensen, A.; Caine, S. B. Modulation of prepulse inhibition through both M(1) and M (4) muscarinic receptors in mice. *Psychopharmacology (Berl)* **2010**, 208, 401–16.

425. Winstock, A. Areca nut-abuse liability, dependence and public health. *Addict Biol* **2002**, 7, 133–8.

426. Chu, N.-S. Neurological aspects of areca and betel chewing. *Addict Biol* **2002**, 7, 111–14.

427. Pedersen, H.; Bräuner-Osborne, H.; Ball, R. G.; Frydenvang, K.; Meier, E.; Bøgesø, K. P.; Krogsgaard-Larsen, P. Synthesis and muscarinic receptor pharmacology of a series of 4,5,6,7-tetrahydroisothiazolo[4,5-c]pyridine bioisosteres of arecoline. *Bioorg Med Chem* **1999**, 7, 795–809.

428. Sauerberg, P.; Olesen, P. H.; Nielsen, S.; Treppendahl, S.; Sheardown, M. J.; Honore, T.; Mitch, C. H.; Ward, J. S.; Pike, A. J.; Bymaster, F. P.; et al. Novel functional M1 selective muscarinic agonists. Synthesis and structure-activity relationships of 3-(1,2,5-thiadiazolyl)-1,2,5,6-tetrahydro-1-methylpyridines. *J Med Chem* **1992**, 35, 2274–83.

429. Ngur, D.; Roknich, S.; Mitch, C. H.; Quimby, S. J.; Ward, J. S.; Merritt, L.; Sauerberg, P.; Messer, W. S., Jr.; Hoss, W. Steric and electronic requirements for muscarinic receptor-stimulated phosphoinositide turnover in the CNS in a series of arecoline bioisosteres. *Biochem Biophys Res Commun* **1992**, 187, 1389–94.

430. Christopoulos, A.; Pierce, T. L.; Sorman, J. L.; El-Fakahany, E. E. On the unique binding and activating properties of xanomeline at the M1 muscarinic acetylcholine receptor. *Mol Pharmacol* **1998**, 53, 1120–30.

431. Jakubik, J.; Tucek, S.; El-Fakahany, E. E. Role of receptor protein and membrane lipids in xanomeline wash-resistant binding to muscarinic M1 receptors. *J Pharmacol Exp Ther* **2004**, 308, 105–10.

432. Kane, B. E.; Grant, M. K.; El-Fakahany, E. E.; Ferguson, D. M. Synthesis and evaluation of xanomeline analogs–probing the wash-resistant phenomenon at the M1 muscarinic acetylcholine receptor. *Bioorg Med Chem* **2008**, 16, 1376–92.

433. Stauffer, S. R. Progress toward positive allosteric modulators of the metabotropic glutamate receptor subtype 5 (mGlu(5)). *ACS Chem Neurosci* **2011**, 2, 450–70.

434. Emmitte, K. A. Recent advances in the design and development of novel negative allosteric modulators of mGlu(5). *ACS Chem Neurosci* **2011**, 2, 411–32.

435. Maccarrone, M.; Rossi, S.; Bari, M.; De Chiara, V.; Rapino, C.; Musella, A.; Bernardi, G.; Bagni, C.; Centonze, D. Abnormal mGlu 5 receptor/endocannabinoid coupling in mice lacking FMRP and BC1 RNA. *Neuropsychopharmacology* **2010**, 35, 1500–9.

436. Monn, J. A.; Valli, M. J.; Massey, S. M.; Wright, R. A.; Salhoff, C. R.; Johnson, B. G.; Howe, T.; Alt, C. A.; Rhodes, G. A.; Robey, R. L.; et al. Design, synthesis, and pharmacological characterization of (+)-2-aminobicyclo[3.1.0]hexane-2,6-dicarboxylic acid (LY354740): a potent, selective, and orally active group 2 metabotropic glutamate receptor agonist possessing anticonvulsant and anxiolytic properties. *J Med Chem* **1997**, 40, 528–37.

437. Klodzinska, A.; Chojnacka-Wojcik, E.; Palucha, A.; Branski, P.; Popik, P.; Pilc, A. Potential anti-anxiety, anti-addictive effects of LY 354740, a selective group II glutamate metabotropic receptors agonist in animal models. *Neuropharmacology* **1999**, 38, 1831–9.

438. Helton, D. R.; Tizzano, J. P.; Monn, J. A.; Schoepp, D. D.; Kallman, M. J. LY354740: a metabotropic glutamate receptor agonist which ameliorates symptoms of nicotine withdrawal in rats. *Neuropharmacology* **1997**, 36, 1511–6.

439. Helton, D. R.; Tizzano, J. P.; Monn, J. A.; Schoepp, D. D.; Kallman, M. J. Anxiolytic and side-effect profile of LY354740: a potent, highly selective, orally active agonist for group II metabotropic glutamate receptors. *J Pharmacol Exp Ther* **1998**, 284, 651–60.

440. Adewale, A. S.; Platt, D. M.; Spealman, R. D. Pharmacological stimulation of group ii metabotropic glutamate receptors reduces cocaine self-administration and cocaine-induced reinstatement of drug seeking in squirrel monkeys. *J Pharmacol Exp Ther* **2006**, 318, 922–31.

441. Imre, G. The preclinical properties of a novel group II metabotropic glutamate receptor agonist LY379268. *CNS Drug Rev* **2007**, 13, 444–64.

442. Ohfune, Y.; Demura, T.; Iwama, S.; Matsuda, H.; Namba, K.; Shimamoto, K.; Shinada, T. Stereocontrolled synthesis of a potent agonist of group II metabotropic glutamate receptors, (+)-LY354740, and its related derivatives. *Tetrahedron Lett* **2003**, 44, 5431–4.

443. Ornstein, P. L.; Bleisch, T. J.; Arnold, M. B.; Kennedy, J. H.; Wright, R. A.; Johnson, B. G.; Tizzano, J. P.; Helton, D. R.; Kallman, M. J.; Schoepp, D. D.; et al. 2-Substituted (2SR)-2-Amino-2-((1SR,2SR)-2-carboxycycloprop-1-yl)glycines as potent and selective antagonists of group II metabotropic glutamate receptors. 2. Effects of aromatic substitution, pharmacological characterization, and bioavailability. *J Med Chem* **1998**, 41, 358–78.

444. Wang, J.-Q.; Zhang, Z.; Kuruppu, D.; Brownell, A.-L. Radiosynthesis of PET radiotracer as a prodrug for imaging group II metabotropic glutamate receptors in vivo. *Bioorg Med Chem Lett* **2012**, 22, 1958–62.

445. Bonnefous, C.; Vernier, J. M.; Hutchinson, J. H.; Gardner, M. F.; Cramer, M.; James, J. K.; Rowe, B. A.; Daggett, L. P.; Schaffhauser, H.; Kamenecka, T. M. Biphenyl-indanones: allosteric potentiators of the metabotropic glutamate subtype 2 receptor. *Bioorg Med Chem Lett* **2005**, 15, 4354–8.

446. Dhanya, R. P.; Sidique, S.; Sheffler, D. J.; Nickols, H. H.; Herath, A.; Yang, L.; Dahl, R.; Ardecky, R.; Semenova, S.; Markou, A.; et al. Design and synthesis of an orally active metabotropic glutamate receptor subtype-2 (mGluR2) positive allosteric modulator (PAM) that decreases cocaine self-administration in rats. *J Med Chem* **2011**, 54, 342–53.

447. Zhang, L.; Brodney, M. A.; Candler, J.; Doran, A. C.; Duplantier, A. J.; Efremov, I. V.; Evrard, E.; Kraus, K.; Ganong, A. H.; Haas, J. A.; et al. 1-[(1-Methyl-1H-imidazol-2-yl)methyl]-4-phenylpiperidines as mGluR2 positive allosteric modulators for the treatment of psychosis. *J Med Chem* **2011**, 54, 1724–39.

448. Cid, J. M.; Duvey, G.; Tresadern, G.; Nhem, V.; Furnari, R.; Cluzeau, P.; Vega, J. A.; de Lucas, A. I.; Matesanz, E.; Alonso, J. M.; et al. Discovery of 1,4-disubstituted 3-cyano-2-pyridones: a new class of positive allosteric modulators of the metabotropic glutamate 2 receptor. *J Med Chem* **2012**, 55, 2388–405.

449. Kinney, G. G.; O'Brien, J. A.; Lemaire, W.; Burno, M.; Bickel, D. J.; Clements, M. K.; Chen, T. B.; Wisnoski, D. D.; Lindsley, C. W.; Tiller, P. R.; et al. A novel selective positive allosteric modulator of metabotropic glutamate receptor subtype 5 has in vivo activity and antipsychotic-like effects in rat behavioral models. *J Pharmacol Exp Ther* **2005**, 313, 199–206.

450. Pecknold, J. C.; McClure, D. J.; Appeltauer, L.; Wrzesinski, L.; Allan, T. Treatment of anxiety using fenobam (a nonbenzodiazepine) in a double-blind standard (diazepam) placebo-controlled study. *J Clin Psychopharmacol* **1982**, 2, 129–33.

451. Keck, T. M.; Yang, H. J.; Bi, G. H.; Huang, Y.; Zhang, H. Y.; Srivastava, R.; Gardner, E. L.; Newman, A. H.; Xi, Z. X. Fenobam sulfate inhibits cocaine-taking and cocaine-seeking behavior in rats: implications for addiction treatment in humans. *Psychopharmacology (Berl)* **2013**, 229, 253–65.

452. Watterson, L. R.; Kufahl, P. R.; Nemirovsky, N. E.; Sewalia, K.; Hood, L. E.; Olive, M. F. Attenuation of reinstatement of methamphetamine-, sucrose-, and food-seeking behavior in rats by fenobam, a metabotropic glutamate receptor 5 negative allosteric modulator. *Psychopharmacology (Berl)* **2013**, 225, 151–9.

453. Cleva, R. M.; Watterson, L. R.; Johnson, M. A.; Olive, M. F. Differential modulation of thresholds for intracranial self-stimulation by mGlu5 positive and negative allosteric modulators: implications for effects on drug self-administration. *Front Pharmacol* **2012**, 2, 93.

454. Chesworth, R.; Brown, R. M.; Kim, J. H.; Lawrence, A. J. The metabotropic glutamate 5 receptor modulates extinction and reinstatement of methamphetamine-seeking in mice. *PLoS One* **2013**, 8, e68371.

455. Pellizer, G.; Rubessa, F. 1H- and 13C-NMR spectra of fenobam. *J Pharm Sci* **1983**, 72, 189–90.

456. Thomas, S. P.; Nagarajan, K.; Row, T. N. G. Polymorphism and tautomeric preference in fenobam and the utility of NLO response to detect polymorphic impurities. *Chem Commun* **2012**, 48, 10559–61.

457. Malherbe, P.; Kratochwil, N.; Muhlemann, A.; Zenner, M. T.; Fischer, C.; Stahl, M.; Gerber, P. R.; Jaeschke, G.; Porter, R. H. Comparison of the binding pockets of two chemically unrelated allosteric antagonists of the mGlu5 receptor and identification of crucial residues involved in the inverse agonism of MPEP. *J Neurochem* **2006**, 98, 601–15.

458. Pagano, A.; Ruegg, D.; Litschig, S.; Stoehr, N.; Stierlin, C.; Heinrich, M.; Floersheim, P.; Prezeau, L.; Carroll, F.; Pin, J. P.; et al. The non-competitive antagonists 2-methyl-6-(phenylethynyl)pyridine and 7-hydroxyiminocyclopropan[b]chromen-1a-carboxylic acid ethyl ester interact with overlapping binding pockets in the transmembrane region of group I metabotropic glutamate receptors. *J Biol Chem* **2000**, 275, 33750–8.

459. McGeehan, A. J.; Olive, M. F. The mGluR5 antagonist MPEP reduces the conditioned rewarding effects of cocaine but not other drugs of abuse. *Synapse* **2003**, 47, 240–2.

460. Cosford, N. D.; Tehrani, L.; Roppe, J.; Schweiger, E.; Smith, N. D.; Anderson, J.; Bristow, L.; Brodkin, J.; Jiang, X.; McDonald, I.; et al. 3-[(2-Methyl-1,3-thiazol-4-yl)ethynyl]-pyridine: a potent and highly selective metabotropic glutamate subtype 5 receptor antagonist with anxiolytic activity. *J Med Chem* **2003**, 46, 204–6.

461. Iso, Y.; Grajkowska, E.; Wroblewski, J. T.; Davis, J.; Goeders, N. E.; Johnson, K. M.; Sanker, S.; Roth, B. L.; Tueckmantel, W.; Kozikowski, A. P. Synthesis and structure-activity relationships of 3-[(2-methyl-1,3-thiazol-4-yl)ethynyl]pyridine analogues as potent, noncompetitive metabotropic glutamate receptor subtype 5 antagonists; search for cocaine medications. *J Med Chem* **2006**, 49, 1080–100.

462. Lindsley, C. W.; Bates, B. S.; Menon, U. N.; Jadhav, S. B.; Kane, A. S.; Jones, C. K.; Rodriguez, A. L.; Conn, P. J.; Olsen, C. M.; Winder, D. G.; et al. (3-Cyano-5-fluorophenyl)biaryl negative allosteric modulators of mGlu(5): Discovery of a new tool compound with activity in the OSS mouse model of addiction. *ACS Chem Neurosci* **2011**, 2, 471–82.

463. Mitsukawa, K.; Yamamoto, R.; Ofner, S.; Nozulak, J.; Pescott, O.; Lukic, S.; Stoehr, N.; Mombereau, C.; Kuhn, R.; McAllister, K. H.; et al. A selective metabotropic glutamate receptor 7 agonist: activation of receptor signaling via an allosteric site modulates stress parameters in vivo. *Proc Natl Acad Sci USA* **2005**, 102, 18712–7.

464. McFarland, K.; Lapish, C. C.; Kalivas, P. W. Prefrontal glutamate release into the core of the nucleus accumbens mediates cocaine-induced reinstatement of drug-seeking behavior. *J Neurosci* **2003**, 23, 3531–7.

465. Schmaal, L.; Veltman, D. J.; Nederveen, A.; van den Brink, W.; Goudriaan, A. E. N-acetylcysteine normalizes glutamate levels in cocaine-dependent patients: a randomized crossover magnetic resonance spectroscopy study. *Neuropsychopharmacology* **2012**, 37, 2143–52.

466. LaRowe, S. D.; Myrick, H.; Hedden, S.; Mardikian, P.; Saladin, M.; McRae, A.; Brady, K.; Kalivas, P. W.; Malcolm, R. Is cocaine desire reduced by N-acetylcysteine? *Am J Psychiat* **2007**, 164, 1115–7.

467. Kupchik, Y. M.; Moussawi, K.; Tang, X. C.; Wang, X.; Kalivas, B. C.; Kolokithas, R.; Ogburn, K. B.; Kalivas, P. W. The effect of N-acetylcysteine in the nucleus accumbens on neurotransmission and relapse to cocaine. *Biol Psychiat* **2012**, 71, 978–86.

468. Micheli, F. Recent advances in the development of dopamine D3 receptor antagonists: a medicinal chemistry perspective. *ChemMedChem* **2011**, 6, 1152–62.

469. Banala, A. K.; Levy, B. A.; Khatri, S. S.; Furman, C. A.; Roof, R. A.; Mishra, Y.; Griffin, S. A.; Sibley, D. R.; Luedtke, R. R.; Newman, A. H. N-(3-fluoro-4-(4-(2-methoxy or 2,3-dichlorophenyl)piperazine-1-yl)butyl)arylcarboxamides as selective dopamine D3 receptor ligands: critical role of the carboxamide linker for D3 receptor selectivity. *J Med Chem* **2011**, 54, 3581–94.

470. Black, K. J.; Hershey, T.; Koller, J. M.; Videen, T. O.; Mintun, M. A.; Price, J. L.; Perlmutter, J. S. A possible substrate for dopamine-related changes in mood and behavior: prefrontal and limbic effects of a D3-preferring dopamine agonist. *Proc Natl Acad Sci USA* **2002**, 99, 17113–8.

471. Lieberman, J. A. Dopamine partial agonists: a new class of antipsychotic. *CNS Drugs* **2004**, 18, 251–67.

472. Lange, J. H.; Reinders, J. H.; Tolboom, J. T.; Glennon, J. C.; Coolen, H. K.; Kruse, C. G. Principal component analysis differentiates the receptor binding profiles of three antipsychotic drug candidates from current antipsychotic drugs. *J Med Chem* **2007**, 50, 5103–8.

473. Coffin, P. O.; Santos, G. M.; Das, M.; Santos, D. M.; Huffaker, S.; Matheson, T.; Gasper, J.; Vittinghoff, E.; Colfax, G. N. Aripiprazole for the treatment of methamphetamine dependence: a randomized, double-blind, placebo-controlled trial. *Addiction* **2013**, 108, 751–61.

474. Holmes, I. P.; Blunt, R. J.; Lorthioir, O. E.; Blowers, S. M.; Gribble, A.; Payne, A. H.; Stansfield, I. G.; Wood, M.; Woollard, P. M.; Reavill, C.; et al. The identification of a selective dopamine D2 partial agonist, D3 antagonist displaying high levels of brain exposure. *Bioorg Med Chem Lett* **2010**, 20, 2013–6.

475. Rothman, R. B.; Baumann, M. H.; Prisinzano, T. E.; Newman, A. H. Dopamine transport inhibitors based on GBR12909 and benztropine as potential medications to treat cocaine addiction. *Bioch Pharmacol* **2008**, 75, 2–16.

476. Newman, A. H.; Kulkarni, S. Probes for the dopamine transporter: new leads toward a cocaine-abuse therapeutic–A focus on analogues of benztropine and rimcazole. *Med Res Rev* **2002**, 22, 429–64.

477. Zhang, A.; Neumeyer, J. L.; Baldessarini, R. J. Recent progress in development of dopamine receptor subtype-selective agents: potential therapeutics for neurological and psychiatric disorders. *Chem Rev* **2007**, 107, 274–302.

478. Newman, A. H.; Kline, R. H.; Allen, A. C.; Izenwasser, S.; George, C.; Katz, J. L. Novel 4′-substituted and 4′,4″-disubstituted 3 alpha-(diphenylmethoxy)tropane analogs as potent and selective dopamine uptake inhibitors. *J Med Chem* **1995**, 38, 3933–40.

479. Bisgaard, H.; Larsen, M. A.; Mazier, S.; Beuming, T.; Newman, A. H.; Weinstein, H.; Shi, L.; Loland, C. J.; Gether, U. The binding sites for benztropines and dopamine in the dopamine transporter overlap. *Neuropharmacology* **2011**, 60, 182–90.

480. Penetar, D. M.; Looby, A. R.; Su, Z.; Lundahl, L. H.; Eros-Sarnyai, M.; McNeil, J. F.; Lukas, S. E. Benztropine pretreatment does not affect responses to acute cocaine administration in human volunteers. *Hum Psychopharmacol* **2006**, 21, 549–59.

481. Walker, J. M.; Bowen, W. D.; Walker, F. O.; Matsumoto, R. R.; De Costa, B.; Rice, K. C. Sigma receptors: biology and function. *Pharmacol Rev* **1990**, 42, 355–402.

482. Cobos, E. J.; Entrena, J. M.; Nieto, F. R.; Cendan, C. M.; Del Pozo, E. Pharmacology and therapeutic potential of sigma(1) receptor ligands. *Curr Neuropharmacol* **2008**, 6, 344–66.

483. Maurice, T.; Su, T. P. The pharmacology of sigma-1 receptors. *Pharmacol Ther* **2009**, 124, 195–206.

484. Hiranita, T.; Soto, P. L.; Tanda, G.; Katz, J. L. Reinforcing effects of sigma-receptor agonists in rats trained to self-administer cocaine. *J Pharmacol Exp Ther* **2010**, 332, 515–24.

485. Matsumoto, R. R.; Liu, Y.; Lerner, M.; Howard, E. W.; Brackett, D. J. Sigma receptors: potential medications development target for anti-cocaine agents. *Eur J Pharmacol* **2003**, 469, 1–12.

486. Cao, J.; Kulkarni, S. S.; Husbands, S. M.; Bowen, W. D.; Williams, W.; Kopajtic, T.; Katz, J. L.; George, C.; Newman, A. H. Dual probes for the dopamine transporter and sigma1 receptors: novel piperazinyl alkyl-bis(4′-fluorophenyl)amine analogues as potential cocaine-abuse therapeutic agents. *J Med Chem* **2003**, 46, 2589–98.

487. Ferris, R. M.; Tang, F. L.; Chang, K. J.; Russell, A. Evidence that the potential antipsychotic agent rimcazole (BW 234U) is a specific, competitive antagonist of sigma sites in brain. *Life Sci* **1986**, 38, 2329–37.

488. Izenwasser, S.; Newman, A. H.; Katz, J. L. Cocaine and several sigma receptor ligands inhibit dopamine uptake in rat caudate-putamen. *Eur J Pharmacol* **1993**, 243, 201–5.

489. Gilmore, D. L.; Liu, Y.; Matsumoto, R. R. Review of the pharmacological and clinical profile of rimcazole. *CNS Drug Rev* **2004**, 10, 1–22.

490. Husbands, S. M.; Izenwasser, S.; Kopajtic, T.; Bowen, W. D.; Vilner, B. J.; Katz, J. L.; Newman, A. H. Structure-activity relationships at the monoamine transporters and sigma receptors for a novel series of 9-[3-(cis-3, 5-dimethyl-1-piperazinyl)propyl]carbazole (rimcazole) analogues. *J Med Chem* **1999**, 42, 4446–55.

491. Smith, T. A.; Yang, X.; Wu, H.; Pouw, B.; Matsumoto, R. R.; Coop, A. Trifluoromethoxyl substituted phenylethylene diamines as high affinity sigma receptor ligands with potent anti-cocaine actions. *J Med Chem* **2008**, 51, 3322–5.

492. Mesangeau, C.; Narayanan, S.; Green, A. M.; Shaikh, J.; Kaushal, N.; Viard, E.; Xu, Y. T.; Fishback, J. A.; Poupaert, J. H.; Matsumoto, R. R.; et al. Conversion of a highly selective sigma-1 receptor-ligand to sigma-2 receptor preferring ligands with anticocaine activity. *J Med Chem* **2008**, 51, 1482–6.

493. Seminerio, M. J.; Robson, M. J.; Abdelazeem, A. H.; Mesangeau, C.; Jamalapuram, S.; Avery, B. A.; McCurdy, C. R.; Matsumoto, R. R. Synthesis and pharmacological characterization of a novel sigma receptor ligand with improved metabolic stability and antagonistic effects against methamphetamine. *AAPS J* **2011**, 14, 43–51.

494. Hiranita, T.; Soto, P. L.; Kohut, S. J.; Kopajtic, T.; Cao, J.; Newman, A. H.; Tanda, G.; Katz, J. L. Decreases in cocaine self-administration with dual inhibition of the dopamine transporter and sigma receptors. *J Pharmacol Exp Ther* **2011**, 339, 662–77.

495. Senda, T.; Matsuno, K.; Okamoto, K.; Kobayashi, T.; Nakata, K.; Mita, S. Ameliorating effect of SA4503, a novel sigma 1 receptor agonist, on memory impairments induced by cholinergic dysfunction in rats. *Eur J Pharmacol* **1996**, 315, 1–10.

496. Matsuno, K.; Nakazawa, M.; Okamoto, K.; Kawashima, Y.; Mita, S. Binding properties of SA4503, a novel and selective sigma 1 receptor agonist. *Eur J Pharmacol* **1996**, 306, 271–9.

497. Lever, J. R.; Gustafson, J. L.; Xu, R.; Allmon, R. L.; Lever, S. Z. Sigma1 and sigma2 receptor binding affinity and selectivity of SA4503 and fluoroethyl SA4503. *Synapse* **2006**, 59, 350–8.

498. Cheng,   K.   1-([4-Methoxy-11C]-3,4-dimethoxyphenethyl)-4-(3-phenylpropyl) piperazinephenoxy). In: *Molecular Imaging and Contrast Agent Database (MICAD)*, National Center for Biotechnology Information (US), 2008.

499. Ishikawa, M.; Ishiwata, K.; Ishii, K.; Kimura, Y.; Sakata, M.; Naganawa, M.; Oda, K.; Miyatake, R.; Fujisaki, M.; Shimizu, E.; et al. High occupancy of sigma-1 receptors in the human brain after single oral administration of fluvoxamine: a positron emission tomography study using [11C]SA4503. *Biol Psychiat* **2007**, 62, 878–83.

500. Mori, T.; Rahmadi, M.; Yoshizawa, K.; Itoh, T.; Shibasaki, M.; Suzuki, T. Inhibitory effects of SA4503 on the rewarding effects of abused drugs. *Addict Biol* **2012**, 19, 362–9.

501. Rodvelt, K. R.; Oelrichs, C. E.; Blount, L. R.; Fan, K. H.; Lever, S. Z.; Lever, J. R.; Miller, D. K. The sigma receptor agonist SA4503 both attenuates and enhances the effects of methamphetamine. *Drug Alcohol Depend* **2011**, 116, 203–10.

502. Fujimura, K.; Matsumoto, J.; Niwa, M.; Kobayashi, T.; Kawashima, Y.; In, Y.; Ishida, T. Synthesis, structure and quantitative structure-activity relationships of sigma receptor ligands, 1-[2-(3,4-dimethoxyphenyl)ethyl]-4-(3-phenylpropyl) piperazines. *Bioorg Med Chem* **1997**, 5, 1675–83.

503. Colfax, G. N.; Santos, G. M.; Das, M.; Santos, D. M.; Matheson, T.; Gasper, J.; Shoptaw, S.; Vittinghoff, E. Mirtazapine to reduce methamphetamine use: a randomized controlled trial. *Arch Gen Psychiat* **2011**, 68, 1168–75.

504. Afshar, M.; Knapp, C. M.; Sarid-Segal, O.; Devine, E.; Colaneri, L. S.; Tozier, L.; Waters, M. E.; Putnam, M. A.; Ciraulo, D. A. The efficacy of mirtazapine in the treatment of cocaine dependence with comorbid depression. *Am J Drug Alcohol Abuse* **2012**, 38, 181–6.

505. Kelder, J. A. N.; Funke, C.; De Boer, T.; Delbressine, L.; Leysen, D.; Nickolson, V. A comparison of the physicochemical and biological properties of mirtazapine and mianserin. *J Pharm Pharmacol* **1997**, 49, 403–11.

506. Cunningham, K. A.; Anastasio, N. C.; Fox, R. G.; Stutz, S. J.; Bubar, M. J.; Swinford, S. E.; Watson, C. S.; Gilbertson, S. R.; Rice, K. C.; Rosenzweig-Lipson, S.; et al. Synergism between a serotonin 5-HT2A receptor (5-HT2AR) antagonist and 5-HT2CR

agonist suggests new pharmacotherapeutics for cocaine addiction. *ACS Chem Neurosci* **2013**, 4, 110–21.

507. Fulton, B.; Brogden, R. Buspirone. *CNS Drugs* **1997**, 7, 68–88.

508. Bergman, J.; Roof, R. A.; Furman, C. A.; Conroy, J. L.; Mello, N. K.; Sibley, D. R.; Skolnick, P. Modification of cocaine self-administration by buspirone (buspar(R)): potential involvement of D3 and D4 dopamine receptors. *Int J Neuropsychopharmacol* **2013**, 16, 445–58.

509. Winhusen, T.; Brady, K. T.; Stitzer, M.; Woody, G.; Lindblad, R.; Kropp, F.; Brigham, G.; Liu, D.; Sparenborg, S.; Sharma, G.; et al. Evaluation of buspirone for relapse-prevention in adults with cocaine dependence: an efficacy trial conducted in the real world. *Contemp Clin Trials* **2012**, 33, 993–1002.

510. Winstanley, E. L.; Bigelow, G. E.; Silverman, K.; Johnson, R. E.; Strain, E. C. A randomized controlled trial of fluoxetine in the treatment of cocaine dependence among methadone-maintained patients. *J Subst Abuse Treat* **2011**, 40, 255–64.

511. Ciraulo, D. A.; Knapp, C.; Rotrosen, J.; Sarid-Segal, O.; Ciraulo, A. M.; LoCastro, J.; Greenblatt, D. J.; Leiderman, D. Nefazodone treatment of cocaine dependence with comorbid depressive symptoms. *Addiction* **2005**, 100 Suppl 1, 23–31.

512. Funk, S. Pharmacological treatment in alcohol-, drug- and benzodiazepine-dependent patients – the significance of trazodone. *Neuropsychopharmacol Hung* **2013**, 15, 85–93.

513. Madras, B. K.; Xie, Z.; Lin, Z.; Jassen, A.; Panas, H.; Lynch, L.; Johnson, R.; Livni, E.; Spencer, T. J.; Bonab, A. A.; et al. Modafinil occupies dopamine and norepinephrine transporters in vivo and modulates the transporters and trace amine activity in vitro. *J Pharmacol Exp Ther* **2006**, 319, 561–9.

514. Volkow, N. D.; Fowler, J. S.; Logan, J.; Alexoff, D.; Zhu, W.; Telang, F.; Wang, G. J.; Jayne, M.; Hooker, J. M.; Wong, C.; et al. Effects of modafinil on dopamine and dopamine transporters in the male human brain: clinical implications. *J Am Med Assoc* **2009**, 301, 1148–54.

515. Morgan, P. T.; Pace-Schott, E.; Pittman, B.; Stickgold, R.; Malison, R. T. Normalizing effects of modafinil on sleep in chronic cocaine users. *Am J Psychiat* **2010**, 167, 331–40.

516. Hart, C. L.; Haney, M.; Vosburg, S. K.; Rubin, E.; Foltin, R. W. Smoked cocaine self-administration is decreased by modafinil. *Neuropsychopharmacology* **2008**, 33, 761–8.

517. Schmitz, J. M.; Rathnayaka, N.; Green, C.; Moeller, F. G.; Dougherty, A. E.; Grabowski, J. Combination of modafinil and d-Amphetamine for the treatment of cocaine dependence: a preliminary investigation. *Front Psychiat* **2012**, 3, Article 77, 1–6.

518. Dackis, C. A.; Kampman, K. M.; Lynch, K. G.; Plebani, J. G.; Pettinati, H. M.; Sparkman, T.; O'Brien, C. P. A double-blind, placebo-controlled trial of modafinil for cocaine dependence. *J Subst Abuse Treat* **2012**, 43, 303–12.

519. Prisinzano, T.; Podobinski, J.; Tidgewell, K.; Luo, M.; Swenson, D. Synthesis and determination of the absolute configuration of the enantiomers of modafinil. *Tetrahedron: Asymmetr* **2004**, 15, 1053–8.

520. Cao, J.; Prisinzano, T. E.; Okunola, O. M.; Kopajtic, T.; Shook, M.; Katz, J. L.; Newman, A. H. SARs at the monoamine transporters for a novel series of modafinil analogues. *ACS Med Chem Lett* **2011**, 2, 48–52.

521. Miller, D. K.; Lever, J. R.; Rodvelt, K. R.; Baskett, J. A.; Will, M. J.; Kracke, G. R. Lobeline, a potential pharmacotherapy for drug addiction, binds to mu opioid receptors

and diminishes the effects of opioid receptor agonists. *Drug Alcohol Depend* **2007**, 89, 282–91.

522. Zheng, G.; Horton, D. B.; Deaciuc, A. G.; Dwoskin, L. P.; Crooks, P. A. Des-keto lobeline analogs with increased potency and selectivity at dopamine and serotonin transporters. *Bioorg Med Chem Lett* **2006**, 16, 5018–21.

523. Flammia, D.; Dukat, M.; Damaj, M. I.; Martin, B.; Glennon, R. A. Lobeline: structure-affinity investigation of nicotinic acetylcholinergic receptor binding. *J Med Chem* **1999**, 42, 3726–31.

524. Crooks, P. A.; Zheng, G.; Vartak, A. P.; Culver, J. P.; Zheng, F.; Horton, D. B.; Dwoskin, L. P. Design, synthesis and interaction at the vesicular monoamine transporter-2 of lobeline analogs: potential pharmacotherapies for the treatment of psychostimulant abuse. *Curr Top Med Chem* **2011**, 11, 1103–27.

525. Harrod, S. B.; Dwoskin, L. P.; Crooks, P. A.; Klebaur, J. E.; Bardo, M. T. Lobeline attenuates d-methamphetamine self-administration in rats. *J Pharmacol Exp Ther* **2001**, 298, 172–9.

526. Harrod, S. B.; Dwoskin, L. P.; Green, T. A.; Gehrke, B. J.; Bardo, M. T. Lobeline does not serve as a reinforcer in rats. *Psychopharmacology (Berl)* **2003**, 165, 397–404.

527. Stead, L. F.; Hughes, J. R. Lobeline for smoking cessation. *Cochrane Database Syst Rev* **2012**, 2, CD000124.

528. Elkashef, A. M.; Rawson, R. A.; Anderson, A. L.; Li, S. H.; Holmes, T.; Smith, E. V.; Chiang, N.; Kahn, R.; Vocci, F.; Ling, W.; et al. Bupropion for the treatment of metham-phetamine dependence. *Neuropsychopharmacology* **2008**, 33, 1162–70.

529. Andersen, K. E.; Braestrup, C.; Groenwald, F. C.; Joergensen, A. S.; Nielsen, E. B.; Sonnewald, U.; Soerensen, P. O.; Suzdak, P. D.; Knutsen, L. J. S. The synthesis of novel GABA uptake inhibitors. 1. Elucidation of the structure-activity studies leading to the choice of (R)-1-[4,4-bis(3-methyl-2-thienyl)-3-butenyl]-3-piperidinecarboxylic acid (Tiagabine) as an anticonvulsant drug candidate. *J Med Chem* **1993**, 36, 1716–25.

530. Dhar, T. G. M.; Borden, L. A.; Tyagarajan, S.; Smith, K. E.; Branchek, T. A.; Weinshank, R. L.; Gluchowski, C. Design, synthesis and evaluation of substituted triarylnipecotic acid derivatives as GABA uptake inhibitors: identification of a ligand with moderate affinity and selectivity for the cloned human GABA transporter GAT-3. *J Med Chem* **1994**, 37, 2334–42.

531. Gonzalez, G.; Sevarino, K.; Sofuoglu, M.; Poling, J.; Oliveto, A.; Gonsai, K.; George, T. P.; Kosten, T. R. Tiagabine increases cocaine-free urines in cocaine-dependent methadone-treated patients: results of a randomized pilot study. *Addiction* **2003**, 98, 1625–32.

532. Bowery, N. G.; Bettler, B.; Froestl, W.; Gallagher, J. P.; Marshall, F.; Raiteri, M.; Bonner, T. I.; Enna, S. J. International Union of Pharmacology. XXXIII. Mammalian gamma-aminobutyric acid(B) receptors: structure and function. *Pharmacol Rev* **2002**, 54, 247–64.

533. Brebner, K.; Childress, A. R.; Roberts, D. C. A potential role for GABA(B) ago-nists in the treatment of psychostimulant addiction. *Alcohol Alcohol* **2002**, 37, 478–84.

534. Shoptaw, S.; Yang, X.; Rotheram-Fuller, E. J.; Hsieh, Y. C.; Kintaudi, P. C.; Charuvastra, V. C.; Ling, W. Randomized placebo-controlled trial of baclofen for cocaine dependence:

preliminary effects for individuals with chronic patterns of cocaine use. *J Clin Psychiat* **2003**, 64, 1440–8.

535. Kahn, R.; Biswas, K.; Childress, A. R.; Shoptaw, S.; Fudala, P. J.; Gorgon, L.; Montoya, I.; Collins, J.; McSherry, F.; Li, S. H.; et al. Multi-center trial of baclofen for abstinence initiation in severe cocaine-dependent individuals. *Drug Alcohol Depend* **2009**, 103, 59–64.

536. Yang, X. F.; Ding, C. H.; Li, X. H.; Huang, J. Q.; Hou, X. L.; Dai, L. X.; Wang, P. J. Regio- and enantioselective palladium-catalyzed allylic alkylation of nitromethane with monosubstituted allyl substrates: synthesis of (R)-rolipram and (R)-baclofen. *J Org Chem* **2012**, 77, 8980–5.

537. Keltner, N. L.; Folks, D. G. *Psychotropic Drugs*, 4th edition. St. Louis, MO: Elsevier Mosby, 2005.

538. Rotheram-Fuller, E.; De La Garza, R., 2nd; Mahoney, J. J., 3rd; Shoptaw, S.; Newton, T. F. Subjective and cardiovascular effects of cocaine during treatment with amantadine and baclofen in combination. *Psychiat Res* **2007**, 152, 205–10.

539. Elkashef, A.; Kahn, R.; Yu, E.; Iturriaga, E.; Li, S. H.; Anderson, A.; Chiang, N.; Ait-Daoud, N.; Weiss, D.; McSherry, F.; et al. Topiramate for the treatment of methamphetamine addiction: a multi-center placebo-controlled trial. *Addiction* **2012**, 107, 1297–306.

540. Jobes, M. L.; Ghitza, U. E.; Epstein, D. H.; Phillips, K. A.; Heishman, S. J.; Preston, K. L. Clonidine blocks stress-induced craving in cocaine users. *Psychopharmacology (Berl)* **2011**, 218, 83–8.

541. Schatzberg, A. F.; Cole, J. O.; DeBattista, C. *Manual of Clinical Psychopharmacology*, 6th edition. Washington, DC: American Psychiatric Pub., 2007.

542. Kampman, K. M.; Dackis, C.; Lynch, K. G.; Pettinati, H.; Tirado, C.; Gariti, P.; Sparkman, T.; Atzram, M.; O'Brien, C. P. A double-blind, placebo-controlled trial of amantadine, propranolol, and their combination for the treatment of cocaine dependence in patients with severe cocaine withdrawal symptoms. *Drug Alcohol Depend* **2006**, 85, 129–37.

543. Schindler, C. W.; Panlilio, L. V.; Gilman, J. P.; Justinova, Z.; Vemuri, V. K.; Makriyannis, A.; Goldberg, S. R. Effects of cannabinoid receptor antagonists on maintenance and reinstatement of methamphetamine self-administration in rhesus monkeys. *Eur J Pharmacol* **2010**, 633, 44–9.

544. De Vries, T. J.; Shaham, Y.; Homberg, J. R.; Crombag, H.; Schuurman, K.; Dieben, J.; Vanderschuren, L. J.; Schoffelmeer, A. N. A cannabinoid mechanism in relapse to cocaine seeking. *Nat Med* **2001**, 7, 1151–4.

545. Moffett, M. C.; Goeders, N. E. CP-154,526, a CRF type-1 receptor antagonist, attenuates the cue-and methamphetamine-induced reinstatement of extinguished methamphetamine-seeking behavior in rats. *Psychopharmacology (Berl)* **2007**, 190, 171–80.

546. Schulz, D. W.; Mansbach, R. S.; Sprouse, J.; Braselton, J. P.; Collins, J.; Corman, M.; Dunaiskis, A.; Faraci, S.; Schmidt, A. W.; Seeger, T.; et al. CP-154,526: a potent and selective nonpeptide antagonist of corticotropin releasing factor receptors. *Proc Natl Acad Sci USA* **1996**, 93, 10477–82.

547. Chen, Y. L.; Mansbach, R. S.; Winter, S. M.; Brooks, E.; Collins, J.; Corman, M. L.; Dunaiskis, A. R.; Faraci, W. S.; Gallaschun, R. J.; Schmidt, A.; et al. Synthesis and oral efficacy of a 4-(butylethylamino)pyrrolo[2,3-d]pyrimidine: a centrally

active corticotropin-releasing factor1 receptor antagonist. *J Med Chem* **1997**, 40, 1749–54.

548. Berry, M. D. Mammalian central nervous system trace amines. Pharmacologic amphetamines, physiologic neuromodulators. *J Neurochem* **2004**, 90, 257–71.

549. Lindemann, L.; Meyer, C. A.; Jeanneau, K.; Bradaia, A.; Ozmen, L.; Bluethmann, H.; Bettler, B.; Wettstein, J. G.; Borroni, E.; Moreau, J. L.; et al. Trace amine-associated receptor 1 modulates dopaminergic activity. *J Pharmacol Exp Ther* **2008**, 324, 948–56.

550. Miller, G. M. Avenues for the development of therapeutics that target trace amine associated receptor 1 (TAAR1). *J Med Chem* **2012**, 55, 1809–14.

551. Revel, F. G.; Moreau, J. L.; Gainetdinov, R. R.; Ferragud, A.; Velazquez-Sanchez, C.; Sotnikova, T. D.; Morairty, S. R.; Harmeier, A.; Groebke Zbinden, K.; Norcross, R. D.; et al. Trace amine-associated receptor 1 partial agonism reveals novel paradigm for neuropsychiatric therapeutics. *Biol Psychiat* **2012**, 72, 934–42.

552. Stalder, H.; Hoener, M. C.; Norcross, R. D. Selective antagonists of mouse trace amine-associated receptor 1 (mTAAR1): discovery of EPPTB (RO5212773). *Bioorg Med Chem Lett* **2011**, 21, 1227–31.

553. Galley, G.; Stalder, H.; Goergler, A.; Hoener, M. C.; Norcross, R. D. Optimisation of imidazole compounds as selective TAAR1 agonists: discovery of RO5073012. *Bioorg Med Chem Lett* **2012**, 22, 5244–8.

554. Cordero, S. L. A.; Stab II, B. R.; Guillen, F.; Montano, E. A. R. A molecular mechanism of ethanol dependence: the influence of the ionotropic glutamate receptor activated by N-methyl-D-aspartate. In: *Addictions – From Pathophysiology to Treatment*, Belin, D., Ed. InTech, 2013; pp. 169–200.

555. Urban, N. B. L.; Kegeles, L. S.; Slifstein, M.; Xu, X.; Martinez, D.; Sakr, E.; Castillo, F.; Moadel, T.; O'Malley, S. S.; Krystal, J. H.; et al. Sex differences in striatal dopamine release in young adults after oral alcohol challenge: a positron emission tomography imaging study with [11C]Raclopride. *Biol Psychiat* **2010**, 68, 689–96.

556. Correa, M.; Salamone, J. D.; Segovia, K. N.; Pardo, M.; Longoni, R.; Spina, L.; Peana, A. T.; Vinci, S.; Acquas, E. Piecing together the puzzle of acetaldehyde as a neuroactive agent. *Neurosci Biobehav Rev* **2012**, 36, 404–30.

557. Deehan, G. A., Jr.; Brodie, M. S.; Rodd, Z. A. What is in that drink: the biological actions of ethanol, acetaldehyde, and salsolinol. *Curr Top Behav Neurosci* **2013**, 13, 163–84.

558. Johansson, B. A review of the pharmacokinetics and pharmacodynamics of disulfiram and its metabolites. *Acta Psychiatr Scand Suppl* **1992**, 369, 15–26.

559. Kitson, T. M. The effect of disulfiram on the aldehyde dehydrogenases of sheep liver. *Biochem J* **1975**, 151, 407–12.

560. Nash, N. G.; Daley, R. D.; Klaus, F. Disulfiram. In: *Analytical Profiles of Drug Substances*. Academic Press, 1975; Vol. 4, pp. 168–91.

561. Spanagel, R.; Kiefer, F. Drugs for relapse prevention of alcoholism: ten years of progress. *Trends Pharmacol Sci* **2008**, 29, 109–15.

562. Harris, B. R.; Prendergast, M. A.; Gibson, D. A.; Rogers, D. T.; Blanchard, J. A.; Holley, R. C.; Fu, M. C.; Hart, S. R.; Pedigo, N. W.; Littleton, J. M. Acamprosate inhibits the binding and neurotoxic effects of trans-ACPD, suggesting a novel site of action at metabotropic glutamate receptors. *Alcohol Clin Exp Res* **2002**, 26, 1779–93.

563. Center for Substance Abuse Treatment. Acamprosate: A new medication for alcohol use disorders. Substance Abuse Treatment Advisory. *Fall.* **2005**, 4.

564. Lam, M. P.; Nurmi, H.; Rouvinen, N.; Kiianmaa, K.; Gianoulakis, C. Effects of acute ethanol on beta-endorphin release in the nucleus accumbens of selectively bred lines of alcohol-preferring AA and alcohol-avoiding ANA rats. *Psychopharmacology (Berl)* **2010**, 208, 121–30.

565. van den Wildenberg, E.; Wiers, R. W.; Dessers, J.; Janssen, R. G.; Lambrichs, E. H.; Smeets, H. J.; van Breukelen, G. J. A functional polymorphism of the mu-opioid receptor gene (OPRM1) influences cue-induced craving for alcohol in male heavy drinkers. *Alcohol Clin Exp Res* **2007**, 31, 1–10.

566. Heilig, M.; Goldman, D.; Berrettini, W.; O'Brien, C. P. Pharmacogenetic approaches to the treatment of alcohol addiction. *Nat Rev Neurosci* **2011**, 12, 670–84.

567. Weerts, E. M.; Kim, Y. K.; Wand, G. S.; Dannals, R. F.; Lee, J. S.; Frost, J. J.; McCaul, M. E. Differences in delta- and mu-opioid receptor blockade measured by positron emission tomography in naltrexone-treated recently abstinent alcohol-dependent subjects. *Neuropsychopharmacology* **2008**, 33, 653–65.

568. Hahn, E. F.; Fishman, J.; Heilman, R. D. Narcotic antagonists. 4. Carbon-6 derivatives of N-substituted noroxymorphones as narcotic antagonists. *J Med Chem* **1975**, 18, 259–62.

569. Ghirmai, S.; Azar, M. R.; Polgar, W. E.; Berzetei-Gurske, I.; Cashman, J. R. Synthesis and biological evaluation of alpha- and beta-6-amido derivatives of 17-cyclopropylmethyl-3, 14beta-dihydroxy-4, 5alpha-epoxymorphinan: potential alcohol-cessation agents. *J Med Chem* **2008**, 51, 1913–24.

570. Gual, A.; He, Y.; Torup, L.; van den Brink, W.; Mann, K. A randomised, double-blind, placebo-controlled, efficacy study of nalmefene, as-needed use, in patients with alcohol dependence. *Eur Neuropsychopharmacol* **2013**, 23, 1432–42.

571. Trial watch: nalmefene reduces alcohol use in phase III trial. *Nat Rev Drug Discov* **2011**, 10, 566.

572. Walker, B. M.; Koob, G. F. Pharmacological evidence for a motivational role of kappa-opioid systems in ethanol dependence. *Neuropsychopharmacology* **2008**, 33, 643–52.

573. Pelotte, A. L.; Smith, R. M.; Ayestas, M.; Dersch, C. M.; Bilsky, E. J.; Rothman, R. B.; Deveau, A. M. Design, synthesis, and characterization of 6-b-naltrexol analogs, and their selectivity for in vitro opioid receptor subtypes. *Bioorg Med Chem Lett* **2009**, 19, 2811–4.

574. Raehal, K. M.; Lowery, J. J.; Bhamidipati, C. M.; Paolino, R. M.; Blair, J. R.; Wang, D.; Sadee, W.; Bilsky, E. J. In vivo characterization of 6beta-naltrexol, an opioid ligand with less inverse agonist activity compared with naltrexone and naloxone in opioid-dependent mice. *J Pharmacol Exp Ther* **2005**, 313, 1150–62.

575. Portoghese, P. S.; Larson, D. L.; Sayre, L. M.; Fries, D. S.; Takemori, A. E. A novel opioid receptor site directed alkylating agent with irreversible narcotic antagonistic and reversible agonistic activities. *J Med Chem* **1980**, 23, 233–4.

576. Wentland, M. P.; Lu, Q.; Lou, R.; Bu, Y.; Knapp, B. I.; Bidlack, J. M. Synthesis and opioid receptor binding properties of a highly potent 4-hydroxy analogue of naltrexone. *Bioorg Med Chem Lett* **2005**, 15, 2107–10.

577. Wentland, M. P.; Lou, R.; Dehnhardt, C. M.; Duan, W.; Cohen, D. J.; Bidlack, J. M. 3-Carboxamido analogues of morphine and naltrexone. Synthesis and opioid receptor binding properties. *Bioorg Med Chem Lett* **2001**, 11, 1717–21.

578. Soyka, M.; Koller, G.; Schmidt, P.; Lesch, O. M.; Leweke, M.; Fehr, C.; Gann, H.; Mann, K. F. Cannabinoid receptor 1 blocker rimonabant (SR 141716) for treatment of alcohol dependence: results from a placebo-controlled, double-blind trial. *J Clin Psychopharmacol* **2008**, 28, 317–24.

579. George, D. T.; Herion, D. W.; Jones, C. L.; Phillips, M. J.; Hersh, J.; Hill, D.; Heilig, M.; Ramchandani, V. A.; Geyer, C.; Spero, D. E.; et al. Rimonabant (SR141716) has no effect on alcohol self-administration or endocrine measures in nontreatment-seeking heavy alcohol drinkers. *Psychopharmacology (Berl)* **2010**, 208, 37–44.

580. Muzyk, A. J.; Rivelli, S. K.; Gagliardi, J. P. Defining the role of baclofen for the treatment of alcohol dependence: a systematic review of the evidence. *CNS Drugs* **2012**, 26, 69–78.

581. Maryanoff, B. E.; Nortey, S. O.; Gardocki, J. F.; Shank, R. P.; Dodgson, S. P. Anticonvulsant O-alkyl sulfamates. 2,3:4,5-Bis-O-(1-methylethylidene)-beta-D-fructopyranose sulfamate and related compounds. *J Med Chem* **1987**, 30, 880–7.

582. Rosenfeld, W. E. Topiramate: a review of preclinical, pharmacokinetic, and clinical data. *Clin Ther* **1997**, 19, 1294–308.

583. Johnson, B. A.; Rosenthal, N.; Capece, J. A.; Wiegand, F.; Mao, L.; Beyers, K.; McKay, A.; Ait-Daoud, N.; Addolorato, G.; Anton, R. F.; et al. Improvement of physical health and quality of life of alcohol-dependent individuals with topiramate treatment: US multisite randomized controlled trial. *Arch Intern Med* **2008**, 168, 1188–99.

584. Florez, G.; Saiz, P. A.; Garcia-Portilla, P.; Alvarez, S.; Nogueiras, L.; Bobes, J. Topiramate for the treatment of alcohol dependence: comparison with naltrexone. *Eur Addict Res* **2011**, 17, 29–36.

585. Del Re, A. C.; Gordon, A. J.; Lembke, A.; Harris, A. H. Prescription of topiramate to treat alcohol use disorders in the Veterans Health Administration. *Addict Sci Clin Pract* **2013**, 8, 12.

586. Gilligan, P. J.; Clarke, T.; He, L.; Lelas, S.; Li, Y. W.; Heman, K.; Fitzgerald, L.; Miller, K.; Zhang, G.; Marshall, A.; et al. Synthesis and structure-activity relationships of 8-(pyrid-3-yl)pyrazolo[1,5-a]-1,3,5-triazines: potent, orally bioavailable corticotropin releasing factor receptor-1 (CRF1) antagonists. *J Med Chem* **2009**, 52, 3084–92.

587. Coric, V.; Feldman, H. H.; Oren, D. A.; Shekhar, A.; Pultz, J.; Dockens, R. C.; Wu, X.; Gentile, K. A.; Huang, S. P.; Emison, E.; et al. Multicenter, randomized, double-blind, active comparator and placebo-controlled trial of a corticotropin-releasing factor receptor-1 antagonist in generalized anxiety disorder. *Depress Anxiety* **2010**, 27, 417–25.

588. Gehlert, D. R.; Cippitelli, A.; Thorsell, A.; Le, A. D.; Hipskind, P. A.; Hamdouchi, C.; Lu, J.; Hembre, E. J.; Cramer, J.; Song, M.; et al. 3-(4-Chloro-2-morpholin-4-yl-thiazol-5-yl)-8-(1-ethylpropyl)-2,6-dimethyl-imidazo [1,2-b]pyridazine: a novel brain-penetrant, orally available corticotropin-releasing factor receptor 1 antagonist with efficacy in animal models of alcoholism. *J Neurosci* **2007**, 27, 2718–26.

589. Tellew, J. E.; Lanier, M.; Moorjani, M.; Lin, E.; Luo, Z.; Slee, D. H.; Zhang, X.; Hoare, S. R.; Grigoriadis, D. E.; St Denis, Y.; et al. Discovery of NBI-77860/GSK561679, a potent corticotropin-releasing factor (CRF1) receptor antagonist with improved pharmacokinetic properties. *Bioorg Med Chem Lett* **2010**, 20, 7259–64.

590. George, D. T.; Gilman, J.; Hersh, J.; Thorsell, A.; Herion, D.; Geyer, C.; Peng, X.; Kielbasa, W.; Rawlings, R.; Brandt, J. E.; et al. Neurokinin 1 receptor antagonism as a possible therapy for alcoholism. *Science* **2008**, 319, 1536–9.

591. Pertwee, R. G.; Howlett, A. C.; Abood, M. E.; Alexander, S. P.; Di Marzo, V.; Elphick, M. R.; Greasley, P. J.; Hansen, H. S.; Kunos, G.; Mackie, K.; et al. International Union of Basic and Clinical Pharmacology. LXXIX. Cannabinoid receptors and their ligands: beyond CB(1) and CB(2). *Pharmacol Rev* **2010**, 62, 588–631.

592. Les, A.; Pucko, W.; Szelejewski, W. Optimization of the reduction of a 5-benzylidenethiazolidine-2,4-dione derivative supported by the reaction response surface analysis: synthesis of pioglitazone hydrochloride. *Org Proc Res Dev* **2004**, 8, 157–62.

593. Sternbach, L. H. The benzodiazepine story. *J Med Chem* **1979**, 22, 1–7.

594. Low, K.; Crestani, F.; Keist, R.; Benke, D.; Brunig, I.; Benson, J. A.; Fritschy, J. M.; Rulicke, T.; Bluethmann, H.; Mohler, H.; et al. Molecular and neuronal substrate for the selective attenuation of anxiety. *Science* **2000**, 290, 131–4.

595. Tan, K. R.; Brown, M.; Laboèbe, G.; Yvon, C.; Creton, C.; Fritschy, J.-M.; Rudolph, U.; Lüscher, C. Neural bases for addictive properties of benzodiazepines. *Nature* **2010**, 463, 769–74.

596. Nielsen, M.; Hansen, E. H.; Gotzsche, P. C. What is the difference between dependence and withdrawal reactions? A comparison of benzodiazepines and selective serotonin re-uptake inhibitors. *Addiction* **2012**, 107, 900–8.

597. Lader, M. Dependence and withdrawal: comparison of the benzodiazepines and selective serotonin re-uptake inhibitors. *Addiction* **2012**, 107, 909–10.

598. Brady, K. Withdrawal or dependence: a matter of context. *Addiction* **2012**, 107, 910–1.

599. Adler, L. E.; Olincy, A.; Waldo, M.; Harris, J. G.; Griffith, J.; Stevens, K.; Flach, K.; Nagamoto, H.; Bickford, P.; Leonard, S.; et al. Schizophrenia, sensory gating, and nicotinic receptors. *Schizophr Bull* **1998**, 24, 189–202.

600. D'Souza, M. S.; Markou, A. Neuronal mechanisms underlying development of nicotine dependence: implications for novel smoking-cessation treatments. *Addict Sci Clin Pract* **2011**, 6, 4–16.

601. Eccles, J. C.; McGeer, P. L. Ionotropic and metabotropic neurotransmission. *Trends Neurosci* **1979**, 2, 39–40.

602. Sugiyama, H.; Ito, I.; Hirono, C. A new type of glutamate receptor linked to inositol phospholipid metabolism. *Nature* **1987**, 325, 531–3.

603. Liechti, M. E.; Markou, A. Role of the glutamatergic system in nicotine dependence: implications for the discovery and development of new pharmacological smoking cessation therapies. *CNS Drugs* **2008**, 22, 705–24.

604. Coe, J. W.; Rollema, H.; O'Neill, B. T. Case history: chantix/champix (varenicline tartrate), a nicotinic acetylcholine receptor partial agonist as a smoking cessation aid. In: *Annual Reports in Medicinal Chemistry,* Academic Press, 2009; Vol. 44, pp. 71–101.

605. Rollema, H.; Coe, J. W.; Chambers, L. K.; Hurst, R. S.; Stahl, S. M.; Williams, K. E. Rationale, pharmacology and clinical efficacy of partial agonists of alpha4beta2 nACh receptors for smoking cessation. *Trends Pharmacol Sci* **2007**, 28, 316–25.

606. FDA Drug Safety Communication: Safety review update of Chantix (varenicline) and risk of neuropsychiatric adverse events. http://www.fda.gov/Drugs/DrugSafety/ucm276737. Last accessed 25 February, 2014.

607. Brandon, T. H.; Drobes, D. J.; Unrod, M.; Heckman, B. W.; Oliver, J. A.; Roetzheim, R. C.; Karver, S. B.; Small, B. J. Varenicline effects on craving, cue reactivity, and smoking reward. *Psychopharmacology (Berl)* **2011**, 218, 391–403.

608. Gould, R. W.; Czoty, P. W.; Nader, S. H.; Nader, M. A. Effects of varenicline on the reinforcing and discriminative stimulus effects of cocaine in rhesus monkeys. *J Pharmacol Exp Ther* **2011**, 339, 678–86.

609. Soroko, F. E.; Mehta, N. B.; Maxwell, R. A.; Ferris, R. M.; Schroeder, D. H. Bupropion hydrochloride ((+/-) alpha-t-butylamino-3-chloropropiophenone HCl): a novel antidepressant agent. *J Pharm Pharmacol* **1977**, 29, 767–70.

610. Musso, D. L.; Mehta, N. B.; Soroko, F. E.; Ferris, R. M.; Hollingsworth, E. B.; Kenney, B. T. Synthesis and evaluation of the antidepressant activity of the enantiomers of bupropion. *Chirality* **1993**, 5, 495–500.

611. Carroll, F. I.; Blough, B. E.; Mascarella, S. W.; Navarro, H. A.; Eaton, J. B.; Lukas, R. J.; Damaj, M. I. Synthesis and biological evaluation of bupropion analogues as potential pharmacotherapies for smoking cessation. *J Med Chem* **2010**, 53, 2204–14.

612. Hesse, L. M.; Venkatakrishnan, K.; Court, M. H.; von Moltke, L. L.; Duan, S. X.; Shader, R. I.; Greenblatt, D. J. CYP2B6 mediates the in vitro hydroxylation of bupropion: potential drug interactions with other antidepressants. *Drug Metab Dispos* **2000**, 28, 1176–83.

613. Damaj, M. I.; Carroll, F. I.; Eaton, J. B.; Navarro, H. A.; Blough, B. E.; Mirza, S.; Lukas, R. J.; Martin, B. R. Enantioselective effects of hydroxy metabolites of bupropion on behavior and on function of monoamine transporters and nicotinic receptors. *Mol Pharmacol* **2004**, 66, 675–82.

614. Lukas, R. J.; Muresan, A. Z.; Damaj, M. I.; Blough, B. E.; Huang, X.; Navarro, H. A.; Mascarella, S. W.; Eaton, J. B.; Marxer-Miller, S. K.; Carroll, F. I. Synthesis and characterization of in vitro and in vivo profiles of hydroxybupropion analogues: aids to smoking cessation. *J Med Chem* **2010**, 53, 4731–48.

615. Damaj, M. I.; Grabus, S. D.; Navarro, H. A.; Vann, R. E.; Warner, J. A.; King, L. S.; Wiley, J. L.; Blough, B. E.; Lukas, R. J.; Carroll, F. I. Effects of hydroxymetabolites of bupropion on nicotine dependence behavior in mice. *J Pharmacol Exp Ther* **2010**, 334, 1087–95.

616. Perrine, D. M.; Ross, J. T.; Nervi, S. J.; Zimmerman, R. H. A short, one-pot synthesis of bupropion (Zyban, Wellbutrin). *J Chem Educ* **2000**, 77, 1479.

617. Fang, Q. K.; Han, Z.; Grover, P.; Kessler, D.; Senanayake, C. H.; Wald, S. A. Rapid access to enantiopure bupropion and its major metabolite by stereospecific nucleophilic substitution on an α-ketotriflate. *Tetrahedron: Asymmetr* **2000**, 11, 3659–63.

618. Mercer, S. L. ACS chemical neuroscience molecule spotlight on contrave. *ACS Chem Neurosci* **2011**, 2, 484–6.

619. Polosa, R.; Benowitz, N. L. Treatment of nicotine addiction: present therapeutic options and pipeline developments. *Trends Pharmacol Sci* **2011**, 32, 281–9.

620. Cahill, K.; Ussher, M. Cannabinoid type 1 receptor antagonists (rimonabant) for smoking cessation. *Cochrane Database Syst Rev* **2007**, CD005353.

621. Micheli, F.; Arista, L.; Bonanomi, G.; Blaney, F. E.; Braggio, S.; Capelli, A. M.; Checchia, A.; Damiani, F.; Di-Fabio, R.; Fontana, S.; et al. 1,2,4-Triazolyl azabicyclo[3.1.0]hexanes: a new series of potent and selective dopamine D(3) receptor antagonists. *J Med Chem* **2010**, 53, 374–91.

622. Mugnaini, M.; Iavarone, L.; Cavallini, P.; Griffante, C.; Oliosi, B.; Savoia, C.; Beaver, J.; Rabiner, E. A.; Micheli, F.; Heidbreder, C.; et al. Occupancy of brain dopamine D3

receptors and drug craving: a translational approach. *Neuropsychopharmacology* **2013**, 38, 302–12.

623. Skolnick, P.; Popik, P.; Janowsky, A.; Beer, B.; Lippa, A. S. Antidepressant-like actions of DOV 21,947: a "triple" reuptake inhibitor. *Eur J Pharmacol* **2003**, 461, 99–104.

624. Pletcher, M. J.; Vittinghoff, E.; Kalhan, R.; Richman, J.; Safford, M.; Sidney, S.; Lin, F.; Kertesz, S. Association between marijuana exposure and pulmonary function over 20 years. *J Am Med Assoc* **2012**, 307, 173–81.

625. Meier, M. H.; Caspi, A.; Ambler, A.; Harrington, H.; Houts, R.; Keefe, R. S.; McDonald, K.; Ward, A.; Poulton, R.; Moffitt, T. E. Persistent cannabis users show neuropsychological decline from childhood to midlife. *Proc Natl Acad Sci USA* **2012**, 109, E2657–64.

626. Oresic, M. Obesity and psychotic disorders: uncovering common mechanisms through metabolomics. *Dis Model Mech* **2012**, 5, 614–20.

627. METSY. http://ec.europa.eu/research/health/medical-research/brain-research/projects/tmetsy_en.html. Last accessed 25 February, 2014.

628. Editor Pertwee, R. G. Handbook of experimental pharmacology. In: *Cannabinoids*, Berlin, New York: Springer-Verlag, 2005; Vol. 168, pp 1–757.

629. Onaivi, E. S. *Marijuana and Cannabinoid Research: Methods and Protocols*. Totowa, N.J.: Humana Press, 2006.

630. Reggio, P. H. (ed.) *The Cannabinoid Receptors*. New York: Humana, 2009.

631. Breivogel, C. S.; Childers, S. R.; Deadwyler, S. A.; Hampson, R. E.; Vogt, L. J.; Sim-Selley, L. J. Chronic delta9-tetrahydrocannabinol treatment produces a time-dependent loss of cannabinoid receptors and cannabinoid receptor-activated G proteins in rat brain. *J Neurochem* **1999**, 73, 2447–59.

632. Lichtman, A. H.; Martin, B. R. Cannabinoid tolerance and dependence. *Handb Exp Pharmacol* **2005**, 691–717.

633. Lee, M. C.; Smith, F. L.; Stevens, D. L.; Welch, S. P. The role of several kinases in mice tolerant to delta 9-tetrahydrocannabinol. *J Pharmacol Exp Ther* **2003**, 305, 593–9.

634. Rinaldi-Carmona, M.; Barth, F.; Heaulme, M.; Shire, D.; Calandra, B.; Congy, C.; Martinez, S.; Maruani, J.; Neliat, G.; Caput, D.; et al. SR141716A, a potent and selective antagonist of the brain cannabinoid receptor. *FEBS Lett* **1994**, 350, 240–4.

635. Seltzman, H. H.; Carroll, F. I.; Burgess, J. P.; Wyrick, C. D.; Burch, D. F. Synthesis, spectral studies and tritiation of the cannabinoid antagonist SR141716A. *J Chem Soc Chem Commun* **1995**, 1549–50.

636. Murray, W. V.; Wachter, M. P. A simple regioselective synthesis of ethyl 1,5-diarylpyrazole-3-carboxylates. *J Heterocycl Chem* **1989**, 26, 1389–92.

637. Pertwee, R. G. Inverse agonism and neutral antagonism at cannabinoid CB1 receptors. *Life Sci* **2005**, 76, 1307–24.

638. Sim-Selley, L. J.; Brunk, L. K.; Selley, D. E. Inhibitory effects of SR141716A on G-protein activation in rat brain. *Eur J Pharmacol* **2001**, 414, 135–43.

639. Gorelick, D. A.; Goodwin, R. S.; Schwilke, E.; Schwope, D. M.; Darwin, W. D.; Kelly, D. L.; McMahon, R. P.; Liu, F.; Ortemann-Renon, C.; Bonnet, D.; et al. Antagonist-elicited cannabis withdrawal in humans. *J Clin Psychopharmacol* **2011**, 31, 603–12.

640. Haney, M.; Hart, C. L.; Vosburg, S. K.; Nasser, J.; Bennett, A.; Zubaran, C.; Foltin, R. W. Marijuana withdrawal in humans: effects of oral THC or divalproex. *Neuropsychopharmacology* **2004**, 29, 158–70.

641. Cooper, Z. D.; Haney, M. Opioid antagonism enhances marijuana's effects in heavy marijuana smokers. *Psychopharmacology (Berl)* **2010**, 211, 141–8.

642. Levin, F. R.; Mariani, J. J.; Brooks, D. J.; Pavlicova, M.; Cheng, W.; Nunes, E. V. Dronabinol for the treatment of cannabis dependence: a randomized, double-blind, placebo-controlled trial. *Drug Alcohol Depend* **2011**, 116, 142–50.

643. Archer, R. A.; Blanchard, W. B.; Day, W. A.; Johnson, D. W.; Lavagnino, E. R.; Ryan, C. W.; Baldwin, J. E. Cannabinoids. 3. Synthetic approaches to 9-ketocannabinoids. Total synthesis of nabilone. *J Org Chem* **1977**, 42, 2277–84.

644. Gareau, Y.; Dufresne, C.; Gallant, M.; Rochette, C.; Sawyer, N.; Slipetz, D. M.; Tremblay, N.; Weech, P. K.; Metters, K. M.; Labelle, M. Structure activity relationships of tetrahydrocannabinol analogues on human cannabinoid receptors. *Bioorg Med Chem Lett* **1996**, 6, 189–94.

645. Mechoulam, R. *Cannabinoids as Therapeutic Agents*. Boca Raton, FL: CRC Press, 1986; p. 186.

646. Bedi, G.; Cooper, Z. D.; Haney, M. Subjective, cognitive and cardiovascular dose-effect profile of nabilone and dronabinol in marijuana smokers. *Addict Biol* **2013**, 18, 872–81.

647. Mechoulam, R.; Peters, M.; Murillo-Rodriguez, E.; Hanuš, L. O. Cannabidiol – recent advances. *Chem Biodivers* **2007**, 4, 1678–92.

648. Nanjayya, S. B.; Shivappa, M.; Chand, P. K.; Murthy, P.; Benegal, V. Baclofen in cannabis dependence syndrome. *Biol Psychiat* **2010**, 68, e9–10.

649. Haney, M.; Hart, C. L.; Vosburg, S. K.; Comer, S. D.; Reed, S. C.; Cooper, Z. D.; Foltin, R. W. Effects of baclofen and mirtazapine on a laboratory model of marijuana withdrawal and relapse. *Psychopharmacology (Berl)* **2010**, 211, 233–44.

650. Dooley, D. J.; Donovan, C. M.; Pugsley, T. A. Stimulus-dependent modulation of [3H] norepinephrine release from rat neocortical slices by gabapentin and pregabalin. *J Pharmacol Exp Ther* **2000**, 295, 1086–93.

651. Mason, B. J.; Crean, R.; Goodell, V.; Light, J. M.; Quello, S.; Shadan, F.; Buffkins, K.; Kyle, M.; Adusumalli, M.; Begovic, A.; et al. A proof-of-concept randomized controlled study of gabapentin: effects on cannabis use, withdrawal and executive function deficits in cannabis-dependent adults. *Neuropsychopharmacology* **2012**, 37, 1689–98.

652. Thorpe, A. J.; Taylor, C. P.; Calcium channel α2δ ligands: gabapentin and pregabalin. In: Taylor, J. B.; Triggle, D. J. (eds.), *Comprehensive Medicinal Chemistry II*, Oxford: Elsevier, 2007; pp. 227–46.

653. Yuen, P.-W. α2δ Ligands: Neurontin® (Gabapentin) and Lyrica® (Pregabalin). In: Johnson, D. S.; Li, J. J. (eds.), *The Art of Drug Synthesis*. John Wiley & Sons, Inc., 2006; pp. 225–40.

654. Lichtman, A. H.; Fisher, J.; Martin, B. R. Precipitated cannabinoid withdrawal is reversed by Delta(9)-tetrahydrocannabinol or clonidine. *Pharmacol Biochem Behav* **2001**, 69, 181–8.

655. Stewart, J. L.; McMahon, L. R. Rimonabant-induced Delta9-tetrahydrocannabinol withdrawal in rhesus monkeys: discriminative stimulus effects and other withdrawal signs. *J Pharmacol Exp Ther* **2010**, 334, 347–56.

656. Haney, M.; Hart, C. L.; Vosburg, S. K.; Comer, S. D.; Reed, S. C.; Foltin, R. W. Effects of THC and lofexidine in a human laboratory model of marijuana withdrawal and relapse. *Psychopharmacology (Berl)* **2008**, 197, 157–68.

657. Carter, A. J. Hippocampal noradrenaline release in awake, freely moving rats is regulated by alpha-2 adrenoceptors but not by adenosine receptors. *J Pharmacol Exp Ther* **1997**, 281, 648–54.

658. Cravatt, B. F.; Giang, D. K.; Mayfield, S. P.; Boger, D. L.; Lerner, R. A.; Gilula, N. B. Molecular characterization of an enzyme that degrades neuromodulatory fatty-acid amides. *Nature* **1996**, 384, 83–7.

659. Rakers, C.; Zoerner, A. A.; Engeli, S.; Batkai, S.; Jordan, J.; Tsikas, D. Stable isotope liquid chromatography-tandem mass spectrometry assay for fatty acid amide hydrolase activity. *Anal Biochem* **2012**, 421, 699–705.

660. Bracey, M. H.; Hanson, M. A.; Masuda, K. R.; Stevens, R. C.; Cravatt, B. F. Structural adaptations in a membrane enzyme that terminates endocannabinoid signaling. *Science* **2002**, 298, 1793–6.

661. Seierstad, M.; Breitenbucher, J. G. Discovery and development of fatty acid amide hydrolase (FAAH) inhibitors. *J Med Chem* **2008**, 51, 7327–43.

662. Otrubova, K.; Ezzili, C.; Boger, D. L. The discovery and development of inhibitors of fatty acid amide hydrolase (FAAH). *Bioorg Med Chem Lett* **2011**, 21, 4674–85.

663. Vandevoorde, S. Overview of the chemical families of fatty acid amide hydrolase and monoacylglycerol lipase inhibitors. *Curr Top Med Chem* **2008**, 8, 247–67.

664. Khanna, I. K.; Alexander, C. W. Fatty acid amide hydrolase inhibitors–progress and potential. *CNS Neurol Disord Drug Targets* **2011**, 10, 545–58.

665. Roques, B. P.; Fournie-Zaluski, M. C.; Wurm, M. Inhibiting the breakdown of endogenous opioids and cannabinoids to alleviate pain. *Nat Rev Drug Discov* **2012**, 11, 292–310.

666. Feledziak, M.; Lambert, D. M.; Marchand-Brynaert, J.; Muccioli, G. G. Inhibitors of the endocannabinoid-degrading enzymes, or how to increase endocannabinoid's activity by preventing their hydrolysis. *Recent Pat CNS Drug Discov* **2012**, 7, 49–70.

667. Minkkila, A.; Saario, S.; Nevalainen, T. Discovery and development of endocannabinoid-hydrolyzing enzyme inhibitors. *Curr Top Med Chem* **2010**, 10, 828–58.

668. Ezzili, C.; Otrubova, K.; Boger, D. L. Fatty acid amide signaling molecules. *Bioorg Med Chem Lett* **2010**, 20, 5959–68.

669. Carroll, F. I.; Lewin, A. H.; Mascarella, S. W.; Seltzman, H. H.; Reddy, P. A. Designer drugs: a medicinal chemistry perspective. *Ann N Y Acad Sci* **2012**, 1248, 18–38.

670. Numan, N. The green leaf: Khat. *World J Med Sci* **2012**, 7, 210–23.

671. Kalix, P. Khat: a plant with amphetamine effects. *J Subst Abuse Treat* **1988**, 5, 163–9.

672. Al-Motarreb, A.; Baker, K.; Broadley, K. J. Khat: pharmacological and medical aspects and its social use in Yemen. *Phytother Res* **2002**, 16, 403–13.

673. Szendrei. Khat. http://www.unodc.org/unodc/en/data-and-analysis/bulletin/bulletin_1980-01-01_3_page003.html. Last accessed 25 January, 2014.

674. Berrang, B. D.; Lewin, A. H.; Carroll, F. I. Enantiomeric alpha-aminopropiophenones (cathinone): preparation and investigation. *J Org Chem* **1982**, 47, 2643–7.

675. Kalix, P.; Khan, I. Khat: an amphetamine-like plant material. *Bull World Health Organ* **1984**, 62, 681–6.

676. Baumann, M. H.; Ayestas, M. A.; Jr.; Partilla, J. S.; Sink, J. R.; Shulgin, A. T.; Daley, P. F.; Brandt, S. D.; Rothman, R. B.; Ruoho, A. E.; Cozzi, N. V. The designer methcathinone

analogs, mephedrone and methylone, are substrates for monoamine transporters in brain tissue. *Neuropsychopharmacology* **2012**, 37, 1192–203.

677. Dybdal-Hargreaves, N. F.; Holder, N. D.; Ottoson, P. E.; Sweeney, M. D.; Williams, T. Mephedrone: public health risk, mechanisms of action, and behavioral effects. *Eur J Pharmacol* **2013**, 714, 32–40.

678. Parrott, A. C. Human psychobiology of MDMA or 'Ecstasy': an overview of 25 years of empirical research. *Hum Psychopharmacol* **2013**, 28, 289–307.

679. Hanks, J. B.; Gonzalez-Maeso, J. Animal models of serotonergic psychedelics. *ACS Chem Neurosci* **2013**, 4, 33–42.

680. United States Drug Enforcement Administration press release July 26, 2012. http://www.justice.gov/dea/pubs/pressrel/pr072612p.html. Last accessed 15 June, 2014.

681. Wermouth, C. G. G.; C.R.; Lindberg, P.; Mitscher, L.A. Glossary of terms used in medicinal chemistry. *Pure Appl Chem* **1998**, 70, 1129–43.

682. Buckle, D. R.; Erhardt, P. W.; Ganellin, C. R.; Kobayashi, T.; Perun, T. J.; Proudfoot, J.; Senn-Bilfinger, J. Glossary of terms used in medicinal chemistry. Part II (IUPAC recommendations 2013). *Pure Appl Chem* **2013**, 85, 1725–58.

683. Tronson, N. C.; Taylor, J. R. Addiction: a drug-induced disorder of memory reconsolidation. *Curr Opin Neurobiol* **2013**, 23, 573–80.

684. Motulsky, H. J.; Neubig, R. R. Analyzing binding data. *Curr Protoc Neurosci* **2010**, Chapter 7, Unit 7.5.

685. Smith, P. A.; Selley, D. E.; Sim-Selley, L. J.; Welch, S. P. Low dose combination of morphine and Δ9-tetrahydrocannabinol circumvents antinociceptive tolerance and apparent desensitization of receptors. *Eur J Pharmacol* **2007**, 571, 129–37.

INDEX

Drug Discovery for the Treatment of Addiction: Medicinal Chemistry Strategies, First Edition. Brian S. Fulton.
© 2014 John Wiley & Sons, Inc. Published 2014 by John Wiley & Sons, Inc.

Printed in the USA
J073579SCl090814   077332